MEAN-FIELD MAGNETOHYDRODYNAMICS
AND DYNAMO THEORY

Some other Pergamon Titles of Interest:

MEAN-FIELD MAGNETOHYDRODYNAMICS AND DYNAMO THEORY

by

F. KRAUSE and K.-H. RÄDLER

Zentralinstitut für Astrophysik der
Akademie der Wissenschaften der DDR

PERGAMON PRESS

Oxford · New York · Toronto · Sydney · Paris · Frankfurt

U.K.	Pergamon Press Ltd., Headington Hill Hall, Oxford OX3 OBW, England
U.S.A.	Pergamon Press Inc., Maxwell House, Fairview Park, Elmsford, New York, 10523, U.S.A.
CANADA	Pergamon of Canada, Suite 104, 150 Consumers Road, Willowdale, Ontario M2J 1P9, Canada
AUSTRALIA	Pergamon Press (Aust.) Pty. Ltd., P. O. Box 544, Potts Point, N. S. W. 2011, Australia
FRANCE	Pergamon Press SARL, 24 rue des Ecoles, 75240 Paris, Cedex 05, France
FEDERAL REPUBLIC OF GERMANY	Pergamon Press GmbH, 6242 Kronberg-Taunus, Pferdstraße 1, Federal Republic of Germany

First edition 1980

British Library Cataloguing in Publication Data

Krause, F

Mean-field magnetohydrodynamics and dynamo theory.
1. Magnetohydrodynamics
2. Dynamo theory (Cosmic physics)
I. Title II. Rädler, K-H
538'.6 QA920 79-42947

ISBN 0-08-025041-6

Printed in GDR

PREFACE

This book is intended to give a systematic introduction to mean-field magnetohydrodynamics and the dynamo theory which is based on it, and to provide a survey of the results achieved. Our first attempt in this direction, which at the same time constitutes the basis for this book, was an extensive contribution in the hand-book "Ergebnisse der Plasmaphysik und der Gaselektronik" (Krause and Rädler [2]). Since that article first appeared, mean-field magnetohydrodynamics as well as the theory of the turbulent dynamo have been considerably advanced by contributions from scientists in various parts of the world. We have endeavoured to incorporate the essential contents of the latest papers. The fairly comprehensive bibliography also includes works not mentioned in the text.

Chapters 1 to 10 deal with mean-field magnetohydrodynamics. We consider mean-field electrodynamics as the central core of this theory which only comprises the kinematic aspects. This fills the largest portion of the book and it is only after a systematic discussion of it that the back-reaction of the magnetic field on motion is then taken into consideration. The systematic elaboration of the theory requires demanding mathematical formalisms. Therefore, we thought it advisable first to demonstrate in Chapters 1 to 3 a simple approach to the essential results, and to employ the more ambitious means only in later chapters.

Chapters 11 to 17 first give a general introduction to the dynamo problem of magnetohydrodynamics, and then explain the dynamo theory which is based on mean-field magnetohydrodynamics, known as the theory of the turbulent dynamo, and discuss its application to cosmical objects.

We wish to express our sincere gratitude and appreciation to our teacher, Professor M. Steenbeck, with whom we have had the honour to work over many years in establishing the fundamentals of the theories presented here. We are also pleased to thank our colleagues Dr. O. A. Lielausis (Riga), Professor K. H. Moffatt (Bristol) and Professor P. H. Roberts (Newcastle upon Tyne) with whom we have maintained a fruitful exchange of ideas for a long time. We also thank our co-workers Dr. H. Bräuer, Dr. L. Oetken and Dr. G. Rüdiger (Potsdam), with whom we are constantly working.

Finally, we ought to mention that since we did not write this book in our native language we needed the advice of Herr Ch. Uhl (Potsdam), whom we thank very much.

CONTENTS

CHAPTER 1

INTRODUCTION

1.1. Turbulence and large-scale structures

Stimulated by astrophysics and geophysics an increasing interest has developed in the behaviour of electromagnetic fields in electrically conducting fluids carrying out irregular, especially turbulent motions. Above all two prominent phenomena were the foci of investigations: the Earth's magnetic field and the solar activity cycle. Their origin has been shown to be closely connected with turbulent motions in the core of the Earth and in the convection zone of the Sun.

From elementary experience the general conviction has developed that turbulent motions destroy structures of all kinds. The magnetic fields in sunspots, for instance, decay within weeks. According to estimations without considering turbulence they should be able to exist for more than a thousand years. The rapid decay is ascribed to the influence of the turbulent motion of matter in the solar convection zone.

It is a remarkable new finding that turbulence need not always be a cause of destruction, but it can give rise to large-scale structures under certain conditions. As L. Biermann [1] (1951) has shown, it is an anisotropic turbulence which is responsible for the non-uniform rotation of the convection zone of the Sun and of other stellar objects. Similarly, turbulent motions may also lead to the generation of large-scale magnetic fields. Ya. H. Frenkel [1] (1945) and L. E. Gurevich and A. H. Lebedinskij [1] (1945) had the idea that convective motions were responsible for the existence of the magnetic fields of the Earth and in the sunspots. E. N. Parker [2] (1955) was the first to discover a generating mechanism. The crucial point is the induction effect of a "cyclonic turbulence" occurring in convective layers of rotating bodies. If this effect occurs in combination with a non-uniform rotation, it allows for self-excitation of large-scale magnetic fields.

The processes described by Biermann and Parker are two examples of how large-scale structures are brought about by small-scale turbulent motions. The feature they have in common is the so-called "inverse cascade", i.e. the flow of energy from the small scales to the large scales, and this is just the opposite of what is normally expected to happen in a turbulent motion.

Recently, further examples of this kind have been found in hydrodynamics, which are known as "negative viscosity phenomena".

Based on a concept suggested by M. STEENBECK [1] (1961) a rather general theory has been developed which describes the behaviour of mean electromagnetic fields, i.e. of the large-scale parts of the electromagnetic fields, in electrically conducting matter carrying out turbulent motions. This theory is generally called "mean-field electroydnamics". Within this frame the conditions for the generation of large-scale magnetic fields from small-scale turbulent motions are most clearly revealed. A completely disordered, i.e. homogeneous isotropic mirrorsymmetric turbulence only influences the decay rate of the mean magnetic fields, which is enhanced in almost all cases of physical interest. A turbulence with the weakest possible deviation from complete disorder is one which is still homogeneous and isotropic but lacks mirrorsymmetry. In this case, right-handed and left-handed helical motions do not occur with the same probability. A homogeneous isotropic non-mirrorsymmetric turbulence

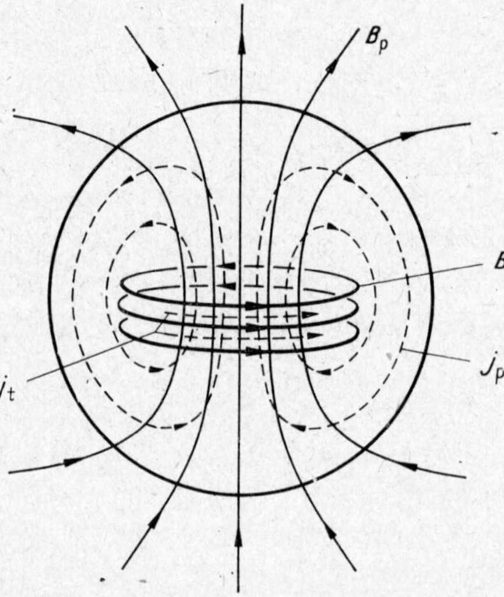

Fig. 1.1. A possible self-maintaining magnetic field configuration in an electrically conducting sphere with α-effect, i.e. Ohm's law reads $\boldsymbol{j} = \sigma(\boldsymbol{E} + \alpha\boldsymbol{B})$. The conducting sphere is embedded in the empty, insulating space. The magnetic field, $\boldsymbol{B}$, is axisymmetric. It is composed of a poloidal part, $\boldsymbol{B}_p$, with the field lines in the meridional planes with respect to the axis of symmetry, and a toroidal part, $\boldsymbol{B}_t$, with the field lines encircling the axis of symmetry. Because of the α-effect $\boldsymbol{B}_t$ drives a toroidal current, $\boldsymbol{j}_t$ (dashed lines). This current $\boldsymbol{j}_t$ is accompanied by the poloidal magnetic field $\boldsymbol{B}_p$ which, in turn, drives a poloidal current, $\boldsymbol{j}_p$ (dashed lines), accompanied by the toroidal magnetic field $\boldsymbol{B}_t$

provides for the quite unusual effect of the appearance of a mean electromotive force parallel to the magnetic field, the so-called α-effect. It is easy to understand that, for sufficiently strong α-effect, a magnetic field as depicted in figure 1.1 can be maintained against Ohmic decay.

In convective layers of rotating bodies, the Coriolis forces provide for a helical structure of the motions. There are regions in which one kind of helical motions dominate, thus causing the α-effect. This opens a direct approach to the explanation of the origin of cosmical magnetic fields.

Another theory of magnetic field generation due to more or less irregular motions has been initiated and elaborated by S. I. Braginskij [1, 2, 3] (1964). Since the large-scale fields are assumed to deviate only weakly from symmetry with respect to the rotational axis of the conducting body this theory has become known as the theory of the "nearly symmetric dynamo". This theory has been widely discussed with respect to the magnetic field of the Earth.

1.2. On the general concept of mean-field magnetohydrodynamics

In this book we shall deal with mean-field magnetohydrodynamics and its applications to the dynamo problem, known as the theory of the turbulent dynamo.

Mean-field magnetohydrodynamics is concerned with the behaviour of mean electromagnetic and hydrodynamic fields in turbulently moving electrically conducting media. It covers the mean-field electrodynamics which is restricted to the investigation of the mean electromagnetic fields in the case of prescribed hydrodynamic fields, i.e. without considering the influence of the electromagnetic fields on the motions.

This theory shall be developed on the basis of usual magnetohydrodynamics and all suppositions apply which are necessary for its justification. In all the following considerations a Euclidian space is supposed. The matter is considered to be a continuous fluid medium with sufficiently high electrical conductivity. Its velocity shall be small compared with the velocity of light in vacuum.

Let us give a short outline of the general concept and some special features of mean-field magnetohydrodynamics. We start from the basic equations of magnetohydrodynamics, which mainly consist of Maxwell's and the corresponding constitutive equations and the Navier-Stokes equation. From these equations corresponding equations for the mean fields shall be derived by an appropriate averaging operation, in the course of which non-linear terms give rise to new terms describing the mean effects of the fluctuating fields on the mean fields. With this we follow Reynolds's treatment of the Navier-

Stokes equation by which Reynolds's stresses were introduced. The corresponding quantity in mean-field electrodynamics is the average of the electromotive force due to the fluctuations of the velocity and the magnetic field, in the following called turbulent electromotive force. One of the major problems is to find representations of the turbulent electromotive force as a functional of the mean magnetic field in order to obtain a closed system of equations.

With this procedure detailed results can only be gained if restricting assumptions are introduced. In almost all deductions a two-scale turbulence shall be considered, i.e. the characteristic time and length scales of the fluctuations are small compared with those of the mean fields. The two-scale property does not only provide a simplified functional dependence of the turbulent electromotive force on the mean magnetic field, it guarantees that the basic rules of averaging operations hold for averages over space and time coordinates too.

In addition, the second order correlation approximation is often applied, in which the turbulent electromotive force, and other quantities of that kind, are calculated by taking into account only the statistical moments up to the second order, and the statistical moments of the third and higher orders are neglected. It has, however, to be noted that the second order correlation approximation is not a primary constituent of mean-field magnetohydrodynamics, it is rather a first step in a convergent approximation procedure. Even this first step provides for a fairly comprehensive insight into the physical processes under consideration. Higher order approximations require far more mathematical effort.

1.3. Technical remarks

The international system of units shall be used throughout. In addition to the unit T (tesla) for the magnetic flux density also the unit G (gauss) is used, where $1\,\mathrm{T} = 10^4\,\mathrm{G}$.

As far as the representation of mathematical relations is concerned some stipulations shall apply.

Vector equations will not only be written in the usual symbolic notation but also in component representations. Unless defined otherwise these representations are related to a Cartesian coordinate system and the summation convention is used.

Integrals without indication of limits are to extend over all space and time.

As usual the complex conjugate of a quantity shall be denoted by an asterisk.

Averages shall be denoted by an overbar. In some of the later chapters the overbar will again be omitted since only averaged quantities are considered.

CHAPTER 2

BASIC IDEAS OF MEAN-FIELD ELECTRODYNAMICS

2.1. Basic equations

Our considerations are based on non-relativistic magnetohydrodynamics. Let $\boldsymbol{B}$ be the magnetic flux density, $\boldsymbol{H}$ the magnetic field, $\boldsymbol{E}$ the electric field, $\boldsymbol{j}$ the electric current density, μ the permeability, σ the electrical conductivity, and $\boldsymbol{u}$ the fluid velocity; μ and σ shall be supposed to be constant.

We shall adopt the magnetohydrodynamic approximation, i.e. suppose that these fields are related by the equations

$$\operatorname{curl} \boldsymbol{E} = -\frac{\partial \boldsymbol{B}}{\partial t}, \quad \operatorname{curl} \boldsymbol{H} = \boldsymbol{j}, \quad \operatorname{div} \boldsymbol{B} = 0, \tag{2.1}$$

$$\boldsymbol{B} = \mu \boldsymbol{H}, \quad \boldsymbol{j} = \sigma(\boldsymbol{E} + \boldsymbol{u} \times \boldsymbol{B}). \tag{2.2}$$

These equations determine $\boldsymbol{B}$, $\boldsymbol{H}$, $\boldsymbol{E}$ and $\boldsymbol{j}$ from an assigned $\boldsymbol{u}$. They show that

$$\frac{\partial \boldsymbol{B}}{\partial t} = \operatorname{curl}\,(\boldsymbol{u} \times \boldsymbol{B}) + \frac{1}{\mu\sigma} \Delta \boldsymbol{B}, \quad \operatorname{div} \boldsymbol{B} = 0. \tag{2.3}$$

Once $\boldsymbol{B}$ has been determined from this equation, $\boldsymbol{H}$, $\boldsymbol{E}$ and $\boldsymbol{j}$ may be obtained from (2.1) and (2.2).

Here, and in most of the following chapters, we shall consider $\boldsymbol{u}$ to be given. In this way we shall avoid the problems raised by the reaction of the magnetic field on the motion, but occasionally discuss this effect qualitatively. Finally, in chapter 10, we shall give a survey of the results obtained in mean-field magnetohydrodynamics up to now.

2.2. Averaging operations

In a turbulent medium, all the fields so far introduced vary irregularly in space and time. Let F be such a fluctuating field, considered as a random function. The corresponding mean field, $\overline{F}$, is then defined to be the expectation value of F in an ensemble of identical systems, and F' shall be used to denote the difference, $F - \overline{F}$, between F and $\overline{F}$. The following relations, the Reynolds relations, hold:

$$F = \overline{F} + F', \quad \overline{\overline{F}} = \overline{F}, \quad \overline{F'} = 0, \tag{2.4}$$

$$\overline{F + G} = \overline{F} + \overline{G}, \quad \overline{\overline{F}G} = \overline{F}\,\overline{G}, \quad \overline{\overline{F}G'} = 0,$$

where G is another fluctuating field. The averaging operator commutes with the differentiation and integration operators in both space and time.

Instead of averaging over an ensemble, we may also define mean values by integration over space or time. Some of the relations (2.4) are then only approximate, although they will be the more accurate the less the means vary over the integration range concerned. We shall not need to refer to the nature of the averaging operation in our work, but only to the properties (2.4) and to the commutation rules mentioned.

Observations of the solar surface, for example, give evidence for both velocity field and magnetic field, showing fluctuations of different kind with different scales. There are: (i) the granules with diameters of about 1,000 km and a life time of some minutes, (ii) the supergranules with diameters of about 30,000 km and life times of about one day, (iii) the giant cells with diameters of the order of the solar radius and life times of some months, (iv) the sunspots with diameters ranging from some hundreds to some ten thousands of kilometers and life times from some hours to some months. However, the mean magnetic field of the Sun is responsible for the solar cycle, consequently it has a time scale of 22 years. Naturally, the length scale is of the order of the solar radius only. From these data we can infer that it is possible to find a proper time scale for the averaging operation with respect to the time, e.g. one year, but the same does not hold for averaging with respect to the space coordinate.

It is worth remarking that the Sun offers the possibility of constructing an ensemble by taking the Sun during one cycle as one member of the ensemble. Then one can take the average over a number of cycles.

2.3. The equations for the mean fields

The problem of mean-field electrodynamics is that of determining $\overline{\boldsymbol{B}}$, $\overline{\boldsymbol{H}}$, $\overline{\boldsymbol{E}}$ and $\overline{\boldsymbol{j}}$ when $\overline{\boldsymbol{u}}$ and some of the properties of $\boldsymbol{u}'$ are known.

On averaging (2.1) and (2.2) we obtain

$$\operatorname{curl} \overline{\boldsymbol{E}} = -\frac{\partial \overline{\boldsymbol{B}}}{\partial t}, \quad \operatorname{curl} \overline{\boldsymbol{H}} = \overline{\boldsymbol{j}}, \quad \operatorname{div} \overline{\boldsymbol{B}} = 0, \tag{2.5}$$

$$\overline{\boldsymbol{B}} = \mu \overline{\boldsymbol{H}}, \quad \overline{\boldsymbol{j}} = \sigma(\overline{\boldsymbol{E}} + \overline{\boldsymbol{u}} \times \overline{\boldsymbol{B}} + \overline{\boldsymbol{u}' \times \boldsymbol{B}'}). \tag{2.6}$$

These equations would be identical with (2.1) and (2.2), were it not for the single new term $\overline{\boldsymbol{u}' \times \boldsymbol{B}'}$ arising from a non-linearity, which gives rise to an additional electromotive force in Ohm's law for the mean fields. We shall speak of the "turbulent electromotive force" and, following P. H. ROBERTS [3], introduce the notation

$$\mathfrak{E} = \overline{\boldsymbol{u}' \times \boldsymbol{B}'}. \tag{2.7}$$

In order to determine $\overline{\boldsymbol{B}}$, $\overline{\boldsymbol{H}}$, $\overline{\boldsymbol{E}}$ and $\overline{\boldsymbol{j}}$, we must know not only $\overline{\boldsymbol{u}}$ but also this electromotive force $\mathfrak{E}$.

Let us now consider the electromotive force $\mathfrak{E}$ in more detail. Substituting $\boldsymbol{B} = \overline{\boldsymbol{B}} + \boldsymbol{B}'$ and $\boldsymbol{u} = \overline{\boldsymbol{u}} + \boldsymbol{u}'$ in (2.3), we obtain

$$\begin{gathered} \frac{\partial \boldsymbol{B}'}{\partial t} - \operatorname{curl}\,(\overline{\boldsymbol{u}} \times \boldsymbol{B}') - \operatorname{curl}\,(\boldsymbol{u}' \times \boldsymbol{B}') - \frac{1}{\mu\sigma}\Delta \boldsymbol{B}' \\ = -\frac{\partial \overline{\boldsymbol{B}}}{\partial t} + \operatorname{curl}\,(\overline{\boldsymbol{u}} \times \overline{\boldsymbol{B}}) + \operatorname{curl}\,(\boldsymbol{u}' \times \overline{\boldsymbol{B}}) + \frac{1}{\mu\sigma}\Delta \overline{\boldsymbol{B}}, \quad \operatorname{div}\,(\overline{\boldsymbol{B}} + \boldsymbol{B}') = 0. \end{gathered} \tag{2.8}$$

These equations determine $\boldsymbol{B}'$, if $\overline{\boldsymbol{B}}$, $\overline{\boldsymbol{u}}$ and $\boldsymbol{u}'$ are known. We may therefore consider $\mathfrak{E}$ as a functional of $\overline{\boldsymbol{B}}$, $\overline{\boldsymbol{u}}$ and (since $\mathfrak{E}$ is a mean field) the average statistical properties of $\boldsymbol{u}'$. The solution of (2.8) depends, of course, on initial and boundary conditions on $\boldsymbol{B}'$, but we shall postpone a discussion of these until later.

Let us return to equations (2.5) and (2.6). Insofar that the turbulent electromotive force $\mathfrak{E}$ as a functional of $\overline{\boldsymbol{B}}$, $\overline{\boldsymbol{u}}$ and $\boldsymbol{u}'$ is known, it becomes possible to determine $\overline{\boldsymbol{B}}$, $\overline{\boldsymbol{H}}$, $\overline{\boldsymbol{E}}$ and $\overline{\boldsymbol{j}}$ from a knowledge of $\overline{\boldsymbol{u}}$ and $\boldsymbol{u}'$. Consequently, the determination of that functional is the crucial point in the development of mean-field electrodynamics.

2.4. General properties of the turbulent electromotive force

It is a characteristic of turbulence that quantities like $\boldsymbol{u}'$ and $\boldsymbol{B}'$ at a certain space-time point have a correlation with some other quantity at another space-time point only if the separation with respect to space as well as to time is not too large. Accordingly, for a determination of $\mathfrak{E}$ at a certain space-time point the quantities $\overline{\boldsymbol{B}}$, $\overline{\boldsymbol{u}}$ and $\boldsymbol{u}'$ only need to be known in a certain neighbourhood of the point considered. As a consequence initial and boundary conditions on $\boldsymbol{B}'$ will not influence $\mathfrak{E}$ if we assume the initial time point and the boundaries to lie outside this neighbourhood. The following investigations are based on this assumption.

As already mentioned, $\boldsymbol{B}'$ may be determined from equations (2.8), and an inspection of these shows that $\boldsymbol{B}'$ is a linear functional of the mean magnetic field $\overline{\boldsymbol{B}}$. Obviously the same statement holds for the turbulent electromotive force $\mathfrak{E}$ defined by (2.7). We cannot completely exclude the possibility that there may be non-trivial solutions $\boldsymbol{B}'$ of equations (2.8) when $\overline{\boldsymbol{B}} = 0$. Such a possibility is indicated by results deduced by KAZANTSEV [1], GAILITIS [4] and KRAICHNAN [4]. In this case we have to expect that $\mathfrak{E}$ may be an inhomogeneous linear functional of $\overline{\boldsymbol{B}}$ (RÄDLER [15]). Initial and boundary conditions may also provide for an inhomogeneity but, due to our assumption in the previous paragraph, they are of no influence.

However, in most of the following we adopt the viewpoint that $\boldsymbol{B}'$, and with that the turbulent electromotive force $\mathfrak{E}$, owes its existence to the interaction of the fluctuating velocity field $\boldsymbol{u}'$ with the mean magnetic field $\overline{\boldsymbol{B}}$ and, consequently, that $\mathfrak{E}$ can be assumed to be a linear homogeneous functional of $\overline{\boldsymbol{B}}$.

CHAPTER 3

ELEMENTARY TREATMENT OF A SIMPLE EXAMPLE

3.1. Assumptions

By way of introduction, we now investigate a rather simple case, but one which allows the clarification of fundamental questions and the deduction of certain far-reaching results.

As already mentioned, for the determination of $\mathfrak{E}$ at a special space-time point it is necessary to know the mean field $\overline{\boldsymbol{B}}$ in a proper neighbourhood only. In this region we suppose $\overline{\boldsymbol{B}}$ is so weakly dependent on time and position that a sufficient representation is provided by the components of $\overline{\boldsymbol{B}}$ itself and its first spatial derivatives.

We assume, furthermore, the medium to have no mean motion, i.e. $\overline{\boldsymbol{u}} = 0$, and the turbulence, $\boldsymbol{u}'$, to be homogeneous, isotropic and steady in the statistical sense. Sometimes we shall also consider the turbulence to be mirror-symmetric.

3.2. Homogeneity, isotropy and mirrorsymmetry of turbulent fields

A homogeneous turbulent field is defined by the property that every average derived from it is independent of position, or—in other words—every such quantity is invariant if the field undergoes a translation. For a steady turbulent field the same will hold with respect to the time.

We speak of an isotropic turbulent field if no direction exists which is preferred over any other. We define such a field by the property that every mean quantity derived from it does not vary if the field undergoes an arbitrary rotation about an arbitrary axis.

Finally, we shall speak of a mirrorsymmetric turbulent field if the mean quantities derived from it are invariant if the field is reflected in an arbitrary plane.

It is less restrictive to demand the latter properties in special circumstances only, e.g. a turbulent field can be isotropic at one point only, or it can show mirrorsymmetry with respect to a special plane or a set of planes only. An illustration of the latter is the turbulent motion in the solar convection zone:

The mean properties of this field remain unchanged only it is reflected in the Sun's equatorial plane or any meridional plane.

In a turbulent motion it may happen that left-handed helical motions are more probable than right-handed ones, or vice-versa. We shall speak of a turbulence with a preferred "Schraubensinn" (STEENBECK, KRAUSE and RÄDLER [2]) or—to use the English equivalent introduced by MOFFATT [2]—we say that the turbulence has helicity. A reflection transforms a left-handed screw into a right-handed one, and vice versa. Consequently, if a turbulent field is mirrorsymmetric to at least one plane, it cannot have helicity, i.e. motions associated with a left-handed screw are as probable as motions associated with a right-handed screw.

It is worth noting some relations between homogeneity, isotropy and mirror symmetry. A homogeneous turbulent field need not be isotropic, and an isotropic field need not be mirrorsymmetric. However, if a homogeneous field is isotropic at least at one point, then it is isotropic in all points. Furthermore, if a field which is isotropic at one point is mirrorsymmetric with respect to at least one plane through this point, then it has this property with respect to every plane through this point. Conversely, a turbulent field is homogeneous, if it is isotropic at every point, and it is isotropic at one point if it is mirrorsymmetric with respect to every plane through this point. The latter statements are due to the facts that any translation can be represented by two rotations about two parallel axes, and any rotation can be represented by two reflections in planes containing the axis of rotation.

Insofar as homogeneous isotropic turbulence exists in nature it is difficult to find convincing reasons for such turbulence to be non-mirrorsymmetric. For this reason, in the literature of hydrodynamic turbulence the word "isotropy" is often used to describe the isotropic mirrorsymmetric case. However, by assuming homogeneity and isotropy without demanding mirrorsymmetry, an aspect important for dynamo theory is most clearly identified.

3.3. Symmetry laws

Mirror symmetry is a fundamental property of the basic equations of classical physics. The background of this is the idea that from any real system another real system can be derived by reflecting the former in an arbitrary plane or at an arbitrary point. However, in the definition of some quantities, right-handedness is involved; such quantities are the cross-product, or those defined by the curl-operation and the volume (triple scalar product). Since right-handedness is not invariant with respect to reflections, these quantities show different transformation properties if compared with those which do not need the right-handedness for their definition. We have the relations for a reflection at a certain point

$$\boldsymbol{a}^{\text{ref}} = -\boldsymbol{a}, \quad \boldsymbol{a}^{\text{ref}} \times \boldsymbol{b}^{\text{ref}} = \boldsymbol{a} \times \boldsymbol{b}, \quad \text{curl}\, \boldsymbol{a}^{\text{ref}} = \text{curl}\, \boldsymbol{a}. \tag{3.1}$$

$$v^{\text{ref}} = (\boldsymbol{a}^{\text{ref}} \times \boldsymbol{b}^{\text{ref}})\, \boldsymbol{c}^{\text{ref}} = -(\boldsymbol{a} \times \boldsymbol{b}) \cdot \boldsymbol{c} = -v;$$

$\boldsymbol{a}$, $\boldsymbol{b}$, $\boldsymbol{c}$ are arbitrary vectors and v denotes the volume of the parallelepiped defined by the vectors $\boldsymbol{a}$, $\boldsymbol{b}$, $\boldsymbol{c}$. The superscript "ref" denotes the reflected quantity.

Our basic equations (2.1), (2.2) relate quantities which have different transformation properties if the system under consideration is reflected. Let $\boldsymbol{E}, \boldsymbol{j}, \boldsymbol{H}, \boldsymbol{B}$ be a solution of (2.1), (2.2) for a given $\boldsymbol{u}$. If we carry out a reflection at a point, obviously $\boldsymbol{u}$ as a motion of particles has to change sign and with it $\boldsymbol{j}$ and $\boldsymbol{E}$, whereas $\boldsymbol{H}$ and $\boldsymbol{B}$ according to (3.1) and (2.1), remain unchanged. We have, if reflecting at the point $\boldsymbol{x} = 0$,

$$\boldsymbol{u}^{\text{ref}}(\boldsymbol{x}, t) = -\boldsymbol{u}(-\boldsymbol{x}, t), \quad \boldsymbol{E}^{\text{ref}}(\boldsymbol{x}, t) = -\boldsymbol{E}(-\boldsymbol{x}, t),$$
$$\boldsymbol{j}^{\text{ref}}(\boldsymbol{x}, t) = -\boldsymbol{j}(-\boldsymbol{x}, t), \tag{3.2}$$

but

$$\boldsymbol{H}^{\text{ref}}(\boldsymbol{x}, t) = \boldsymbol{H}(-\boldsymbol{x}, t), \quad \boldsymbol{B}^{\text{ref}}(\boldsymbol{x}, t) = \boldsymbol{B}(-\boldsymbol{x}, t). \tag{3.3}$$

Vectors which behave like the quantities (3.2) are usually called polar vectors, those like $\boldsymbol{H}$ and $\boldsymbol{B}$ axial vectors. Other synonyms are "pseudo" and "skew" for quantities which are dependent on using right-handedness, e.g. a scalar like the volume v is called a pseudo-scalar.

For turbulence a pseudo-scalar of interest is defined by the average

$$h = \overline{\boldsymbol{u}' \cdot \text{curl}\, \boldsymbol{u}'}. \tag{3.4}$$

Obviously, $h \neq 0$ is an indicator that the turbulence has helicity. In the mirrorsymmetric case h has on the one hand to be unchanged since all statistical properties of $\boldsymbol{u}'$ are unchanged under a reflection, on the other hand h has to change the sign since it is skew. Hence, $h = 0$ for mirror-symmetric turbulence.

The symmetry laws often prove to be a useful tool for simplifying a general ansatz for a solution of the basic equations. If we know that $F + G = 0$ is an integral of the basic equations and G is skew but F not, then $F - G = 0$ is a solution of the basic equations too. Consequently, $F = 0$ and $G = 0$ must be valid.

It is possible that we are concerned with products of different quantities. As we can see from (3.1), (3.2), (3.3) the scalar product of a polar and an axial vector is a pseudoscalar, whereas the cross-product of these quantities is a polar vector; in particular, the turbulent electromotive force $\mathfrak{E}$ is, of course, a polar vector.

3.4. The structure of the turbulent electromotive force

As already mentioned, $\mathfrak{E}$ is a functional of $\overline{\boldsymbol{B}}$, $\overline{\boldsymbol{u}}$ and $\boldsymbol{u}'$. We consider this quantity at a certain space-time point. On the assumptions introduced in 3.1. it may depend on $\overline{\boldsymbol{B}}$ and its first spatial derivatives at this point and on $\boldsymbol{u}'$ at the neighbourhood of this point only. Naturally, only mean quantities derived from $\boldsymbol{u}'$ are relevant.

We notice that $\mathfrak{E}$ varies with position only as far as $\overline{\boldsymbol{B}}$ and its first spatial derivatives do. This is simply a consequence of the homogeneity assumed for $\boldsymbol{u}'$.

Now we have to take into account the fact that $\mathfrak{E}$ is a vector quantity. The only vectors available to construct it are $\overline{\boldsymbol{B}}$, curl $\overline{\boldsymbol{B}}$, $\overline{\boldsymbol{B}} \times \text{curl}\, \overline{\boldsymbol{B}}$, $(\overline{\boldsymbol{B}} \cdot \nabla)\, \overline{\boldsymbol{B}}, \ldots$, where the dots stand for vector products of higher order in $\overline{\boldsymbol{B}}$. No contribution can be furnished by $\boldsymbol{u}'$, as is easily seen: Since $\boldsymbol{u}'$ is isotropic any mean quantity derived from it must be invariant under an arbitrary rotation. The only vector with this property is the zero-vector.

Noting finally that $\mathfrak{E}$ is a linear functional of $\overline{\boldsymbol{B}}$ we find as the only possible representation

$$\mathfrak{E} = \overline{\boldsymbol{u}' \times \boldsymbol{B}'} = \alpha \overline{\boldsymbol{B}} - \beta \,\text{curl}\, \overline{\boldsymbol{B}}. \tag{3.5}$$

α and β are constant mean quantities determined by the turbulent velocity field $\boldsymbol{u}'$.

Relation (3.5) is deduced from our basic equations (2.1), (2.2) and consequently the symmetry laws discussed in the foregoing paragraph will apply. In this way we find that β is scalar whereas α is pseudo-scalar.

We now assume, in addition, that $\boldsymbol{u}'$ is mirrorsymmetric, i.e. in the ensemble of fields one realisation $\boldsymbol{u}'$ and every other derived from it by a reflection have the same probability. Then, on the one hand, the quantities α and β do not change if a reflection is carried out, since the ensemble is not altered. On the other hand, however, α must change its sign since it is skew. Consequently $\alpha = 0$ for homogeneous isotropic mirrorsymmetric turbulence. A connection with the helicity h defined by (3.4) is thus indicated.

3.5. Ohm's law

Returning now to the equations for the mean fields (2.5) and (2.6) and substituting (3.5) we find the relation

$$\overline{\boldsymbol{j}} = \sigma_T(\overline{\boldsymbol{E}} + \alpha \overline{\boldsymbol{B}}) \tag{3.6}$$

as Ohm's law for the mean fields, where σ_T, the turbulent conductivity, is given by

$$\sigma_T = \frac{\sigma}{1 + \mu\sigma\beta}. \tag{3.7}$$

Using (3.6) instead of Ohm's law in (2.6) the equations (2.5) and (3.6) form a complete set of equations to determine $\overline{\boldsymbol{B}}$, $\overline{\boldsymbol{H}}$, $\overline{\boldsymbol{E}}$ and $\overline{\boldsymbol{j}}$ when α and β are known.

In the mirrorsymmetric case, $\alpha = 0$, our equations for the mean fields, i.e. (2.5) and (2.6) with Ohm's law from (3.6), are in formal conformity with the equations for the unaveraged fields, i.e. (2.1) and (2.2) with $\boldsymbol{u} = 0$. The replacement of the molecular conductivity, σ ,by the turbulent conductivity, σ_T, is the only difference. The ratio σ_T/σ can be determined if β is known.

In the non-mirrorsymmetric case an electromotive force $\alpha\overline{\boldsymbol{B}}$ appears in Ohm's law for the mean fields, which is parallel ($\alpha > 0$) or antiparallel ($\alpha < 0$) to the mean magnetic field $\overline{\boldsymbol{B}}$. This new effect due to the turbulent motion, called α-effect, is of great importance in dynamo theory as we shall see later on.

3.6. Preliminary steps for a determination of α and β on special assumptions

We start from equations (2.8) assuming $\overline{\boldsymbol{u}} = 0$. If we subtract the equations obtained from (2.3) by taking the average we find

$$\frac{\partial \boldsymbol{B}'}{\partial t} - \frac{1}{\mu\sigma}\Delta\boldsymbol{B}' - \operatorname{curl}(\boldsymbol{u}'\times\boldsymbol{B}' - \overline{\boldsymbol{u}'\times\boldsymbol{B}'}) = \operatorname{curl}(\boldsymbol{u}'\times\overline{\boldsymbol{B}}),$$

$$\operatorname{div}\boldsymbol{B}' = 0. \tag{3.8}$$

It is our intention to determine $\boldsymbol{B}'$ from these equations. However, in order to do so, further assumptions are needed to exclude the non-linear terms. In a certain sense, which is specified later, we demand the turbulent velocity field to be small, and omit all terms of higher order in $\boldsymbol{u}'$ and $\boldsymbol{B}'$. Thus, we replace (3.8) by

$$\frac{\partial \boldsymbol{B}'}{\partial t} - \frac{1}{\mu\sigma}\Delta\boldsymbol{B}' = \operatorname{curl}(\boldsymbol{u}'\times\overline{\boldsymbol{B}}), \quad \operatorname{div}\boldsymbol{B}' = 0. \tag{3.9}$$

Furthermore, we restrict our investigations to two limiting cases which allow additional mathematical simplification, although they are of high physical importance. Dissipative effects are neglected in the first case, i.e. we replace (3.9) by

$$\frac{\partial \boldsymbol{B}'}{\partial t} = \operatorname{curl}(\boldsymbol{u}'\times\overline{\boldsymbol{B}}), \quad \operatorname{div}\boldsymbol{B}' = 0. \tag{3.10}$$

In the second case dissipative effects are assumed to be so strong that $\boldsymbol{B}'$ is stationary at any time. Hence we can omit the derivative with respect to time in (3.9) and arrive at the equations

$$\Delta \boldsymbol{B}' = -\mu\sigma \operatorname{curl} (\boldsymbol{u}' \times \overline{\boldsymbol{B}}), \quad \operatorname{div} \boldsymbol{B}' = 0. \tag{3.11}$$

Before the conditions are formulated under which the simplifications just introduced are warranted, it is worth remarking that they are not a fundamental ingredient of mean-field magnetohydrodynamics. The remarkable relation (3.5), for example, is not based on them. However, these simplifications enable us to calculate certain models in a rather simple manner.

We now introduce two scales characterizing the variation of $\boldsymbol{u}'$ with time and position, i.e. the correlation time, τ_{cor}, and the correlation length, λ_{cor}. Since $\boldsymbol{B}'$ is assumed to be caused by the interaction of $\boldsymbol{u}'$ with $\overline{\boldsymbol{B}}$, we conclude that τ_{cor} and λ_{cor} are the scales of $\boldsymbol{B}'$ too, since $\overline{\boldsymbol{B}}$ is assumed to depend weakly on time and position.

Our first simplification—the discard of the terms of second order in the fluctuations—is justified if at least one of the terms, $\dfrac{\partial \boldsymbol{B}'}{\partial t}$, $\dfrac{1}{\mu\sigma} \Delta \boldsymbol{B}'$, is large compared with the term $\operatorname{curl} (\boldsymbol{u}' \times \boldsymbol{B}' - \overline{\boldsymbol{u}' \times \boldsymbol{B}'})$. This is the case if

$$\min (\mathrm{Rm}, \mathrm{S}) \ll 1, \tag{3.12}$$

where Rm is the magnetic Reynolds number

$$\mathrm{Rm} = \mu\sigma u' \lambda_{\text{cor}}, \tag{3.13}$$

and S the Strouhal number

$$\mathrm{S} = u' \frac{\tau_{\text{cor}}}{\lambda_{\text{cor}}}. \tag{3.14}$$

When defining Rm and S we have already used the notation u' for the r.m.s. velocity fluctuation, i.e.

$$u' = \sqrt{\overline{\boldsymbol{u}'^2}}. \tag{3.15}$$

The theory which is based on equation (3.9) and is justified if the inequality (3.12) holds, shall be referred to here as the "second order correlation approximation", other descriptions are "first order smoothing theory" or "the discard of third order cumulants". The first and the last terminologies become clearer if one realizes that $\overline{\boldsymbol{u}' \times \boldsymbol{B}'}$ (rather than $\boldsymbol{B}'$) is the quantity of interest.

Equation (3.9) was further simplified by directing our interest to two limiting cases, by assuming that one of the terms of the left-hand side will dominate. Obviously, the first case is characterized by the inequality

$$\tau_{\text{cor}} \ll \mu\sigma \lambda_{\text{cor}}^2, \tag{3.16}$$

—we shall speak of the high-conductivity limit—and the second one by

$$\tau_{\mathrm{cor}} \gg \mu\sigma\lambda_{\mathrm{cor}}^2, \tag{3.17}$$

the low-conductivity limit.

From (3.13) and (3.14) we find the relation

$$\mathrm{Rm} = \frac{\mu\sigma\lambda_{\mathrm{cor}}^2}{\tau_{\mathrm{cor}}}\,\mathrm{S}. \tag{3.18}$$

Taking into account (3.12) it becomes obvious that the applicability of the second order correlation approximation implies $\mathrm{S} \ll 1$ in the high-conductivity limit, but $\mathrm{Rm} \ll 1$ in the low-conductivity limit. In the former case there is no restriction on Rm, especially, Rm can be very large as is often the case for cosmical objects. Since one generally expects $\mathrm{S} \approx 1$ the application of results deduced on the basis of the second order correlation approximation in the high-conductivity limit is an extrapolation from $\mathrm{S} \ll 1$ to $\mathrm{S} \approx 1$ only. That may, perhaps, explain the sometimes convincing quantitative agreement with the observations as we shall see later.

3.7. The high-conductivity limit

The governing equation for this case, (3.10), can be solved by the expression

$$\boldsymbol{B}'(\boldsymbol{x}, t) = \boldsymbol{B}'(\boldsymbol{x}, t_0) + \int_{t_0}^{t} \mathrm{curl}\,(\boldsymbol{u}'(\boldsymbol{x}, t') \times \overline{\boldsymbol{B}}(\boldsymbol{x}, t'))\, \mathrm{d}t', \quad \mathrm{div}\, \boldsymbol{B}'(\boldsymbol{x}, t_0) = 0. \tag{3.19}$$

t_0 is arbitrary. According to earlier assumptions $t - t_0$, however, is to be chosen so large that no correlation between $\boldsymbol{u}'(\boldsymbol{x}, t)$ and $\boldsymbol{B}'(\boldsymbol{x}, t_0)$ exists. It is, moreover, convenient to assume the initial time point so distant that any initial fluctuating magnetic field has disappeared. Thus we assume

$$t - t_0 \gg \mu\sigma\lambda_{\mathrm{cor}}^2 \gg \tau_{\mathrm{cor}}, \tag{3.20}$$

and the appropriate solution of (3.10) takes the form

$$\boldsymbol{B}'(\boldsymbol{x}, t) = \int_{t_0}^{t} \mathrm{curl}\,(\boldsymbol{u}'(\boldsymbol{x}, t') \times \overline{\boldsymbol{B}}(\boldsymbol{x}, t'))\, \mathrm{d}t'. \tag{3.21}$$

We shall now derive the turbulent electromotive force $\mathfrak{E}$ by multiplying (3.21) with $\boldsymbol{u}'(\boldsymbol{x}, t)$ and taking the average. The resulting expression does not depend on t_0 as far as (3.20) holds. Hence we can put $t_0 = -\infty$ and obtain

$$\mathfrak{E}(\boldsymbol{x}, t) = \int_{-\infty}^{t} \overline{\boldsymbol{u}'(\boldsymbol{x}, t) \times \mathrm{curl}\,(\boldsymbol{u}'(\boldsymbol{x}, t') \times \overline{\boldsymbol{B}}(\boldsymbol{x}, t'))}\, \mathrm{d}t'. \tag{3.22}$$

The integrand is significantly different from zero only if $t - t' < \tau_{cor}$. According to the assumptions introduced in 3.1. $\overline{\boldsymbol{B}}$ does not vary in this time interval and, consequently, $\overline{\boldsymbol{B}}(\boldsymbol{x}, t)$ can be substituted for $\overline{\boldsymbol{B}}(\boldsymbol{x}, t')$. If we finally replace t' by $t - \tau$ we derive

$$\mathfrak{E}(\boldsymbol{x}, t) = \int_0^\infty \overline{\boldsymbol{u}'(\boldsymbol{x}, t) \times \operatorname{curl}\,(\boldsymbol{u}'(\boldsymbol{x}, t-\tau) \times \overline{\boldsymbol{B}}(\boldsymbol{x}, t))}\, \mathrm{d}\tau. \tag{3.23}$$

We will now write (3.23) in a representation by vector components with respect to a right-handed cartesian coordinate system x_1, x_2, x_3. Considering homogeneous isotropic turbulence we already know that (3.23) is of the form (3.5). Thus for $\mathfrak{E}_1$, we have only to write down the terms with $\overline{B}_1$, $\dfrac{\partial \overline{B}_3}{\partial x_2}$ and $\dfrac{\partial \overline{B}_2}{\partial x_3}$; the others $\overline{B}_2$, $\overline{B}_3$, $\dfrac{\partial \overline{B}_1}{\partial x_1}$, ..., can be dropped from the beginning. In this way we obtain

$$\begin{aligned}
\mathfrak{E}_1 &= -(w_{312} - w_{213})\, \overline{B}_1 - w_{22} \frac{\partial \overline{B}_3}{\partial x_2} + w_{33} \frac{\partial \overline{B}_2}{\partial x_3}, \\
\mathfrak{E}_2 &= -(w_{123} - w_{321})\, \overline{B}_2 - w_{33} \frac{\partial \overline{B}_1}{\partial x_3} + w_{11} \frac{\partial \overline{B}_3}{\partial x_1}, \\
\mathfrak{E}_3 &= -(w_{231} - w_{132})\, \overline{B}_3 - w_{11} \frac{\partial \overline{B}_2}{\partial x_1} + w_{22} \frac{\partial \overline{B}_1}{\partial x_2},
\end{aligned} \tag{3.24}$$

where the notations

$$w_{ijk} = \int_0^\infty \overline{u_i'(\boldsymbol{x}, t) \frac{\partial u_k'(\boldsymbol{x}, t-\tau)}{\partial x_j}}\, \mathrm{d}\tau, \tag{3.25}$$

$$w_{ij} = \int_0^\infty \overline{u_i'(\boldsymbol{x}, t)\, u_j'(\boldsymbol{x}, t-\tau)}\, \mathrm{d}\tau, \tag{3.26}$$

have been used. Comparing (3.24) with (3.5) we immediately find that

$$\begin{aligned}
\alpha &= -(w_{312} - w_{213}) = -(w_{123} - w_{321}) = -(w_{231} - w_{132}) \\
&= -\frac{1}{3}(w_{123} + w_{231} + w_{312} - w_{132} - w_{321} - w_{213}),
\end{aligned} \tag{3.27}$$

and

$$\beta = w_{11} = w_{22} = w_{33} = \frac{1}{3}(w_{11} + w_{22} + w_{33}). \tag{3.28}$$

Inserting (3.25) and (3.26) into (3.27) and (3.28) we find finally

$$\alpha = -\frac{1}{3} \int_0^\infty \overline{\boldsymbol{u}'(\boldsymbol{x}, t) \cdot \operatorname{curl}\, \boldsymbol{u}'(\boldsymbol{x}, t-\tau)}\, \mathrm{d}\tau, \tag{3.29}$$

and

$$\beta = \frac{1}{3} \int_0^\infty \overline{\boldsymbol{u}'(\boldsymbol{x}, t) \cdot \boldsymbol{u}'(\boldsymbol{x}, t - \tau)}\, d\tau . \tag{3.30}$$

Deducing these formulae we used the knowledge of (3.5). However, without this knowledge, (3.24) has only at first a more complicated form because for reasons of isotropy, relations between the coefficients w_{ijk}, w_{ij} are easily established. These coefficients have to be unchanged if undergoing an arbitrary rotation. If, for example, a 90°-rotation about the x_3-axis is carried out, x_1 is transformed into x_2, u_1' into u_2'; but x_2 is transformed into $-x_1$ and u_2' into $-u_1'$; x_3 and u_3' remain unchanged. Inspecting (3.25) and (3.26) we find that $w_{123} = -w_{213}$, and $w_{222} = -w_{111}$ but also $w_{111} = w_{222}$ which leads to $w_{111} = w_{222} = 0$. Similarly we conclude $w_{11} = w_{22}$; or $w_{13} = w_{23}$ but $w_{23} = -w_{13}$, thus $w_{12} = w_{13} = 0$. In this way a direct derivation of (3.24), (3.29) and (3.30) is provided.

An interesting property of α, which is a pseudo-scalar, becomes obvious from (3.29). The integrand is related to the quantity h defined by (3.4). In the case of a turbulence where motions corresponding to a right-handed screw are more probable than those corresponding to a left-handed screw, α takes negative values, in the opposite case positive values.

For an estimate of the order of magnitude we evaluate the integrals in (3.29) and (3.30) such that

$$\alpha = -\frac{1}{3} \overline{\boldsymbol{u}' \cdot \operatorname{curl} \boldsymbol{u}'}\, \tau_{\text{cor}} = -\frac{1}{3} h\, \tau_{\text{cor}} , \tag{3.31}$$

and (Rädler (1966) [2], Kazantsev (1967) [1])

$$\beta = \frac{1}{3} u'^2 \tau_{\text{cor}} . \tag{3.32}$$

Inserting (3.32) into (3.7) we find for the turbulent conductivity

$$\sigma_T = \frac{\sigma}{1 + \frac{1}{3} \mu \sigma u'^2 \tau_{\text{cor}}} , \tag{3.33}$$

which shows that this quantity is smaller than the molecular conductivity.

3.8. Applications to the solar convection zone

The solar convection zone is an interesting example. Considering the granules we can take as characteristic scales, according to the observations,

$$\lambda_{\text{cor}} = 10^6\ \text{m},\ \tau_{\text{cor}} \equiv 3 \cdot 10^2\ \text{s};\ u' = 3 \cdot 10^2\, \frac{\text{m}}{\text{s}} ; \tag{3.34}$$

furthermore, the conductivity is generally assumed to be $\sigma = 3 \cdot 10^3\,\Omega^{-1}\,\text{m}^{-1}$ and for the permeability we take the vacuum value, i.e. $\mu = \mu_0 = 4\pi \cdot 10^{-7}$ Vs $\text{A}^{-1}\,\text{m}^{-1}$. The decay time of a magnetic field having the length scale λ_{cor} is thus of the order $3 \cdot 10^9$ s = 100 years, hence the condition (3.16) for the high-conductivity limit is well satisfied. From the data given in (3.34) and below, we find furthermore

$$\text{Rm} = 10^6, \quad \text{S} = 10^{-1}, \tag{3.35}$$

so that (3.12) is at least reasonably well satisfied.

If we apply (3.33) despite the fact that S is smaller than 1 but not very small compared with it, we find

$$\frac{1}{3}\,\mu\sigma u'^2\tau_{\text{cor}} \approx 10^4, \tag{3.36}$$

and thus

$$\sigma_T = 10^{-4}\,\sigma, \tag{3.37}$$

for the solar convection zone.

This result must be seen in relation to processes in the Sun connected with magnetic fields. The large scale magnetic field of the Sun is an alternating field with a period of about 22 years, twice the period of the solar activity cycle for which it is responsible. The characteristic length scale, $\bar{\lambda}$, of this field is the skin depth, which must be at least of the order of the diameters of the largest sunspots, i.e. some ten thousands of kilometers. Taking $\bar{\lambda} \approx 50 \cdot 10^3$ km we obtain for the time scale, $\bar{\tau}$, if using the turbulent conductivity, $\tau = \mu\sigma_T\bar{\lambda}^2 \approx 30$ years, which is just the correct order of magnitude. Using the molecular conductivity this time scale will be, obviously, too large by a factor 10^4. Thus we find a good quantitative agreement.

One is tempted to apply (3.37) to the decay of sunspots and will find again a good quantitative agreement. In sunspots the convective motion must, however, be expected to be strongly influenced by the magnetic field, otherwise their darkness can hardly be understood. This influence was neglected in the derivation of (3.37) so that we have to discuss the decay of sunspots later.

3.9. The low-conductivity limit

Now we are concerned with equation (3.11). We consider the whole space, but assume at first that $\boldsymbol{u}'$ is different from zero in a bounded region only, for this reason $\boldsymbol{B}'$ tends to zero if the argument tends to infinity. The solution of (3.11) is then given by

$$\boldsymbol{B}'(\boldsymbol{x}, t) = \frac{\mu\sigma}{4\pi} \int \frac{\operatorname{curl}' \left(\boldsymbol{u}'(\boldsymbol{x}', t) \times \overline{\boldsymbol{B}}(\boldsymbol{x}', t)\right)}{|\boldsymbol{x}' - \boldsymbol{x}|} \, \mathrm{d}\boldsymbol{x}'. \tag{3.38}$$

We multiply this expression by $\boldsymbol{u}'(\boldsymbol{x}, t)$ and average. After inserting $\boldsymbol{\xi} = \boldsymbol{x}' - \boldsymbol{x}$ one finds for the turbulent electromotive force

$$\mathfrak{E} = \frac{\mu\sigma}{4\pi} \int \overline{\boldsymbol{u}'(\boldsymbol{x}, t) \times \operatorname{curl} \left(\boldsymbol{u}'(\boldsymbol{x} + \boldsymbol{\xi}, t) \times \overline{\boldsymbol{B}}(\boldsymbol{x} + \boldsymbol{\xi}, t)\right)} \, \frac{\mathrm{d}\boldsymbol{\xi}}{\xi}. \tag{3.39}$$

Now we can withdraw the assumption made above that $\boldsymbol{u}'$ has to be different from zero in a bounded region only, since $\overline{\boldsymbol{u}'(\boldsymbol{x}, t) \cdot \boldsymbol{u}'(\boldsymbol{x} + \boldsymbol{\xi}, t)} \to 0$ if $\boldsymbol{\xi} \to \infty$.

Noting that $\overline{\boldsymbol{B}}$ varies weakly over distances of the order of λ_{cor} we insert

$$\overline{\boldsymbol{B}}(\boldsymbol{x} + \boldsymbol{\xi}, t) = \overline{\boldsymbol{B}}(\boldsymbol{x}, t) + (\boldsymbol{\xi} \cdot \operatorname{grad}_x) \, \overline{\boldsymbol{B}}(\boldsymbol{x}, t) \tag{3.40}$$

into (3.39) and, by carrying out the vector operations as in the previous section, we find

$$\begin{aligned}
\mathfrak{E}_1 &= -(w_{312} - w_{213}) \, \overline{B}_1 \\
&\quad - (w_{22} + w_{3223} + w_{2211} + w_{2222}) \frac{\partial \overline{B}_3}{\partial x_2} + (w_{33} + w_{2332} + w_{3311} + w_{3333}) \frac{\partial \overline{B}_2}{\partial x_3}, \\
\mathfrak{E}_2 &= -(w_{123} - w_{321}) \, \overline{B}_2 \\
&\quad - (w_{33} + w_{1331} + w_{3322} + w_{3333}) \frac{\partial \overline{B}_1}{\partial x_3} + (w_{11} + w_{3113} + w_{1122} + w_{1111}) \frac{\partial \overline{B}_1}{\partial x_1}, \\
\mathfrak{E}_3 &= -(w_{231} - w_{132}) \, \overline{B}_3 \\
&\quad - (w_{11} + w_{2112} + w_{1133} + w_{1111}) \frac{\partial \overline{B}_2}{\partial x_1} + (w_{22} + w_{1221} + w_{2233} + w_{2222}) \frac{\partial \overline{B}_1}{\partial x_2},
\end{aligned} \tag{3.41}$$

where the quantities w_{ij}, w_{ijk}, w_{ijkl} are defined by

$$w_{ij} = \frac{\mu\sigma}{4\pi} \int \overline{u_i'(\boldsymbol{x}, t) \, u_j'(\boldsymbol{x} + \boldsymbol{\xi}, t)} \, \frac{\mathrm{d}\boldsymbol{\xi}}{\xi}, \tag{3.42}$$

$$w_{ijk} = \frac{\mu\sigma}{4\pi} \int \overline{u_i'(\boldsymbol{x}, t) \, \frac{\partial u_k'(\boldsymbol{x} + \boldsymbol{\xi}, t)}{\partial \xi_j}} \, \frac{\mathrm{d}\boldsymbol{\xi}}{\xi}, \tag{3.43}$$

$$w_{ijkl} = \frac{\mu\sigma}{4\pi} \int \overline{u_i'(\boldsymbol{x}, t) \, \xi_j \, \frac{\partial u_k'(\boldsymbol{x} + \boldsymbol{\xi}, t)}{\partial \xi_l}} \, \frac{\mathrm{d}\boldsymbol{\xi}}{\xi}. \tag{3.44}$$

A similarity between (3.41) and the analogous relation in the high-conductivity limit, (3.24), only exists for the terms proportional to $\overline{\boldsymbol{B}}$; for those with partial derivatives the more complicated additional quantities w_{ijkl} appear.

By comparing (3.41) with (3.5) we obtain

$$\alpha = -(w_{312} - w_{213}) = -(w_{123} - w_{321}) = -(w_{231} - w_{132})$$
$$= -\frac{1}{3}(w_{123} - w_{132} + w_{231} - w_{213} + w_{312} - w_{321}), \tag{3.45}$$

and, furthermore,

$$\beta = (w_{22} + w_{3223} + w_{2211} + w_{2222}) = (w_{33} + w_{2332} + w_{3311} + w_{3333})$$
$$= (w_{33} + w_{1331} + w_{3322} + w_{3333}) = (w_{11} + w_{3113} + w_{1122} + w_{1111})$$
$$= (w_{11} + w_{2112} + w_{1133} + w_{1111}) = (w_{33} + w_{1221} + w_{2211} + w_{2222})$$
$$= \frac{1}{3}(w_{11} + w_{22} + w_{33} + w_{1111} + w_{1133} + w_{2211} + w_{2222} \tag{3.46}$$
$$+ w_{1331} + w_{2112} + w_{3322} + w_{3223} + w_{3333}).$$

Hence we find

$$\alpha = -\frac{\mu\sigma}{12\pi}\int \overline{\boldsymbol{u}'(\boldsymbol{x}, t) \cdot \operatorname{curl} \boldsymbol{u}'(\boldsymbol{x} + \boldsymbol{\xi}, t)}\,\frac{d\boldsymbol{\xi}}{\xi}, \tag{3.47}$$

and

$$\beta = \frac{\mu\sigma}{12\pi}\int \overline{(\boldsymbol{\xi} \cdot \boldsymbol{u}'(\boldsymbol{x}, t))\,(\boldsymbol{\xi} \cdot \boldsymbol{u}'(\boldsymbol{x} + \boldsymbol{\xi}, t)}\,\frac{d\boldsymbol{\xi}}{\xi^3}. \tag{3.48}$$

An alternative derivation of relations (3.47) and (3.48) for α and β can be given with the help of a representation of the turbulent velocity field by vector potentials. If we define the fields $\boldsymbol{a}_1$, $\boldsymbol{a}_2$, φ by

$$\boldsymbol{u}' = \operatorname{curl} \boldsymbol{a}_1 - \nabla\varphi, \quad \operatorname{div} \boldsymbol{a}_1 = 0, \tag{3.49}$$

and

$$\boldsymbol{a}_1 = \operatorname{curl} \boldsymbol{a}_2, \quad \varphi = \operatorname{div} \boldsymbol{a}_2, \tag{3.50}$$

we have

$$\boldsymbol{u}' = -\Delta \boldsymbol{a}_2. \tag{3.51}$$

Restricting our attention firstly to a derivation of the quantity α we regard $\overline{\boldsymbol{B}}$ as constant, thus having

$$\operatorname{curl}(\Delta \boldsymbol{a}_2 \times \overline{\boldsymbol{B}}) = \operatorname{curl} \Delta(\boldsymbol{a}_2 \times \overline{\boldsymbol{B}}) = \Delta \operatorname{curl}(\boldsymbol{a}_2 \times \overline{\boldsymbol{B}}). \tag{4.52}$$

For equations (3.11) we can now write

$$\Delta \boldsymbol{B}' = \mu\sigma \Delta \operatorname{curl}(\boldsymbol{a}_2 \times \overline{\boldsymbol{B}}), \quad \operatorname{div} \boldsymbol{B}' = 0, \tag{3.53}$$

and solve them by the relation

$$\boldsymbol{B}' = \mu\sigma \operatorname{curl} (\boldsymbol{a}_2 \times \overline{\boldsymbol{B}}) + \boldsymbol{c}, \tag{3.54}$$

where $\boldsymbol{c}$ is constant with zero mean, if the same is true for the potential $\boldsymbol{a}_2$. Hence, we can put $\boldsymbol{c} = 0$. Multiplying (3.54) by the turbulent velocity field $\boldsymbol{u}'$ we obtain

$$\mathfrak{E} = \mu\sigma \overline{\boldsymbol{u}' \times \operatorname{curl} (\boldsymbol{a}_2 \times \overline{\boldsymbol{B}})}, \tag{3.55}$$

and with the arguments leading to (3.29) we find

$$\alpha = -\frac{\mu\sigma}{3} \overline{\boldsymbol{u}' \cdot \operatorname{curl} \boldsymbol{a}_2}, \tag{3.56}$$

or, taking (3.49) and (3.50) into account,

$$\alpha = -\frac{\mu\sigma}{3} (\overline{\boldsymbol{a}_1 \cdot \operatorname{curl} \boldsymbol{a}_1} - \overline{\boldsymbol{a}_1 \cdot \nabla\varphi}). \tag{3.57}$$

The latter summand vanishes, since it is equal to div $(\overline{\boldsymbol{a}_1\varphi})$ because of (3.49), and the quantity $\overline{\boldsymbol{a}_1\varphi}$ is constant since $\boldsymbol{u}'$ is assumed to be homogeneous. Thus we finally find

$$\alpha = -\frac{\mu\sigma}{3} \overline{\boldsymbol{a}_1 \cdot \operatorname{curl} \boldsymbol{a}_1}. \tag{3.58}$$

The equivalence of the formulae just derived with (3.47) is easily seen if $\boldsymbol{a}_2$ as the solution of the Poisson equation (3.51) is represented by the well-known integral expression and then inserted in (3.56).

A determination of β in the same way is possible if $\overline{\boldsymbol{B}}$ is assumed to depend on position. One will find then

$$\beta = \frac{\mu\sigma}{3} (\overline{\boldsymbol{a}_1^2} - \overline{\varphi^2}). \tag{3.59}$$

It is notable that

$$\beta = -\frac{\mu\sigma}{3} \overline{\varphi^2} < 0, \tag{3.60}$$

if

$$\boldsymbol{u}' = -\operatorname{grad} \varphi, \tag{3.61}$$

i.e. if $\boldsymbol{u}'$ is a potential random field (Krause and Roberts [13]). It can easily be deduced from (3.48). If we substitute (3.61) we have

$$\beta = \frac{\mu\sigma}{12\pi} \int \xi_i \overline{\frac{\partial\varphi(\boldsymbol{x}, t)}{\partial x_i} \xi_j \frac{\partial\varphi(\boldsymbol{x}+\boldsymbol{\xi}, t)}{\partial \xi_j}} \frac{d\boldsymbol{\xi}}{\xi^3}$$

$$= \frac{\mu\sigma}{12\pi} \int \xi_i\xi_j \left[\frac{\partial}{\partial x_i}\left(\overline{\varphi(\boldsymbol{x}, t)\frac{\partial\varphi(\boldsymbol{x}+\boldsymbol{\xi}, t)}{\partial \xi_j}}\right) - \overline{\frac{\partial^2\varphi(\boldsymbol{x}, t)\,\varphi(\boldsymbol{x}+\boldsymbol{\xi}, t)}{\partial\xi_i\,\partial\xi_j}}\right] \frac{d\boldsymbol{\xi}}{\xi^3}.$$

Here, and in the following, the summation convention is adopted. The first summand in the brackets vanishes since $\boldsymbol{u}'$, and thus φ too, is homogeneous. Introducing the correlation function,

$$\Phi(\boldsymbol{\xi}, \tau) = \overline{\varphi(\boldsymbol{x}, t)\,\varphi(\boldsymbol{x}+\boldsymbol{\xi}, t+\tau)}, \tag{3.62}$$

which is a function of $\xi = \sqrt{\boldsymbol{\xi}^2}$ only, we obtain

$$\beta = -\frac{\mu\sigma}{12\pi}\int \frac{\xi_i\xi_j}{\xi^3}\frac{\partial^2\Phi(\xi, 0)}{\partial\xi_i\,\partial\xi_j}\,d\boldsymbol{\xi} = -\frac{\mu\sigma}{12\pi}\int \frac{1}{\xi}\frac{\partial^2\Phi}{\partial\xi^2}\,d\boldsymbol{\xi}$$

$$= -\frac{\mu\sigma}{3}\int_0^\infty \xi\frac{\partial^2\Phi}{\partial\xi^2}\,d\xi = -\frac{\mu\sigma}{3}\Phi(0, 0) = -\frac{\mu\sigma}{3}\overline{\varphi'^2}. \tag{3.63}$$

From (3.7) we find with (3.60) in this case

$$\sigma_{\mathrm{T}} = \frac{\sigma}{1 - \frac{1}{3}\overline{(\mu\sigma\varphi)^2}}, \tag{3.64}$$

i.e. the conductivity with respect to the mean fields is larger than the molecular conductivity. A realization of such a turbulence is provided by a random superposition of sound waves.

It is worth noting that

$$\overline{(\mu\sigma\varphi)^2} \approx (\mu\sigma u'\lambda_{\mathrm{cor}})^2 = \mathrm{Rm}^2 \ll 1 \tag{3.65}$$

according to (3.12), (3.17) and (3.18). For this reason no singularity of σ_T is indicated by the result (3.64).

For reasons of the results just deduced we restrict our attention to incompressible turbulence in estimating the order of magnitude of β from (3.48). Since $\overline{(\boldsymbol{\xi u}')\,(\boldsymbol{\xi u}')} \approx \frac{1}{3}\xi^2 u'^2$ we find

$$\beta = \frac{1}{9}\mu\sigma u'^2\lambda_{\mathrm{cor}}^2, \tag{3.66}$$

and

$$\sigma_{\mathrm{T}} = \frac{\sigma}{1 + \frac{1}{9} (\mu\sigma u' \lambda_{\mathrm{cor}})^2}, \tag{3.67}$$

a result which was first published by STEENBECK [2] in 1963.

3.10. Illustration of the α-effect and the α-experiment

The turbulent electromotive force $\alpha\overline{\boldsymbol{B}}$ which is proportional to the mean magnetic field clearly represents an effect of quite a new quality. For this reason it is advisable to illuminate the physical background by illustrating examples. Moreover, there is an experimental verification which inspires reliable confidence in this effect.

For a first illustration we introduce a cartesian coordinate system (fig. 3.1) where the x-axis and the z-axis are lying in the plane of the drawing, and the y-direction perpendicular to it. Let $\overline{\boldsymbol{B}}$ be a mean magnetic field parallel to the

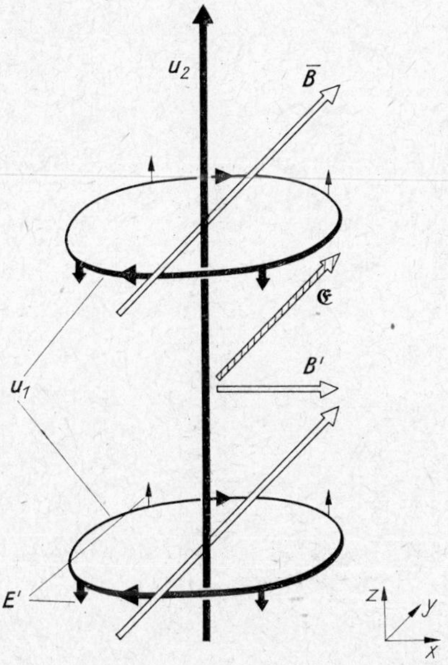

Fig. 3.1. Schematic drawing which shows that a helical motion, $\boldsymbol{u} = \boldsymbol{u}_1 + \boldsymbol{u}_2$, in a homogeneous magnetic field, $\overline{\boldsymbol{B}}$, provides for an electromotive force, $\mathfrak{E}$, parallel to the magnetic field (α-effect)

y-direction and $\boldsymbol{u}'$ a velocity corresponding to a left-handed helical motion. We represent $\boldsymbol{u}'$ by the sum $\boldsymbol{u}' = \boldsymbol{u}_1 + \boldsymbol{u}_2$, where $\boldsymbol{u}_1$ is a rotational motion about the z-axis and $\boldsymbol{u}_2$ is the motion parallel to the z-direction. The interaction of the rotational motion $\boldsymbol{u}_1$ with the mean magnetic field provides for an electrical field $\boldsymbol{E}' = \boldsymbol{u}_1 \times \overline{\boldsymbol{B}}$, which is directed parallel to the z-direction, downwards in front of the plane of the drawing and upwards behind it. $\boldsymbol{E}'$ drives a current $\boldsymbol{j}'$, in the same manner directed downwards in front of the plane of the drawing and upwards behind it. This current is combined with a magnetic field $\boldsymbol{B}'$ in the positive x-direction. The crossproduct of $\boldsymbol{u}_2$ with this magnetic field $\boldsymbol{B}'$ is directed in the positive y-direction, i.e. parallel to the applied mean magnetic field $\overline{\boldsymbol{B}}$.

This statement can also be established if the interaction of $\boldsymbol{u}_2$ with the mean magnetic field is considered first. It is also true if the axis of the helical motion is turned within the (x, z)-plane. A helical motion with arbitrary direction of its axis can be considered as the sum of two helical motions, one with its axis within the (x, z)-plane and one with its axis parallel to the mean magnetic field. All helical motions of the first type provide for an electromotive force $\boldsymbol{u}' \times \boldsymbol{B}'$ parallel to the mean magnetic field. In this way we clearly see that randomly distributed left-handed helical motions produce a mean electromotive force which is directed parallel to the mean magnetic field. This electromotive force will be antiparallel to the mean magnetic field for randomly distributed right-handed helical motions.

Having just illustrated the α-effect in the low-conductivity limit we can also do so in the high-conductivity limit. Under this condition a flux rope of the mean magnetic field is deformed to a twisted Ω by a local helical motion (fig. 3.2). The current $\boldsymbol{j}'$ flowing through the loop of the Ω has, because of the twist, a component which is antiparallel to the original magnetic field for a right-handed helical motion as depicted here. A left-handed helical motion gives rise to a current parallel to the mean magnetic field. So we again find

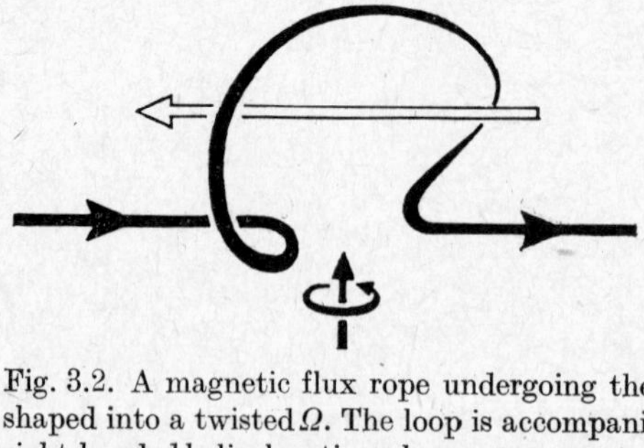

Fig. 3.2. A magnetic flux rope undergoing the influence of a helical motion is shaped into a twisted Ω. The loop is accompanied by a current which, in case of right-handed helical motions, has a component anti-parallel to the magnetic field

that such local helical motions randomly appearing in a conducting fluid have a common property, i.e. to drive a current parallel or antiparallel to an applied mean magnetic field.

This illustration very much resembles that given by PARKER [2] for the effect of the so-called cyclonic turbulence. We shall see later that Parker's effect can be, indeed, interpreted as a special case of the α-effect.

Let us now consider (fig. 3.3) a box in which we assume an electrically conducting fluid carrying out turbulent motions. The turbulence should be homogeneous and isotropic but have helicity, i.e. one kind of helical motions shall be more probable than the other one. Let us assume we have a higher probability of left-handed helical motions. Let a magnetic field $\bar{\boldsymbol{B}}$ be directed from the

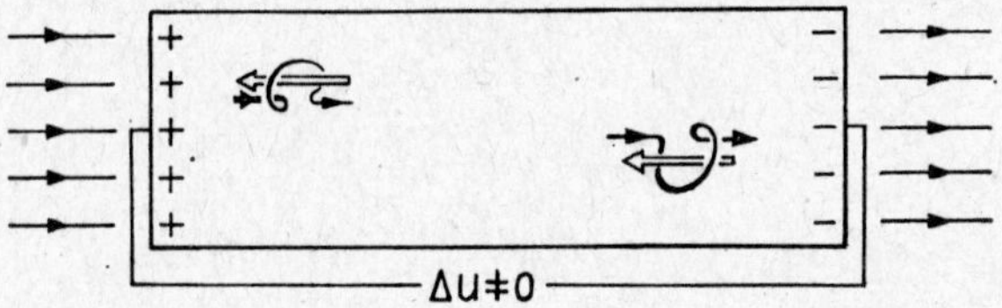

Fig. 3.3. Small-scale helical motions in a box provide a voltage along a magnetic field

left to the right. With these conditions the α-effect must be expected and, consequently, an electrical field parallel to the magnetic field. It provides for positive electrical charges at the right wall of the box and for negative at the left. From the outside we can measure a voltage parallel to the applied magnetic field.

Experimental evidence for the existence of the α-effect was given by an experiment (STEENBECK, KIRKO, GAILITIS, KLAWINA, KRAUSE, LAUMANIS and LIELAUSIS [1, 2]) which is depicted in fig. 3.4. Liquid sodium was used as the working fluid, its flow was compelled to be non-mirrorsymmetric. A system of copper plates created two channels which wound about each other orthogonally (cf. fig. 3.4a). In this way, the direction of flow was twisted in steps about the direction of the magnetic field, $\bar{\boldsymbol{B}}$; this type of motion is clearly not mirrorsymmetric. The conditions drafted in fig. 3.4 give rise to right-handed helical motions. The liquid sodium flowed through a non-corroding steel box. Figure 3.5 is a photograph of this box. A section of the outer wall being removed we are able to see the copper walls inside the box, which force the motion to take the desired helical structure.

An estimation of the electromotive force $\mathfrak{E}$ caused by this motion shall be provided by considering a simplified model. Let $u(x)$, $0 \leqq x \leqq l$, be the velocity

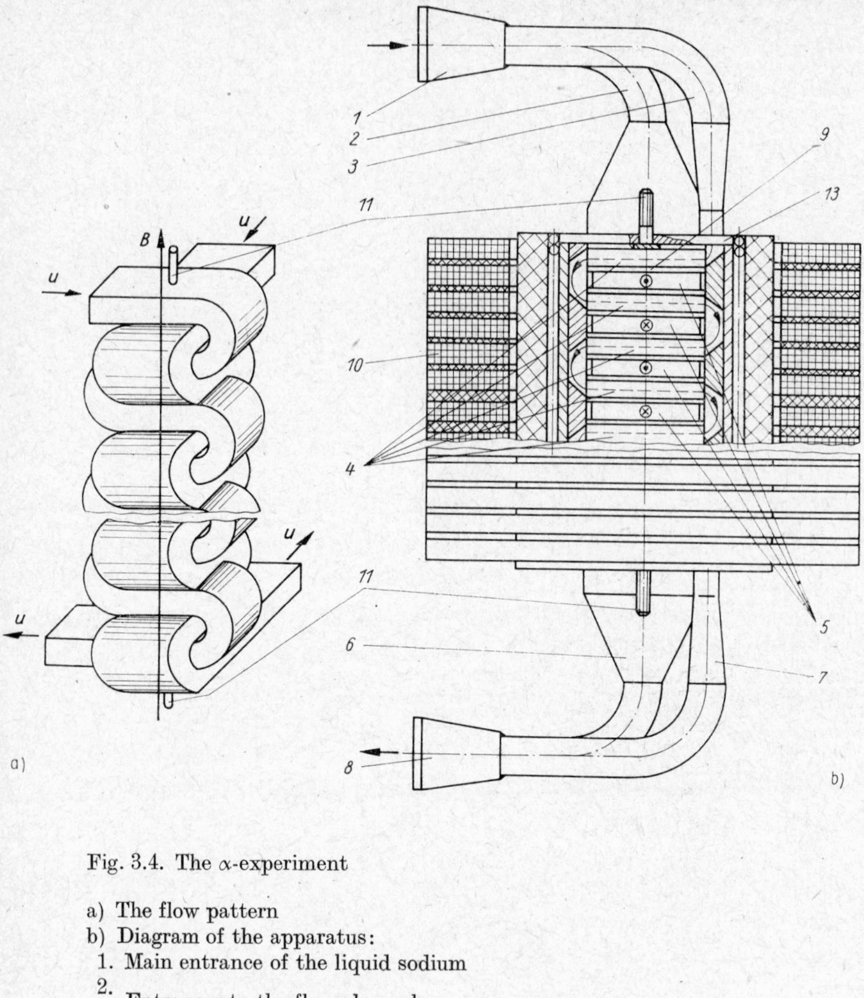

Fig. 3.4. The α-experiment

a) The flow pattern
b) Diagram of the apparatus:
 1. Main entrance of the liquid sodium
 2. 3. Entrances to the flow channels
 4. Flow, in the plane of the paper
 5. Flow, perpendicular to the plane of the paper ($\odot$ out and $\otimes$ in)
 6. 7. Exits from the two flow channels
 8. Main exit of the liquid sodium
 9. Copper plates separating the channels
10. The coil generating the applied magnetic field
11. 12. Electrodes between which the voltage created by the α-effect was measured
13. Walls of non-corroding steel

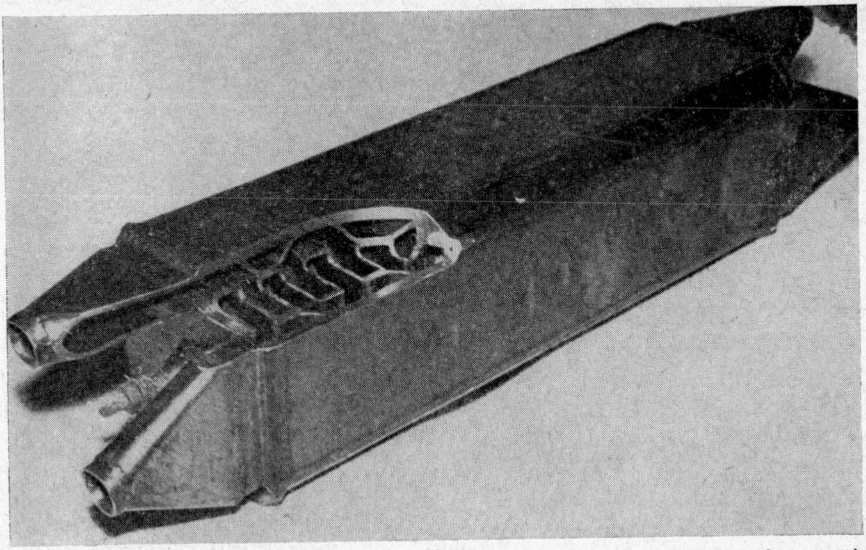

Fig. 3.5. Photograph of the "α-yashchik", i.e., of the box which is the essential piece of the α-experiment. A section of the outer wall is removed and we can see the copper walls inside the box, which force the motion of liquid sodium to take the helical structure

profile in one channel, and

$$u' = \frac{1}{l} \int_0^l u(x)\, \mathrm{d}x\,. \tag{3.68}$$

We define the function $\varphi(x)$ to be periodic with the period $4l$ and

$$\varphi(x) = \begin{cases} \dfrac{u(x)}{u'}, & \text{if } 0 \leqq x \leqq l, \\ 0, & \text{if } l \leqq x \leqq 2l, \\ -\dfrac{u(x-2l)}{u'}, & \text{if } 2l \leqq x \leqq 3l, \\ 0, & \text{if } 3l \leqq x \leqq 4l. \end{cases} \tag{3.69}$$

By definition we have

$$\int_0^l \varphi(x)\, \mathrm{d}x = 1\,. \tag{3.70}$$

The velocity field of the experiment we represent by

$$\boldsymbol{u}' = u'(0, \varphi(x), \pm\varphi(x-l))\,. \tag{3.71}$$

The positive sign of the z-component clearly provides for a left-handed helical motion if $\varphi(x) \geqq 0$ for $0 \leqq x \leqq l$ (cf. fig. 3.6). Function f defined by

$$f(x) = \int_0^x \varphi(\xi)\,\mathrm{d}\xi - \frac{1}{2}, \tag{3.72}$$

we have

$$\int_0^{4l} f(x)\,\mathrm{d}x = 0, \tag{3.73}$$

i.e. the average of f taken over the interval $0 \leqq x \leqq 4l$ is zero.

The fluctuating magnetic field shall be derived from equation (3.11). With (3.71), (3.72), (3.73) and on the assumption $\overline{\boldsymbol{B}} = (\overline{B}, 0, 0)$, $\overline{B} =$ const we find

$$\begin{aligned} B'_x &= 0, \\ B'_y &= -\mu\sigma u'\overline{B}f(x), \\ B'_z &= \mp\mu\sigma u'\overline{B}f(x-l). \end{aligned} \tag{3.74}$$

Hence we obtain for the cross product

$$(\boldsymbol{u}'\times\boldsymbol{B}')_x = \pm\mu\sigma u'^2\overline{B}\{f(x)\,\varphi(x-l) - \varphi(x)\,f(x-l)\}. \tag{3.75}$$

In figure 3.6 this electromotive force is depicted where the positive sign is taken for the z-component in (3.71). It has a unique direction, parallel to $\overline{\boldsymbol{B}}$ for the positive sign in (3.71) and antiparallel to $\overline{\boldsymbol{B}}$ for the negative one. From (3.75) we obtain for the voltage over one channel

$$\varDelta U = \int_0^l (\boldsymbol{u}'\times\boldsymbol{B}')_x\,\mathrm{d}x = \pm\frac{1}{2}\mu\sigma u'^2l^2\overline{B}, \tag{3.76}$$

and for the voltage over the whole box with n channels

$$U = \pm\frac{1}{2}\mu\sigma u'^2l^2\overline{B}n. \tag{3.77}$$

The two signs reflect that we can have either left-handed or right-handed helical motions.

The result of the experiment is best seen in figure 3.7, where the ratio of the measured voltage U_m and the theoretical value U given in (3.77) is plotted as a function of the Stuart number

$$N = \frac{\overline{B}^2\sigma l}{\varrho u'}, \tag{3.78}$$

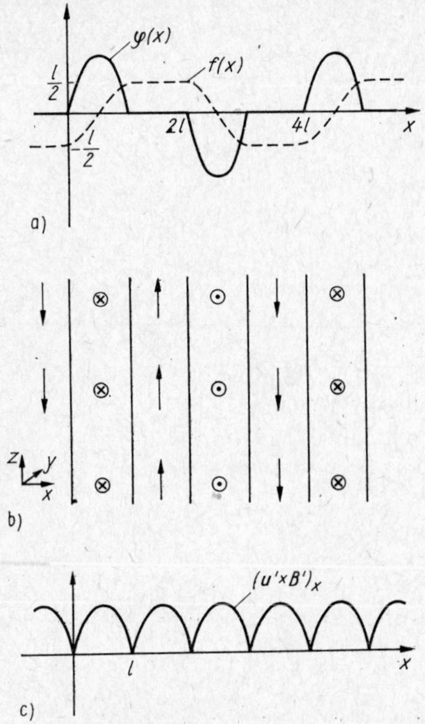

Fig. 3.6. Simplified model of the α-experiment
a) Representation of the profile of the y-component of the velocity field in dependence on x
b) The flow pattern
c) Profile of the x-component of the induced electromotive force

ϱ denoting the mass density. We observe that the two are related in a unique way. For small N, when the fluid velocity is large and the magnetic field is not, the measured voltage U_m was about 20% of that given in (3.77). This is the result of boundary effects, because we assumed the model to have infinite extension in the y- and z-directions. It was also found that, if the Stuart number is increased, the voltage decreases. This is clearly due to the influence of the magnetic field on the velocity field. It is of interest that the ratio U_m/U was found to be independent of the magnetic Reynolds number $\mathrm{Rm} = \mu\sigma l u'$.
Two predictions of (3.77), crucial to the validity of the theory, were verified by the experiment; (i) the sign of the voltage difference U_m between the electrodes is independent of the sense of the flow; (ii) it reverses if the applied magnetic field is reversed. Figure 3.8 shows that, for several values of the magnetic

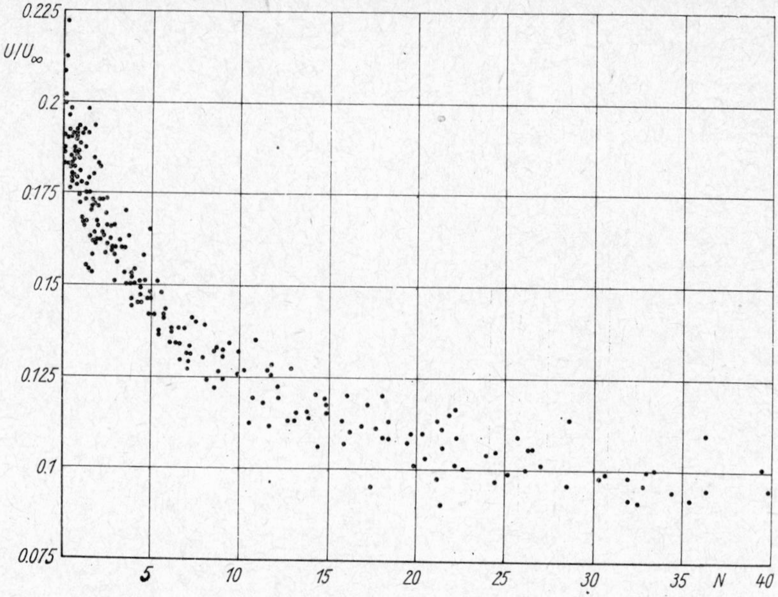

Fig. 3.7. Representation of the voltage measured along the magnetic field in dependence on the Stuart number

field, U_m is indeed proportional to u'^2. Figure 3.9 demonstrates that, when the magnetic field is small, U_m is proportional to it. For large fields, however, the effect is nonlinear in the $\overline{\boldsymbol{B}}$ applied.

It is also easy to illustrate that the α-effect is closely connected with dynamo excitation, as was mentioned above. Let us connect the two ends of the box in figure 3.3 by a wire which is wound around the box in such a way, that it produces the applied $\overline{\boldsymbol{B}}$. Then, for a sufficiently strong α-effect, this device will act as a self-excited dynamo. Another illustration is depicted in figure 3.10. The two rings shall contain an electrically conducting medium with α-effect. Let us assume a magnetic field $\boldsymbol{B}_0$ in ring (I). It drives a current $\boldsymbol{j}_1$, because of the α-effect. This current $\boldsymbol{j}_1$ is combined with a magnetic field $\boldsymbol{B}_1$, in ring (II). $\boldsymbol{B}_1$ drives a current $\boldsymbol{j}_2$ which is combined with a magnetic field $\boldsymbol{B}_2$ in ring (I). $\boldsymbol{B}_2$ clearly supports $\boldsymbol{B}_0$. Consequently, we can expect that the fields are maintained for sufficiently strong α-effect in spite of Ohmic losses. The energy of the Joule heat produced by the currents comes from the energy of the turbulent motion which has helicity.

We finally mention that a more realistic model of that kind was described in section 1.2.

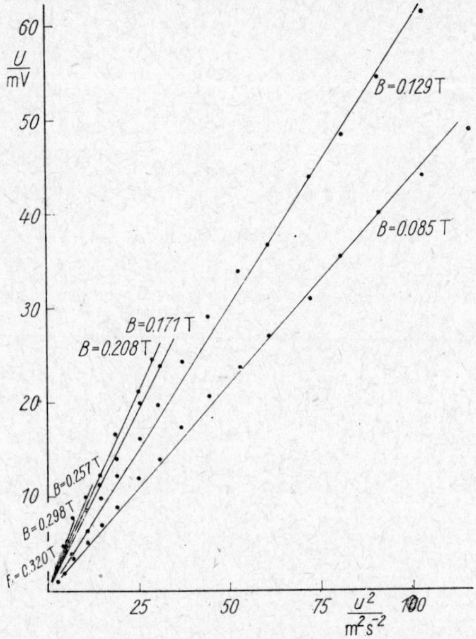

Fig. 3.8. The voltage created by the α-effect as a function of the square of the velocity, for a number of different applied fields

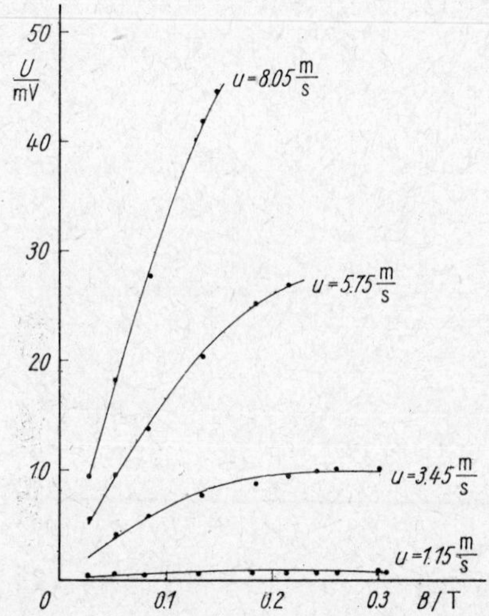

Fig. 3.9. The voltage created by the α-effect as a function of the applied magnetic field strength, for a number of different flow velocities

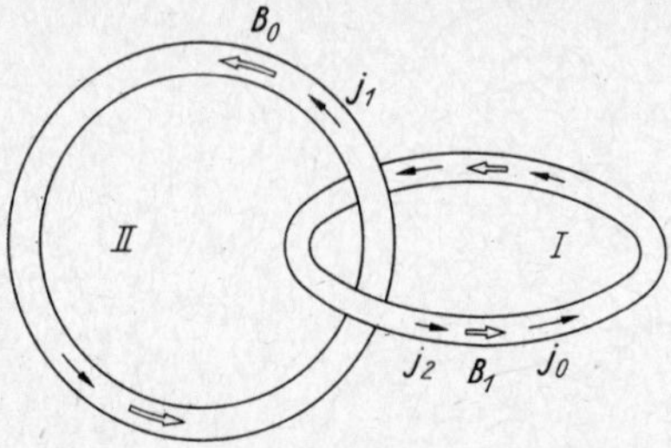

Fig. 3.10. Model illustrating the self-excited generation of a magnetic field in a medium with Ohm's law $\boldsymbol{j} = \sigma(\boldsymbol{E} + \alpha\boldsymbol{B})$

3.11. The mean square of the fluctuations

For a treatment of the high-conductivity limit it is necessary to describe the physical background in more detail. We adopt the standpoint that $\boldsymbol{B}'$ is produced by the interaction of the velocity field $\boldsymbol{u}'$ with the mean magnetic field $\overline{\boldsymbol{B}}$. Within the framework of the second order correlation approximation this production works over a time of order τ_{cor}, producing a field element $\boldsymbol{B}'_{\text{el}}$ of order

$$B'_{\text{el}} \approx \frac{u'\tau_{\text{cor}}}{\lambda_{\text{cor}}} \overline{B} \tag{3.79}$$

according to equation (3.10). After that the velocity field $\boldsymbol{u}'$ changes nearly completely and by interaction with $\overline{\boldsymbol{B}}$ a new field element $\boldsymbol{B}'_{\text{el}}$ is produced, being uncorrelated to the former, and so on. However, such a field element, once produced, will exist over the time $\mu\sigma\lambda^2_{\text{cor}}$, which is large compared with τ_{cor} according to (3.16). Hence the really existing field $\boldsymbol{B}'$ is the superposition of such uncorrelated field elements $\boldsymbol{B}'_{\text{el}}$, the number of them, n, will approximately be given by

$$n \approx \frac{\mu\sigma\lambda^2_{\text{cor}}}{\tau_{\text{cor}}}. \tag{3.80}$$

Since $\boldsymbol{B}'$ is the statistical sum of the uncorrelated field elements $\boldsymbol{B}'_{\text{el}}$, its order of magnitude is given by the square root of n times the order of one element, hence according to (3.79) and (3.80)

$$B' \approx \sqrt{n}\, B'_{\text{el}} \approx \sqrt{\mu\sigma u'^2\tau_{\text{cor}}}\, \overline{B} \tag{3.81}$$

or

$$\overline{\boldsymbol{B}'^2} \approx \mu\sigma u'^2\tau_{\text{cor}}\overline{\boldsymbol{B}}^2. \tag{3.82}$$

This derivation is due to STEENBECK and KRAUSE [6]. A similar result was derived by PARKER [14],

$$\overline{B'^2} \approx \mathrm{Rm}\, \overline{\boldsymbol{B}}^2, \tag{3.83}$$

which is in agreement with (3.82) if $S \approx 1$ as assumed by PARKER. Finally, a result due to MOSS [1] may be added. Carrying out numerical experiments with two-dimensional turbulence MOSS found

$$\overline{B'^2} \approx \mathrm{Rm}^{0.7}\, \overline{\boldsymbol{B}}^2. \tag{3.84}$$

A direct comparison with (3.82) is not possible since the range of the parameter S considered by MOSS is different.

We can derive the estimate (3.82) directly from (3.21) if $t - t_0$ is taken as the lifetime of one field element, i.e. $t = t_0 - \mu\sigma\lambda_{\mathrm{cor}}^2$. We obtain

$$\overline{B'^2} = \int\limits_{t-\mu\sigma\lambda_{\mathrm{cor}}^2}^{t} \int\limits_{t-\mu\sigma\lambda_{\mathrm{cor}}^2}^{t} \overline{\mathrm{curl}\,(\boldsymbol{u}'(\boldsymbol{x}, t') \times \overline{\boldsymbol{B}}) \cdot \mathrm{curl}\,(\boldsymbol{u}'(\boldsymbol{x}, t'') \times \overline{\boldsymbol{B}})}\, \mathrm{d}t'\, \mathrm{d}t''. \tag{3.85}$$

The integrand is significantly different from zero only in a strip of width τ_{cor} along the line $t' = t''$ in the (t', t'')-plane. Thus we find in the isotropic case

$$\overline{B'^2} \approx (\mu\sigma\lambda_{\mathrm{cor}}^2\tau_{\mathrm{cor}}) \left(\frac{\overline{u'^2}}{\lambda_{\mathrm{cor}}^2}\, \overline{\boldsymbol{B}}^2\right)$$

in agreement with (3.82).

In our derivation of the latter results we have used rather qualitative considerations. It is, therefore, important to note that the results (3.82), and the analogous one for the low-conductivity case are confirmed by a rigorous deduction in chapter 7.

In the low-conductivity limit the estimate of $\overline{B'^2}$ analogue to (3.82) is immediately derived from (3.54) if we are confined to the incompressible case. We obtain from (3.54)

$$\overline{B'^2} = \mu^2\sigma^2 \overline{(\overline{\boldsymbol{B}} \cdot \mathrm{grad})\, \boldsymbol{a}_2 \cdot (\overline{\boldsymbol{B}} \cdot \mathrm{grad})\, \boldsymbol{a}_2}$$

$$= \mu^2\sigma^2 \left\{\overline{B}_1^2 \left[\overline{\left(\frac{\partial a_{21}}{\partial x_1}\right)^2} + \overline{\left(\frac{\partial a_{22}}{\partial x_1}\right)^2} + \overline{\left(\frac{\partial a_{23}}{\partial x_1}\right)^2}\right] + \overline{B}_2^2 \left[\overline{\left(\frac{\partial a_{21}}{\partial x_2}\right)^2} + \dots\right] + \dots\right\},$$

which simplifies in the isotropic case to

$$\overline{B'^2} = q\overline{\boldsymbol{B}}^2, \tag{3.86}$$

where

$$\frac{q}{\mu^2\sigma^2} = \overline{\left(\frac{\partial a_{21}}{\partial x_1}\right)^2} + \overline{\left(\frac{\partial a_{22}}{\partial x_1}\right)^2} + \overline{\left(\frac{\partial a_{23}}{\partial x_1}\right)^2} = \overline{\left(\frac{\partial a_{21}}{\partial x_2}\right)^2} + \overline{\left(\frac{\partial a_{22}}{\partial x_2}\right)^2} + \overline{\left(\frac{\partial a_{23}}{\partial x_2}\right)^2}$$

$$= \overline{\left(\frac{\partial a_{21}}{\partial x_3}\right)^2} + \overline{\left(\frac{\partial a_{22}}{\partial x_3}\right)^2} + \overline{\left(\frac{\partial a_{23}}{\partial x_3}\right)^2}$$

$$= \frac{1}{3}\left\{\overline{\left(\frac{\partial a_{21}}{\partial x_1}\right)^2} + \overline{\left(\frac{\partial a_{22}}{\partial x_1}\right)^2} + \overline{\left(\frac{\partial a_{23}}{\partial x_1}\right)^2} + \overline{\left(\frac{\partial a_{21}}{\partial x_2}\right)^2} + \overline{\left(\frac{\partial a_{22}}{\partial x_2}\right)^2}\right.$$

$$\left. + \overline{\left(\frac{\partial a_{23}}{\partial x_2}\right)^2} + \overline{\left(\frac{\partial a_{21}}{\partial x_3}\right)^2} + \overline{\left(\frac{\partial a_{22}}{\partial x_3}\right)^2} + \overline{\left(\frac{\partial a_{23}}{\partial x_3}\right)^2}\right\}. \tag{3.87}$$

Hence we find as an estimate of the order of magnitude

$$\frac{q}{\mu^2\sigma^2} = \frac{1}{3}\,3\,\frac{\overline{\boldsymbol{a}_2^2}}{\lambda_{\rm cor}^2} = \frac{\overline{\boldsymbol{a}_2^2}}{\lambda_{\rm cor}^2} \approx u'^2\lambda_{\rm cor}^2, \tag{3.88}$$

and obtain finally in the low-conductivity limit

$$\overline{\boldsymbol{B}'^2} \approx \mathrm{Rm}^2\overline{\boldsymbol{B}}^2. \tag{3.89}$$

CHAPTER 4

GENERAL METHODS FOR A CALCULATION OF THE TURBULENT ELECTROMOTIVE FORCE

4.1. Introductory remarks. Definitions

The treatment of the two limiting cases in the foregoing chapter makes it clear that for a general investigation more efficient mathematical methods are needed. We now make wide use of tensor calculus. Often the isotropic tensors of the Euclidean space will appear, i.e. the Kronecker-tensor,

$$\delta_{ij} = \begin{cases} 1 \text{ if } i = j \\ 0 \text{ if } i \neq j \end{cases}, \tag{4.1}$$

and the Levi-Civita skew-symmetric tensor,

$$\varepsilon_{ijk} = \begin{cases} 1 \text{ if } i, j, k = 1, 2, 3, \text{ or } 3, 1, 2, \text{ or } 2, 3, 1 \\ 0 \text{ if } i = j \text{ or } i = k \text{ or } j = k \\ -1 \text{ if } i, j, k = 2, 1, 3, \text{ or } 3, 2, 1 \text{ or } 1, 3, 2. \end{cases} \tag{4.2}$$

The averages of products of vector quantities lead to the introduction of correlation tensors. We have the two-point-two-time correlation tensor of the velocity field,

$$Q_{ik}(\boldsymbol{x}, \boldsymbol{\xi}, t, \tau) = \overline{u'_i(\boldsymbol{x}, t)\, u'_k(\boldsymbol{x} + \boldsymbol{\xi}, t + \tau)}, \tag{4.3}$$

the mixed two-point-two-time correlation tensor of the velocity field and the magnetic field,

$$P_{ik}(\boldsymbol{x}, \boldsymbol{\xi}, t, \tau) = \overline{u'_i(\boldsymbol{x}, t)\, B'_k(\boldsymbol{x} + \boldsymbol{\xi}, t + \tau)}, \tag{4.4}$$

and the two-point-two-time correlation tensor of the magnetic field,

$$B_{ik}(\boldsymbol{x}, \boldsymbol{\xi}, t, \tau) = \overline{B'_i(\boldsymbol{x}, t)\, B'_k(\boldsymbol{x} + \boldsymbol{\xi}, t + \tau)}. \tag{4.5}$$

If we again look at the deduction in the previous chapter we see that the quantities just introduced have been used before. For the quantity h, for example, defined by (3.4) we have

$$h = \varepsilon_{ijk} \left. \frac{\partial Q_{ik}}{\partial \xi_j} \right|_{\xi=0,\tau=0}, \tag{4.6}$$

if (4.2) and (4.3) are taken into account. Furthermore, the quantities w_{ik}, w_{ijk}, defined by (3.25) and (3.26) are closely connected with the tensor Q_{ik} too. We find

$$w_{ij} = \int_0^\infty Q_{ij}(\boldsymbol{x}, 0, t, -\tau)\, \mathrm{d}\tau, \tag{4.7}$$

$$w_{ijk} = \int_0^\infty \frac{\partial Q_{ik}(\boldsymbol{x}, 0, t, -\tau)}{\partial x_j}\, \mathrm{d}\tau. \tag{4.8}$$

Analogous relations exist for the similar quantities defined by (3.42), (3.43) and (3.44).

Moreover, relations (3.27) and (3.28) between the components of the tensors w_{ij} and w_{ijk} reflect a relation between these tensors and the isotropic tensors. One easily confirms the relations

$$w_{ijk} = -\frac{\alpha}{2}\varepsilon_{ijk}, \quad w_{ij} = \beta\, \delta_{ij}. \tag{4.9}$$

Finally, we remark in this connection that the turbulent electromotive force $\mathfrak{E}$ can be represented by

$$\mathfrak{E}_i = \varepsilon_{ijk} P_{jk}(\boldsymbol{x}, 0, t, 0), \tag{4.10}$$

and the mean square of the magnetic field fluctuations by

$$\overline{\boldsymbol{B}'^2} = B_{ii}(\boldsymbol{x}, 0, t, 0). \tag{4.11}$$

There are good reasons for introducing correlation tensors of higher rank, i.e. the velocity correlation tensor of third rank

$$Q_{ijk}(\boldsymbol{x}, \boldsymbol{\xi}, \boldsymbol{\eta}, t, \tau, \sigma) = \overline{u_i'(\boldsymbol{x}, t)\, u_j'(\boldsymbol{x} + \boldsymbol{\xi}, t + \tau)\, u_k'(\boldsymbol{x} + \boldsymbol{\xi} + \boldsymbol{\eta}, t + \tau + \sigma)}, \tag{4.12}$$

the mixed tensor of third rank

$$P_{ijk}(\boldsymbol{x}, \boldsymbol{\xi}, \boldsymbol{\eta}, t, \tau, \sigma) = \overline{u_i'(\boldsymbol{x}, t)\, u_j'(\boldsymbol{x} + \boldsymbol{\xi}, t + \tau)\, B_k'(\boldsymbol{x} + \boldsymbol{\xi} + \boldsymbol{\eta}, t + \tau + \sigma)}, \tag{4.13}$$

and the corresponding quantities of fourth rank

$$\begin{aligned} &Q_{ijkl}(\boldsymbol{x}, \boldsymbol{\xi}, \boldsymbol{\eta}, \boldsymbol{\zeta}, t, \tau, \sigma, \varrho) \\ &= \overline{u_i'(\boldsymbol{x}, t)\, u_j'(\boldsymbol{x} + \boldsymbol{\xi}, t + \tau)\, u_k'(\boldsymbol{x} + \boldsymbol{\xi} + \boldsymbol{\eta}, t + \tau + \sigma)} \\ &\quad \overline{u_l'(\boldsymbol{x} + \boldsymbol{\xi} + \boldsymbol{\eta} + \boldsymbol{\zeta}, t + \tau + \sigma + \varrho)}, \end{aligned} \tag{4.14}$$

$$\begin{aligned} &P_{ijkl}(\boldsymbol{x}, \boldsymbol{\xi}, \boldsymbol{\eta}, \boldsymbol{\zeta}, t, \tau, \sigma, \varrho) \\ &= \overline{u_i'(\boldsymbol{x}, t)\, u_j'(\boldsymbol{x} + \boldsymbol{\xi}, t + \tau)\, u_k'(\boldsymbol{x} + \boldsymbol{\xi} + \boldsymbol{\eta}, t + \tau + \sigma)} \\ &\quad \overline{B_l'(\boldsymbol{x} + \boldsymbol{\xi} + \boldsymbol{\eta} + \boldsymbol{\zeta}, t + \tau + \sigma + \varrho)}. \end{aligned} \tag{4.15}$$

In the following the correlation tensors P_{ij}, P_{ijk}, P_{ijkl} shall appear where the last set of coordinates is equal to zero. Therefore we shall introduce, in addition, the tensors p_{ij}, p_{ijk}, p_{ijkl} defined by

$$p_{ij}(\boldsymbol{x}, t) = P_{ij}(\boldsymbol{x}, 0, t, 0), \quad p_{ijk}(\boldsymbol{x}, \boldsymbol{\xi}, t, \tau) = P_{ijk}(\boldsymbol{x}, \boldsymbol{\xi}, 0, t, \tau, 0),$$
$$p_{ijkl}(\boldsymbol{x}, \boldsymbol{\xi}, \boldsymbol{\eta}, t, \tau, \sigma) = P_{ijkl}(\boldsymbol{x}, \boldsymbol{\xi}, \boldsymbol{\eta}, 0, t, \tau, \sigma, 0). \tag{4.16}$$

According to these definitions the turbulent electromotive force shall be given by

$$\mathfrak{E}_i = \varepsilon_{ijk} p_{jk}. \tag{4.17}$$

For reasons of convenience we introduce the operators

$$\mathfrak{D}_{jn} = \left(\frac{\partial}{\partial t} - \eta \Delta\right) \delta_{jn} - \varepsilon_{jkl}\varepsilon_{lmn} \frac{\partial}{\partial x_k} \bar{u}_m \tag{4.18}$$

and

$$\mathfrak{D}_{jmn} = \varepsilon_{jkl}\varepsilon_{lmn} \frac{\partial}{\partial x_k}, \tag{4.19}$$

which are derived from the induction equation (2.3).

In the formulae (4.18) and (4.19) defining the operators $\mathfrak{D}_{jn}$ and $\mathfrak{D}_{jmn}$, $\boldsymbol{x}$ and t are written as the coordinates with respect to which the differentiation is to be carried out. However, applying these operators to one of the correlation tensors our understanding will always be that the differentiation is to be carried out with respect to the last set of space or time coordinates, i.e., for example, if written explicitly

$$\mathfrak{D}_{jn} P_{in} = \left[\left(\frac{\partial}{\partial \tau} - \eta \frac{\partial^2}{\partial \xi_p \, \partial \xi_p}\right) \delta_{jn} - \varepsilon_{jkl}\varepsilon_{lmn} \frac{\partial}{\partial \xi_k} \bar{u}_m(\boldsymbol{x} + \boldsymbol{\xi}, t + \tau)\right] P_{in}(\boldsymbol{x}, \boldsymbol{\xi}, t, \tau)$$

and

$$\mathfrak{D}_{jn} P_{ipn} = \left[\left(\frac{\partial}{\partial \sigma} - \eta \frac{\partial^2}{\partial \eta_r \, \partial \eta_r}\right) \delta_{jn} - \varepsilon_{jkl}\varepsilon_{lmn} \frac{\partial}{\partial \eta_k} \bar{u}_m(\boldsymbol{x} + \boldsymbol{\xi} + \boldsymbol{\eta}, t + \tau + \sigma)\right] P_{ipn}(\boldsymbol{x}, \boldsymbol{\xi}, \boldsymbol{\eta}, t, \tau, \sigma),$$

but

$$\mathfrak{D}_{jn} p_{ipn} = \left[\left(\frac{\partial}{\partial \tau} - \eta \frac{\partial^2}{\partial \xi_r \, \partial \xi_r}\right) \delta_{jn} - \varepsilon_{jkl}\varepsilon_{lmn} \frac{\partial}{\partial \xi_k} \bar{u}_m(\boldsymbol{x} + \boldsymbol{\xi}, t + \tau)\right] p_{ipn}(\boldsymbol{x}, \boldsymbol{\xi}, t, \tau).$$

The argument of the mean velocity field is always the sum of all the coordinate sets, as indicated here.

4.2. The hierarchy of equations for the correlation tensors

We start from the induction equation (2.3), which we now write in the form

$$\mathfrak{D}_{jn} B_n = \mathfrak{D}_{jmn} u'_m B_n. \tag{4.20}$$

Taking the average we arrive at the equation

$$\mathfrak{D}_{jn}\overline{B}_n = \mathfrak{D}_{jmn}p_{mn}, \tag{4.21}$$

where the correlation tensor p_{mn} appears on the right.

We now exchange in the equation (4.20) the arguments $\boldsymbol{x}$, t for $\boldsymbol{x} + \xi$, $t + \tau$ and take the derivatives with respect to ξ and τ. In addition we replace the subscripts m and n on the right by n and p. Multiplying the resulting equation by $u_m'(\boldsymbol{x}, t)$ and taking the average we find

$$\mathfrak{D}_{jn}P_{mn} = \mathfrak{D}_{jnp}(Q_{mn}\overline{B}_p) + \mathfrak{D}_{jnp}p_{mn}. \tag{4.22}$$

In equation (4.21) for the mean field—a first order statistical moment—a correlation tensor of second rank appears, i.e. a second order statistical moment. In equation (4.22) for the correlation tensor P_{mn} a correlation tensor of third rank appears, i.e. a third order statistical moment. This behaviour continues. Thus we find a hierarchy of equations for the unknown quantities $\overline{B}_n$, P_{mn}, P_{mnp}, ..., where the first equations are given by

$$\begin{aligned} \mathfrak{D}_{jn}\overline{B}_n &= \mathfrak{D}_{jmn}p_{mn}, \\ \mathfrak{D}_{jn}P_{mn} &= \mathfrak{D}_{jnp}(Q_{mn}\overline{B}_p) + \mathfrak{D}_{jnp}p_{mnp}, \\ \mathfrak{D}_{jp}P_{mnp} &= \mathfrak{D}_{jpq}(Q_{mnp}\overline{B}_q) - \mathfrak{D}_{jp}(Q_{mn}\overline{B}_p) + \mathfrak{D}_{jpq}p_{mnpq}, \end{aligned} \tag{4.23}$$

The system (4.23) represents an infinite set of equations. For deriving a finite one some closure technique is needed. This is a well known general problem of turbulence theory, which is not yet satisfactorily solved. It is not our intention to discuss this sort of problem.

We derive a finite system of equations out of (4.23) by taking into account the perturbing turbulent velocity field $\boldsymbol{u}'$ up to a certain order. Thus we deal—in a certain sense—with weak turbulence only. However, effects deduced in the frame of our theory can, indeed, be very large without breaking up the scales, as already indicated in the foregoing chapter.

4.3. Second order correlation approximation

In this case we develop a theory which takes into account the perturbing turbulent velocity field $\boldsymbol{u}'$ up to the second order only. Since the second term on the right-hand side of the second equation of the system (4.23) is of third order, we find

$$\begin{aligned} \mathfrak{D}_{jn}\overline{B}_n(\boldsymbol{x}, t) &= \mathfrak{D}_{jmn}p_{mn}(\boldsymbol{x}, t), \\ \mathfrak{D}_{jn}P_{mn}(\boldsymbol{x}, \xi, t, \tau) &= \mathfrak{D}_{jnp}(Q_{mn}(\boldsymbol{x}, \xi, t, \tau)\, \overline{B}_p(\boldsymbol{x} + \xi, t + \tau)), \end{aligned} \tag{4.24}$$

In addition, we have equations which express the fact that the magnetic field is source-free, i.e.

$$\frac{\partial B_n(\boldsymbol{x}, t)}{\partial x_n} = 0, \quad \frac{\partial P_{mn}(\boldsymbol{x}, \boldsymbol{\xi}, t, \tau)}{\partial \xi_n} = 0, \tag{4.25}$$

and, finally, the initial condition

$$P_{mn}(\boldsymbol{x}, \boldsymbol{\xi}, t, \tau) \to 0 \quad \text{if} \quad \tau \to -\infty. \tag{4.26}$$

The latter is obvious from the definition (4.4) of P_{mn}: There is no correlation between the velocity field at time t and the magnetic field at a much earlier time.

(4.24), (4.25) and (4.26) are the basic equations of mean-field electrodynamics within the framework of second order correlation approximation. For a justification (Krause and Roberts [4]) of the neglect of the third order cumulant in the second equation of (4.23) we remark that

$$p_3 \leqq u' p_2, \tag{4.27}$$

if p_2 is a measure of the order of magnitude of P_{mn} and p_3 the same for P_{mnp}. Equality holds in the completely correlated case only. Accordingly, the order of magnitude of the third order term $\mathfrak{D}_{jnp} P_{mnp}$ can be estimated by

$$|\mathfrak{D}_{jnp} P_{mnp}| \approx \frac{p_3}{\lambda_{\text{cor}}} \leqq \frac{u' p_2}{\lambda_{\text{cor}}}. \tag{4.28}$$

On the other hand the order of magnitude of the second order term $\mathfrak{D}_{jn} P_{mn}$ can be estimated by

$$|\mathfrak{D}_{jn} P_{mn}| \gtrapprox \max \left\{ \frac{p_2}{\tau_{\text{cor}}}, \frac{p_2}{\mu\sigma\lambda_{\text{cor}}^2} \right\}, \tag{4.29}$$

where the "$>$" sign is included only to avoid discussion of the term connected with the mean velocity field in the operator $\mathfrak{D}_{jn}$.

Comparing now (4.28) and (4.29) we find that the second order correlation approximation is applicable if either $u' p_2/\lambda_{\text{cor}} \ll p_2/\tau_{\text{cor}}$, or $u' p_2/\lambda_{\text{cor}} \ll p_2/\mu\sigma\lambda_{\text{cor}}^2$ or both. Hence, we have the condition

$$\min(\text{Rm}, \text{S}) \ll 1 \tag{4.30}$$

already mentioned in the foregoing chapter, which is a guarantee for the applicability of second order correlation approximation. The Strouhal number, S, and the magnetic Reynolds number, Rm, are defined by (3.13) and (3.14).

4.4. Higher order correlation approximation

As in the second order case similar developments of higher order theories are possible. If we neglect the fourth order terms in the third equation of (4.23) we arrive at the system

$$\mathfrak{D}_{jn}\overline{B}_n(\boldsymbol{x}, t) = \mathfrak{D}_{jmn}P_{mn}(\boldsymbol{x}, t), \tag{4.31}$$

$$\mathfrak{D}_{jn}P_{mn}(\boldsymbol{x}, \boldsymbol{\xi}, t, \tau) = \mathfrak{D}_{jnp}(Q_{mn}(\boldsymbol{x}, \boldsymbol{\xi}, t, \tau)\, \overline{B}_p(\boldsymbol{x} + \boldsymbol{\xi}, t + \tau)) + \mathfrak{D}_{jnp}P_{mnp}(\boldsymbol{x}, \boldsymbol{\xi}, t, \tau),$$

$$\mathfrak{D}_{jp}P_{mnp}(\boldsymbol{x}, \boldsymbol{\xi}, \boldsymbol{\eta}, t, \tau, \sigma) = \mathfrak{D}_{jpq}(Q_{mnp}(\boldsymbol{x}, \boldsymbol{\xi}, \boldsymbol{\eta}, t, \tau, \sigma)\, \overline{B}_q(\boldsymbol{x} + \boldsymbol{\xi} + \boldsymbol{\eta}, t + \tau + \sigma)).$$

In addition, we have the conditions

$$\frac{\partial \overline{B}_n(\boldsymbol{x}, t)}{\partial x_n} = 0, \quad \frac{\partial P_{mn}(\boldsymbol{x}, \boldsymbol{\xi}, t, \tau)}{\partial \xi_n} = 0, \quad \frac{\partial P_{mnp}(\boldsymbol{x}, \boldsymbol{\xi}, \boldsymbol{\eta}, t, \tau, \sigma)}{\partial \eta_p} = 0, \tag{4.32}$$

and the initial conditions

$$P_{mn}(\boldsymbol{x}, \boldsymbol{\xi}, t, \tau) \to 0 \text{ if } \tau \to -\infty, \tag{4.33}$$

$$P_{mnp}(\boldsymbol{x}, \boldsymbol{\xi}, \boldsymbol{\eta}, t, \tau, \sigma) \to 0 \text{ if } \sigma \to -\infty.$$

(4.31), (4.32) and (4.33) form the system of basic equations of mean-field electrodynamics in the frame of third order correlation approximation.

Finally, we give the analogous set of equations and conditions for the fourth order correlation approximation:

$$\mathfrak{D}_{jn}\overline{B}_n(\boldsymbol{x}, t) = \mathfrak{D}_{jmn}P_{mn}(\boldsymbol{x}, t), \tag{4.34}$$

$$\mathfrak{D}_{jn}P_{mn}(\boldsymbol{x}, \boldsymbol{\xi}, t, \tau) = \mathfrak{D}_{jnp}(Q_{mn}(\boldsymbol{x}, \boldsymbol{\xi}, t, \tau)\, \overline{B}_p(\boldsymbol{x} + \boldsymbol{\xi}, t + \tau)) + \mathfrak{D}_{jnp}P_{mnp}(\boldsymbol{x}, \boldsymbol{\xi}, t, \tau),$$

$$\mathfrak{D}_{jp}P_{mnp}(\boldsymbol{x}, \boldsymbol{\xi}, \boldsymbol{\eta}, t, \tau, \sigma) = \mathfrak{D}_{jpq}(Q_{mnp}(\boldsymbol{x}, \boldsymbol{\xi}, \boldsymbol{\eta}, t, \tau, \sigma)\, \overline{B}_q(\boldsymbol{x} + \boldsymbol{\xi} + \boldsymbol{\eta}, t + \tau + \sigma))$$

$$- Q_{mn}(\boldsymbol{x}, \boldsymbol{\xi}, t, \tau)\, \mathfrak{D}_{jp}\overline{B}_p(\boldsymbol{x} + \boldsymbol{\xi} + \boldsymbol{\eta}, t + \tau + \sigma) + \mathfrak{D}_{jpq}P_{mnpq}(\boldsymbol{x}, \boldsymbol{\xi}, \boldsymbol{\eta}, t, \tau, \sigma),$$

$$\mathfrak{D}_{jq}P_{mnpq}(\boldsymbol{x}, \boldsymbol{\xi}, \boldsymbol{\eta}, \boldsymbol{\zeta}, t, \tau, \sigma, \varrho) = \mathfrak{D}_{jqr}(Q_{mnpq}(\boldsymbol{x}, \boldsymbol{\xi}, \boldsymbol{\eta}, \boldsymbol{\zeta}, t, \tau, \sigma, \varrho)$$

$$\overline{B}_r(\boldsymbol{x} + \boldsymbol{\xi} + \boldsymbol{\eta} + \boldsymbol{\zeta}, t + \tau + \sigma + \varrho));$$

$$\frac{\partial \overline{B}_n(\boldsymbol{x}, t)}{\partial x_n} = 0, \quad \frac{\partial P_{mn}(\boldsymbol{x}\ \boldsymbol{\xi}\ t\ \tau)}{\partial \xi_n} = 0, \tag{4.35}$$

$$\frac{\partial P_{mnp}(\boldsymbol{x}, \boldsymbol{\xi}, \boldsymbol{\eta}, t, \tau\ \sigma)}{\partial \eta_p} = 0, \quad \frac{\partial P_{mnpq}(\boldsymbol{x}, \boldsymbol{\xi}, \boldsymbol{\eta}, \boldsymbol{\zeta}, t, \tau, \sigma, \varrho)}{\partial \zeta_q} = 0;$$

$$P_{mn}(\boldsymbol{x}, \boldsymbol{\xi}, t, \tau) \to 0 \quad \text{if} \quad \tau \to -\infty;$$

$$P_{mnp}(\boldsymbol{x}, \boldsymbol{\xi}, \boldsymbol{\eta}, t, \tau, \sigma) \to 0 \quad \text{if} \quad \sigma \to -\infty; \tag{4.36}$$

$$P_{mnpq}(\boldsymbol{x}, \boldsymbol{\xi}, \boldsymbol{\eta}, \boldsymbol{\zeta}, t, \tau, \sigma, \varrho) \to 0 \quad \text{if} \quad \varrho \to -\infty.$$

4.5. Green's function tensor of the induction equation

Looking for general solutions of the equations deduced in the foregoing paragraphs we now introduce the Green's function tensor $G_{jp}(\boldsymbol{x}, \boldsymbol{\xi}, t, \tau)$ of the induction equation which is defined by the following conditions:

(i) $G_{jp}(\boldsymbol{x}, \boldsymbol{\xi}, t, \tau) = 0$ if $t < \tau$;

(ii) $G_{jp}(\boldsymbol{x}, \boldsymbol{\xi}, t, \tau) \to 0$ if $|\boldsymbol{x} - \boldsymbol{\xi}| \to \infty$;

(iii) $$\mathfrak{D}_{jk} G_{kp}(\boldsymbol{x}, \boldsymbol{\xi}, t, \tau) = \left\{\left(\frac{\partial}{\partial t} - \eta \Delta\right) \delta_{jk} - \varepsilon_{jlm}\varepsilon_{mnk} \frac{\partial}{\partial x_l} \bar{u}_k(\boldsymbol{x}, t)\right\} G_{kp}(\boldsymbol{x}, \boldsymbol{\xi}, t, \tau)$$
$$= \delta_{jp}\, \delta(\boldsymbol{x} - \boldsymbol{\xi}, t - \tau) \quad \text{if} \quad t > \tau.$$

$\delta(\boldsymbol{x} - \boldsymbol{\xi}, t - \tau)$ denotes the Dirac δ-function.

It is easily shown that the field

$$B_k(\boldsymbol{x}, t) = \int_0^\infty \int G_{kp}(\boldsymbol{x}, \boldsymbol{\xi}, t, \tau)\, q_p(\boldsymbol{\xi}, \tau)\, \mathrm{d}\boldsymbol{\xi}\, \mathrm{d}\tau \tag{4.37}$$

is the solution of the equation

$$\mathfrak{D}_{jk} B_k = q_j \tag{4.38}$$

for $t > t_0$ with the initial condition $B_k(\boldsymbol{x}, t_0) = 0$, since we will find by applying the operator $\mathfrak{D}_{jk}$ to the field (4.37) and taking into account condition (iii)

$$\begin{aligned} \mathfrak{D}_{jk} B_k &= \int_{t_0}^\infty \int \mathfrak{D}_{jk} G_{kp}(\boldsymbol{x}, \boldsymbol{\xi}, t, \tau)\, q_p(\boldsymbol{\xi}, \tau)\, \mathrm{d}\boldsymbol{\xi}\, \mathrm{d}\tau \\ &= \int_{t_0}^\infty \int \delta_{jp} \delta(\boldsymbol{x} - \boldsymbol{\xi}, t - \tau)\, q_p(\boldsymbol{\xi}, \tau)\, \mathrm{d}\boldsymbol{\xi}\, \mathrm{d}\tau = q_j(\boldsymbol{x}, t). \end{aligned} \tag{4.39}$$

Furthermore, we will show that the initial value problem

$$\mathfrak{D}_{jk} B_k = 0 \quad \text{if} \quad t > \tau, \quad B_k(\boldsymbol{x}, \tau) = B_k^{(0)}(\boldsymbol{x}) \tag{4.40}$$

is solved by the expression

$$B_k(\boldsymbol{x}, t) = \int G_{kp}(\boldsymbol{x}, \boldsymbol{\xi}, t, \tau)\, B_p^{(0)}(\boldsymbol{\xi})\, \mathrm{d}\boldsymbol{\xi}. \tag{4.41}$$

In order to prove this we remark first that

$$\begin{aligned} \mathfrak{D}_{jk} B_k &= \int \mathfrak{D}_{jk} G_{kp}(\boldsymbol{x}, \boldsymbol{\xi}, t, \tau)\, B_p^{(0)}(\boldsymbol{\xi})\, \mathrm{d}\boldsymbol{\xi} \\ &= \int \delta_{jp} \delta(\boldsymbol{x} - \boldsymbol{\xi}, t - \tau)\, B_p^{(0)}(\boldsymbol{\xi})\, \mathrm{d}\boldsymbol{\xi} = B_j^{(0)}(\boldsymbol{x})\, \delta(t - \tau), \end{aligned} \tag{4.42}$$

and hence the differential equation (4.40) is satisfied for $t > \tau$. Secondly, we find that

$$\int_{\tau-\varepsilon}^{\tau+\varepsilon} \mathfrak{D}_{jk} B_k\, \mathrm{d}t = B_j^{(0)}(\boldsymbol{x}) \tag{4.43}$$

for any positive number ε. In the limit $\varepsilon \to 0$ and with the explicit definition of the operator $\mathfrak{D}_{jk}$ it follows that

$$B_j(\boldsymbol{x}, \tau + \varepsilon) - B_j(\boldsymbol{x}, \tau - \varepsilon) + O(\varepsilon) = B_j^{(0)}(\boldsymbol{x}).$$

Since, by definition, $B_j(\boldsymbol{x}, \tau - \varepsilon) = 0$, we have

$$\lim_{\varepsilon \to 0} B_j(\boldsymbol{x}, \tau + \varepsilon) = B_j^{(0)}(\boldsymbol{x}),$$

thus showing that (4.41), indeed, solves the initial value problem of the induction equation.

A general discussion of the existence of the tensor G_{jp} and related problems will not be our concern. In cases of interest in connection with the following deductions we shall restrict ourselves to special mean velocity fields where we can represent G_{jp} by explicit expressions. However, one point of interest has to be mentioned: The tensor G_{jp} can have the property

$$G_{jp} \to 0 \quad \text{if} \quad t - \tau \to \infty, \tag{4.44}$$

but need not. If not, the initial value problem (4.40) of the induction equation has non-decaying solutions. In physical terms, this means that the initial magnetic field does not decay by Ohmic dissipation. The induction action of the mean velocity field overcomes the losses of magnetic energy by Ohmic dissipation and, consequently, the magnetic field either remains stationary or grows indefinitely. In both these cases the model under discussion is referred to as a self-excited dynamo.

By assumption we exclude the possibility that a self-excited dynamo is provided by the mean velocity field. We confine ourselves to those mean velocity fields for which condition (4.44) holds. We can now, under this restriction, solve the inhomogeneous problem (4.38) where $\boldsymbol{q}(\boldsymbol{x}, t)$ is given and bounded for $-\infty < t < +\infty$ by the expression

$$B_k(\boldsymbol{x}, t) = \iint G_{kp}(\boldsymbol{x}, \boldsymbol{\xi}, t, \tau)\, q_p(\boldsymbol{\xi}, \tau)\, \mathrm{d}\boldsymbol{\xi}\, \mathrm{d}\tau. \tag{4.45}$$

If $\boldsymbol{B}$, as defined by (4.37) or (4.45), represents a magnetic field it has to be source-free. We shall now prove for both representations that $\operatorname{div} \boldsymbol{B} = 0$ if $\operatorname{div} \boldsymbol{q} = 0$.

From the property (iii) of the Green's tensor and the definition (4.18) of the operator $\mathfrak{D}_{jk}$ we obtain

$$\frac{\partial}{\partial x_j} \mathfrak{D}_{jk} G_{kp} = \left(\frac{\partial}{\partial t} - \eta \Delta\right) \frac{\partial G_{kp}}{\partial x_k} = -\frac{\partial \delta(\boldsymbol{x} - \boldsymbol{\xi}, t - \tau)}{\partial \xi_p}, \tag{4.46}$$

and subsequently from (4.37)

$$\left(\frac{\partial}{\partial t} - \eta \Delta\right) \operatorname{div} \boldsymbol{B} = \int_{t_0}^{\infty}\!\!\int \left(\frac{\partial}{\partial t} - \eta \Delta\right) \frac{\partial G_{kp}}{\partial x_k} q_p(\boldsymbol{\xi}, \tau)\, \mathrm{d}\boldsymbol{\xi}\, \mathrm{d}\tau$$

$$= -\int_{t_0}^{\infty}\!\!\int \frac{\partial \delta(\boldsymbol{x} - \boldsymbol{\xi}, t - \tau)}{\partial \xi_p} q_p(\boldsymbol{\xi}, \tau)\, \mathrm{d}\boldsymbol{\xi}\, \mathrm{d}\tau = \int_{t_0}^{\infty}\!\!\int \delta(\boldsymbol{x} - \boldsymbol{\xi}, t - \tau) \frac{\partial q_p(\boldsymbol{\xi}, \tau)}{\partial \xi_p} \mathrm{d}\boldsymbol{\xi}\, \mathrm{d}\tau$$

$$= \operatorname{div} \boldsymbol{q}\,.$$

Thus we find $\left(\frac{\partial}{\partial t} - \eta \Delta\right) \operatorname{div} \boldsymbol{B} = 0$ if $\operatorname{div} \boldsymbol{q} = 0$.

It is well known that a solution $U(\boldsymbol{x}, t)$ of the equation

$$\left(\frac{\partial}{\partial t} - \eta \Delta\right) U(\boldsymbol{x}, t) = 0 \tag{4.47}$$

identically vanishes for $t > t_0$ if it is zero for $t = t_0$.

The field $\boldsymbol{B}$ defined by (4.37) vanishes for $t = t_0$, hence $\operatorname{div} \boldsymbol{B} = 0$ for $t = t_0$. Consequently, we have $\operatorname{div} \boldsymbol{B} = 0$ for all $t \geqq t_0$ if $\operatorname{div} \boldsymbol{q} = 0$. The same holds for the representation (4.45) if the integral is sufficiently converging in the limit $t_0 \to -\infty$.

4.6. Application of the Green's function tensor to the equations of mean-field electrodynamics

With the aid of the Green's function tensor we are able to transform the equations deduced from the hierarchy (4.23) into integro-differential equations. In the case of second order correlation approximation, e.g., we find from the second equation of (4.24)

$$P_{mn}(\boldsymbol{x}, \boldsymbol{\xi}, t, \tau) = \iint G_{np}(\boldsymbol{x} + \boldsymbol{\xi}, \boldsymbol{x} + \boldsymbol{\xi}', t + \tau, t + \tau') \tag{4.48}$$

$$\mathfrak{D}_{pqr}(Q_{mq}(\boldsymbol{x}, \boldsymbol{\xi}', t, \tau')\, \overline{B}_r(\boldsymbol{x} + \boldsymbol{\xi}', t + \tau'))\, \mathrm{d}\boldsymbol{\xi}'\, \mathrm{d}\tau'\,.$$

Noting that $\boldsymbol{x} + \boldsymbol{\xi}$ and $t + \tau$ are the arguments of the operator $\mathfrak{D}_{jp}$ in the second equation of (4.24), one understands the same arguments appearing in the Green's tensor on the right-hand side of (4.48). According to the statements at the end of the foregoing paragraph the quantity P_{mn} given by (4.48) satisfies the second condition (4.25). Moreover, since $Q_{mq} \to 0$ if $\tau' \to -\infty$, the condition (4.26) is likewise satisfied. Hence we can replace the basic equations of the second order correlation approximation (4.24), (4.25) and (4.26) by the equivalent set

$$\mathfrak{D}_{jn}\overline{B}_n(\boldsymbol{x}, t) = \mathfrak{D}_{jmn}p_{mn}(\boldsymbol{x}, t),$$

$$p_{mn}(\boldsymbol{x}, t) = \iint G_{np}(\boldsymbol{x}, \boldsymbol{x} + \boldsymbol{\xi}, t, t + \tau)\, \mathfrak{D}_{pqr}(Q_{mq}(\boldsymbol{x}, \boldsymbol{\xi}, t, \tau)\, \overline{B}_r(\boldsymbol{x} + \boldsymbol{\xi}, t + \tau))\, \mathrm{d}\boldsymbol{\xi}\, \mathrm{d}\tau,$$

$$\frac{\partial \overline{B}_k(\boldsymbol{x}, t)}{\partial x_k} = 0. \tag{4.49}$$

One advantage of (4.49) over (4.25) and the related equations is obvious. The second equation of (4.49) is one that directly yields the quantity, p_{mn}, whereas the second equation of (4.25) governs P_{mn}, a quantity not directly relevant to our problem, the determination of the mean magnetic field. Therefore, we are able to cast the basic equations of the second order correlation approximation in the form of a system of integro-differential equations, i.e.

$$\left\{\left(\frac{\partial}{\partial t} - \eta \Delta\right)\delta_{jn} - \varepsilon_{jkl}\varepsilon_{lmn}\frac{\partial}{\partial x_k}\overline{u}_m(\boldsymbol{x}, t)\right\}\overline{B}_n(\boldsymbol{x}, t)$$

$$= -\varepsilon_{jkl}\varepsilon_{lmn}\varepsilon_{pqr}\varepsilon_{rst}\frac{\partial}{\partial x_k}\iint \frac{\partial G_{np}(\boldsymbol{x}, \boldsymbol{x} + \boldsymbol{\xi}, t, t + \tau)}{\partial \xi_q} \tag{4.50}$$

$$Q_{ms}(\boldsymbol{x}, \boldsymbol{\xi}, t, \tau)\, \overline{B}_t(\boldsymbol{x} + \boldsymbol{\xi}, t + \tau)\, \mathrm{d}\boldsymbol{\xi}\, \mathrm{d}\tau,$$

$$\frac{\partial \overline{B}_k(\boldsymbol{x}, t)}{\partial x_k} = 0.$$

A similar procedure can be applied in the higher order approximations. We thus find in the third order case

$$\left\{\left(\frac{\partial}{\partial t} - \eta \Delta\right)\delta_{jn} - \varepsilon_{jkl}\varepsilon_{lmn}\frac{\partial}{\partial x_k}\overline{u}_m(\boldsymbol{x}, t)\right\}\overline{B}_n(\boldsymbol{x}, t)$$

$$= -\varepsilon_{jkl}\varepsilon_{lmn}\varepsilon_{j'k'l'}\varepsilon_{l'm'n}\frac{\partial}{\partial x_k}\iint \frac{\partial G_{nj'}(\boldsymbol{x}, \boldsymbol{x} + \boldsymbol{\xi}, t, t + \tau)}{\partial \xi_{k'}} \tag{4.51}$$

$$\Big\{Q_{mm'}(\boldsymbol{x}, \boldsymbol{\xi}, t, \tau)\, \overline{B}_{n'}(\boldsymbol{x} + \boldsymbol{\xi}, t + \tau)$$

$$- \varepsilon_{j''k''l''}\varepsilon_{l''m''n''}\iint \frac{\partial G_{n'j''}(\boldsymbol{x} + \boldsymbol{\xi}, \boldsymbol{x} + \boldsymbol{\xi} + \boldsymbol{\eta}, t + \tau, t + \tau + \sigma)}{\partial \eta_{k''}}$$

$$Q_{mm'm''}(\boldsymbol{x}, \boldsymbol{\xi}, \boldsymbol{\eta}, t, \tau, \sigma)\, \overline{B}_{n''}(\boldsymbol{x} + \boldsymbol{\xi} + \boldsymbol{\eta}, t + \tau + \sigma)\, \mathrm{d}\boldsymbol{\eta}\, \mathrm{d}\sigma\Big\}\, \mathrm{d}\boldsymbol{\xi}\, \mathrm{d}\tau,$$

$$\frac{\partial \overline{B}_k(\boldsymbol{x}, t)}{\partial x_k} = 0;$$

and in the case of fourth order correlation approximation

$$\left\{\left(\frac{\partial}{\partial t} - \eta \Delta\right) \delta_{jn} - \varepsilon_{jkl}\varepsilon_{lmn} \frac{\partial}{\partial x_k} \bar{u}_m(\boldsymbol{x}, t)\right\} \overline{B}_n(\boldsymbol{x}, t)$$

$$= -\varepsilon_{jkl}\varepsilon_{lmn}\varepsilon_{j'k'l'}\varepsilon_{l'm'n'} \frac{\partial}{\partial x_k} \iint \frac{\partial G_{nj'}(\boldsymbol{x}, \boldsymbol{x}+\boldsymbol{\xi}, t, t+\tau)}{\partial \xi_{k'}} \tag{4.52}$$

$$\Big\{ Q_{mm'}(\boldsymbol{x}, \boldsymbol{\xi}, t, \tau)\, \overline{B}_{n'}(\boldsymbol{x}+\boldsymbol{\xi}, t+\tau)$$

$$- \varepsilon_{j''k''l''}\varepsilon_{l''m''n''} \iint \frac{G_{n'j''}(\boldsymbol{x}+\boldsymbol{\xi}, \boldsymbol{x}+\boldsymbol{\xi}+\boldsymbol{\eta}, t+\tau, t+\tau+\sigma)}{\partial \eta_{k''}}$$

$$[Q_{mm'm''}(\boldsymbol{x}, \boldsymbol{\xi}, \boldsymbol{\eta}, t, \tau, \sigma)\, \overline{B}_{n''}(\boldsymbol{x}+\boldsymbol{\xi}+\boldsymbol{\eta}, t+\tau+\sigma)$$

$$- \varepsilon_{j'''k'''l'''}\varepsilon_{l'''m'''n'''} \iint \frac{\partial G_{n''j'''}(\boldsymbol{x}+\boldsymbol{\xi}+\boldsymbol{\eta}, \boldsymbol{x}+\boldsymbol{\xi}+\boldsymbol{\eta}+\boldsymbol{\zeta}, t+\tau+\sigma, t+\tau+\sigma+\varrho)}{\partial \zeta_{k'''}}$$

$$(Q_{mm'm''m'''}(\boldsymbol{x}, \boldsymbol{\xi}, \boldsymbol{\eta}, \boldsymbol{\zeta}, t, \tau, \sigma, \varrho)$$

$$- Q_{mm'}(\boldsymbol{x}, \boldsymbol{\xi}, t, \tau)$$

$$Q_{m''m'''}(\boldsymbol{x}+\boldsymbol{\xi}+\boldsymbol{\eta}, \boldsymbol{x}+\boldsymbol{\xi}+\boldsymbol{\eta}+\boldsymbol{\zeta}, t+\tau+\sigma, t+\tau+\sigma+\varrho))$$

$$\overline{B}_{n'''}(\boldsymbol{x}+\boldsymbol{\xi}+\boldsymbol{\eta}+\boldsymbol{\zeta}, t+\tau+\sigma+\varrho)\, \mathrm{d}\boldsymbol{\zeta}\, \mathrm{d}\varrho]\, \mathrm{d}\boldsymbol{\eta}\, \mathrm{d}\sigma \Big\}\, \mathrm{d}\boldsymbol{\xi}\, \mathrm{d}\tau,$$

$$\frac{\partial \overline{B}_k(\boldsymbol{x}, t)}{\partial x_k} = 0.$$

When we now recall that $\overline{\boldsymbol{B}}$ is governed by the induction equation

$$\frac{\partial \overline{\boldsymbol{B}}}{\partial t} - \operatorname{curl}(\overline{\boldsymbol{u}} \times \overline{\boldsymbol{B}}) - \frac{1}{\mu\sigma} \Delta \overline{\boldsymbol{B}} = \operatorname{curl} \mathfrak{E}, \tag{4.53}$$

it becomes obvious that the turbulent electromotive force $\mathfrak{E}$ can be represented by the expression

$$\mathfrak{E}_l = \iint K_{lt}(\boldsymbol{x}, \boldsymbol{\xi}, t, \tau)\, \overline{B}_t(\boldsymbol{x}-\boldsymbol{\xi}, t-\tau)\, \mathrm{d}\boldsymbol{\xi}\, \mathrm{d}\tau. \tag{4.54}$$

The tensor K_{lt} may be obtained by the methods discussed in this chapter where a restriction to a low level of approximation is necessary for technica reasons only: The mathematical effort grows rapidly with the degree of approximation.

From (4.50) we derive in the case of the second order correlation approximation

$$K_{lt}(\boldsymbol{x}, \boldsymbol{\xi}, t, \tau) = \varepsilon_{lmn}\varepsilon_{pqr}\varepsilon_{rst} \frac{\partial G_{np}(\boldsymbol{x}, \boldsymbol{x}-\boldsymbol{\xi}, t, t-\tau)}{\partial \xi_q} Q_{ms}(\boldsymbol{x}, -\boldsymbol{\xi}, t, -\tau), \tag{4.55}$$

and for the third order approximation from (4.51)

$$K_{lt}(\boldsymbol{x}, \xi, t, \tau) = \varepsilon_{lmn}\varepsilon_{j'k'l'}\varepsilon_{l'm'n'} \left[\frac{\partial G_{nj'}(\boldsymbol{x}, \boldsymbol{x} - \xi, t, t - \tau)}{\partial \xi_{k'}} Q_{mm'}(\boldsymbol{x}, -\xi, t, -\tau)\, \delta_{n't} \right.$$
$$+ \varepsilon_{j''k''l''}\varepsilon_{l''m''t} \iint \frac{\partial G_{nj'}(\boldsymbol{x}, \boldsymbol{x} - \xi', t, t - \tau')}{\partial \xi'_{k'}} \frac{\partial G_{n'j''}(\boldsymbol{x} - \xi', \boldsymbol{x} - \xi, t - \tau', t - \tau)}{\partial \xi_{k''}}$$
$$\left. Q_{mm'm''}(\boldsymbol{x}, -\xi', \xi' - \xi, t, -\tau', \tau' - \tau)\, \mathrm{d}\xi'\, \mathrm{d}\tau' \right]. \tag{4.56}$$

We shall not give here the expression of the tensor K_{lt} in the fourth order approximation, which may easily be derived from (4.52).

In section 2.4. we stated already that $\mathfrak{E}$ is a linear functional of $\overline{\boldsymbol{B}}$. Relation (4.54) in combination with (4.55) or (4.56) defining the tensor K_{lt} furnishes now a rather explicit representation of this functional for different approximation levels.

4.7. On the convergence of the correlation approximation

For a mathematical justification of the correlation approximation we first discuss the convergence of a field sequence derived from the induction equation.

In the case of a vanishing velocity field the tensor G_{jp} introduced in section 4.5. is given by

$$G_{jp}(\boldsymbol{x}, \xi, t, \tau) = \delta_{jp} G(\boldsymbol{x} - \xi, t - \tau), \tag{4.57}$$

where

$$G(\boldsymbol{x}, t) = \begin{cases} 0, & \text{if} \quad t < 0, \\ \left(\dfrac{\mu\sigma}{4\pi t}\right)^{3/2} \exp\left(-\dfrac{\mu\sigma x^2}{4t}\right), & \text{if} \quad t \geqq 0. \end{cases} \tag{4.58}$$

Let us consider the induction equation (2.3) in the form

$$\frac{\partial B_j}{\partial t} - \frac{1}{\mu\sigma} \Delta B_j = \varepsilon_{jkl}\varepsilon_{lmn} \frac{\partial u_m B_n}{\partial x_k}, \tag{4.59}$$

which can be replaced by the integral equation

$$B_j(\boldsymbol{x}, t) = \int G(\boldsymbol{x} - \boldsymbol{x}', t - t_0)\, B_j(\boldsymbol{x}', t_0)\, \mathrm{d}\boldsymbol{x}'$$
$$- \varepsilon_{jkl}\varepsilon_{lmn} \int_{t_0}^{t}\!\! \int \frac{\partial G(\boldsymbol{x} - \boldsymbol{x}', t - t')}{\partial x'_k} u_m(\boldsymbol{x}', t')\, B_n(\boldsymbol{x}', t')\, \mathrm{d}\boldsymbol{x}'\, \mathrm{d}t'. \tag{4.60}$$

We define a sequence of fields $\boldsymbol{B}^{(\nu)}(\boldsymbol{x}, t)$, $\nu = 0, 1, 2, \ldots$ by

$$\boldsymbol{B}^{(0)}(\boldsymbol{x}, t) \quad \text{arbitrary}, \quad B_j^{(\nu)}(\boldsymbol{x}, t) = \int G(\boldsymbol{x} - \boldsymbol{x}', t - t_0)\, B_j(\boldsymbol{x}', t_0)\, \mathrm{d}\boldsymbol{x}' \tag{4.61}$$

$$- \varepsilon_{jkl}\varepsilon_{lmn} \int_{t_0}^{t}\int \frac{\partial G(\boldsymbol{x} - \boldsymbol{x}', t - t')}{\partial x_k'} u_m(\boldsymbol{x}', t')\, B_n^{(\nu-1)}(\boldsymbol{x}', t')\, \mathrm{d}\boldsymbol{x}'\, \mathrm{d}t', \quad \nu = 1, 2, \ldots$$

It is obvious that $B_j^{(\nu)}(\boldsymbol{x}, t_0) = B_j(\boldsymbol{x}, t_0)$ for any $\nu \geqq 1$ and that, in the case of convergence, the field

$$\boldsymbol{B}^{(\infty)}(\boldsymbol{x}, t) = \lim_{\nu \to \infty} \boldsymbol{B}^{(\nu)}(\boldsymbol{x}, t)$$

solves the integral equation (4.60) and with it the induction equation (4.59). Consequently, the field $\boldsymbol{B}^{(\infty)}(\boldsymbol{x}, t)$ is a solution of the initial value problem of the induction equation.

We shall not give a detailed proof of convergence here but restrict ourselves to some remarks: We can obtain from sequence (4.61) another one by

$$\boldsymbol{b}^{(0)}(\boldsymbol{x}, t) = \boldsymbol{B}^{(0)}(\boldsymbol{x}, t), \tag{4.62}$$

$$\boldsymbol{b}^{(\nu)}(\boldsymbol{x}, t) = \boldsymbol{B}^{(\nu)}(\boldsymbol{x}, t) - \boldsymbol{B}^{(\nu-1)}(\boldsymbol{x}, t), \quad \nu \geqq 1,$$

and represent $\boldsymbol{B}^{(\infty)}$ by the series

$$\boldsymbol{B}^{(\infty)}(\boldsymbol{x}, t) = \sum_{\nu=0}^{\infty} \boldsymbol{b}^{(\nu)}(\boldsymbol{x}, t). \tag{4.63}$$

It has been proved that the series (4.63) is absolutely convergent for arbitrary finite values of t and t_0 and any bounded velocity field $\boldsymbol{u}$ (Krause [1, 2]). It is also necessary to impose certain conditions concerning the continuity and the differentiability of the functions involved in our problem, which we shall not specify here.

If we now construct a sequence $\boldsymbol{B}^{(\nu)}(\boldsymbol{x}, t)$ by choosing

$$B_j^{(0)}(\boldsymbol{x}, t) = \int G(\boldsymbol{x} - \boldsymbol{x}', t - t_0)\, B_j(\boldsymbol{x}', t_0)\, \mathrm{d}\boldsymbol{x}', \tag{4.64}$$

we can obtain from (4.61) the representation

$$B_j^{(\nu)}(\boldsymbol{x}, t) = \int G_{jp}^{(\nu)}(\boldsymbol{x}, \boldsymbol{x}', t, t_0)\, B_p(\boldsymbol{x}', t_0)\, \mathrm{d}\boldsymbol{x}', \tag{4.65}$$

where the tensors $G_{jp}^{(\nu)}(\boldsymbol{x}, \boldsymbol{x}', t, t_0)$ are defined by

$$G_{jp}^{(0)}(\boldsymbol{x}, \boldsymbol{x}', t, t_0) = \delta_{jp} G(\boldsymbol{x} - \boldsymbol{x}', t - t_0),$$

$$G_{jp}^{(\nu)}(\boldsymbol{x}, \boldsymbol{x}', t, t_0) = \delta_{jp} G(\boldsymbol{x} - \boldsymbol{x}', t - t_0) \tag{4.66}$$

$$+ \varepsilon_{jkl}\varepsilon_{lmn} \int_{t_0}^{t}\int \frac{\partial G(\boldsymbol{x} - \boldsymbol{x}'', t - t'')}{\partial x_k} u_m(\boldsymbol{x}'', t'')\, G_{np}^{(\nu-1)}(\boldsymbol{x}'', \boldsymbol{x}', t'', t_0)\, \mathrm{d}\boldsymbol{x}''\, \mathrm{d}t''.$$

The convergence of the sequence of tensors $G_{jp}^{(\nu)}$ is guaranteed by that of the sequence $\boldsymbol{B}^{(\nu)}$. Consequently, the recurrence relation (4.66) provides the possibility of representing the tensor G_{jp} introduced in section 4.5. by

$$G_{jp}(\boldsymbol{x}, \boldsymbol{x}', t, t_0) = \lim_{\nu \to \infty} G_{jp}^{(\nu)}(\boldsymbol{x}, \boldsymbol{x}', t, t_0). \tag{4.67}$$

With this representation the existence of the tensor G_{jp} is guaranteed for any bounded velocity field.

Let us now consider the equation

$$\frac{\partial B_j}{\partial t} - \varepsilon_{jkl}\varepsilon_{lmn} \frac{\partial \bar{u}_m B_n}{\partial x_k} - \frac{1}{\mu\sigma} \Delta B_j = \varepsilon_{jkl}\varepsilon_{lmn} \frac{\partial u'_m B_n}{\partial x_k}, \tag{4.68}$$

and let $G_{jp}(\boldsymbol{x}, \boldsymbol{x}', t, t')$ be the Green's tensor of the left-hand side of this equation. We can replace equation (4.68) by the integral equation

$$B_j(\boldsymbol{x}, t) = \int G_{jp}(\boldsymbol{x}, \boldsymbol{x}', t, t_0)\, B_p(\boldsymbol{x}', t_0)\, \mathrm{d}\boldsymbol{x}' - \varepsilon_{pkl}\varepsilon_{lmn} \iint \frac{\partial G_{jp}(\boldsymbol{x}, \boldsymbol{x}', t, t')}{\partial x'_k} u'_m(\boldsymbol{x}', t')\, B_n(\boldsymbol{x}', t')\, \mathrm{d}\boldsymbol{x}'\, \mathrm{d}t', \tag{4.69}$$

and define a sequence of fields $\tilde{\boldsymbol{B}}^{(\nu)}(\boldsymbol{x}, t)$ similar to (6.41) by

$\tilde{\boldsymbol{B}}^{(0)}(\boldsymbol{x}, t)$ arbitrary,

$$\tilde{B}_j^{(\nu)}(\boldsymbol{x}, t) = \int G_{jp}(\boldsymbol{x}, \boldsymbol{x}', t, t_0)\, B_p(\boldsymbol{x}', t_0)\, \mathrm{d}\boldsymbol{x}' - \varepsilon_{pkl}\varepsilon_{lmn} \int_{t_0}^{t}\!\! \int \frac{\partial G_{jp}(\boldsymbol{x}, \boldsymbol{x}', t, t')}{\partial x'_k} u'_m(\boldsymbol{x}', t')\, \tilde{B}_n^{(\nu-1)}(\boldsymbol{x}', t')\, \mathrm{d}\boldsymbol{x}'\, \mathrm{d}t'. \tag{4.70}$$

It is not necessary to prove convergence again. We can define the sequence

$$\tilde{\boldsymbol{b}}^{(0)}(\boldsymbol{x}, t) = \tilde{\boldsymbol{B}}^{(0)}(\boldsymbol{x}, t), \quad \tilde{\boldsymbol{b}}^{(\nu)}(\boldsymbol{x}, t) = \tilde{\boldsymbol{B}}^{(\nu)}(\boldsymbol{x}, t) - \tilde{\boldsymbol{B}}^{(\nu-1)}(\boldsymbol{x}, t), \tag{4.71}$$

and represent the field $\boldsymbol{B}^{(\infty)}$ by the series

$$\boldsymbol{B}^{(\infty)}(\boldsymbol{x}, t) = \sum_{\nu=0}^{\infty} \boldsymbol{b}^{(\nu)}(\boldsymbol{x}, t) \tag{4.72}$$

as was done subsequently to (4.61). We can, however, also arrive at the series (4.72) by substituting $\boldsymbol{u} = \overline{\boldsymbol{u}} + \boldsymbol{u}'$ into (4.61) and carrying out some rearrangement. Since the series (4.63) has been proved to be absolutely convergent, the same will hold for the series (4.72).

It is now obvious that the correlation approximation is based on a convergent procedure: We construct a sequence $\tilde{\boldsymbol{B}}^{(\nu)}(\boldsymbol{x}, t)$, $\nu = 0, 1, 2, \ldots$ by choosing in the formula (4.70) $\tilde{\boldsymbol{B}}^{(0)} = \bar{\boldsymbol{B}}$ and determine the fields

$$\mathfrak{E}^{(n)} = \overline{\boldsymbol{u}' \times \tilde{\boldsymbol{B}}^{(n)}}. \tag{4.73}$$

Assuming that the order of the operations of taking the limit and the average can be exchanged, we find that this field sequence will converge as well, and we can represent the turbulent electromotive force by the limit

$$\mathfrak{E}(\boldsymbol{x}, t) = \lim_{n\to\infty} \mathfrak{E}^{(n)}(\boldsymbol{x}, t) = \lim_{n\to\infty} \overline{\boldsymbol{u}'(\boldsymbol{x}, t) \times \tilde{\boldsymbol{B}}^{(n)}(\boldsymbol{x}, t)}. \tag{4.74}$$

CHAPTER 5

TWO-SCALE TURBULENCE

5.1. Introductory remarks

We mentioned in section 2.2. that we shall not have to refer to the nature of the averaging operation, but only to the properties (2.4) and to the commutation rules. If, however, the theory is applied to a certain observation or measurement, in most cases averages with respect to space or time coordinates must be taken. Then in particular the rules

$$\overline{\overline{F}} = \overline{F}, \quad \overline{F'} = 0 \tag{5.1}$$

are only approximately valid. The validity will be better the smaller the characteristic scales of the fluctuations compared with those of the mean fields.

Therefore, having in mind applications of the theory it is appropriate to require that either

$$\lambda_{\text{cor}} \ll \bar{\lambda}, \tag{5.2}$$

or

$$\tau_{\text{cor}} \ll \bar{\tau}, \tag{5.3}$$

or both the relations are fulfilled. Here and in the following, $\bar{\lambda}$ and $\bar{\tau}$ denote the characteristic length and time scales of the mean fields. If one of the relations (5.2), (5.3), or both, are fulfilled we speak of two-scale turbulence.

Already in section 2.4. the turbulent electromotive force was shown to be a functional of $\overline{\boldsymbol{B}}$, $\overline{\boldsymbol{u}}$, $\boldsymbol{u}'$, and we also found that only those of their values contribute to the value of $\mathfrak{E}$, which are taken from a certain neighbourhood of the space-time point under consideration. In the case of two-scale turbulence, i.e. (5.2) and (5.3) being fulfilled, $\overline{\boldsymbol{B}}$ can be represented by Taylor's formula, and, consequently, $\mathfrak{E}$ as a linear functional of $\overline{\boldsymbol{B}}$ by

$$\begin{aligned}\mathfrak{E}_i = \overline{(\boldsymbol{u}' \times \boldsymbol{B}')_i} &= g_{ij}^{(00)} \overline{B}_j + g_{ijk}^{(10)} \frac{\partial \overline{B}_j}{\partial x_k} + g_{ijkl}^{(20)} \frac{\partial^2 \overline{B}_j}{\partial x_k \, \partial x_l} + \cdots \\ &+ g_{ij}^{(01)} \frac{\partial \overline{B}_j}{\partial t} + g_{ijk}^{(11)} \frac{\partial^2 \overline{B}_j}{\partial x_k \, \partial t} + g_{ijkl\ldots n}^{(\varkappa\nu)} \frac{\partial^{(\varkappa+\nu)} \overline{B}_j}{\partial x_k \, \partial x_l \ldots \partial x_n \, \partial t}.\end{aligned} \tag{5.4}$$

Now the tensor quantities $g_{i\ldots n}^{(\varkappa\nu)}$ appear, which are functionals of $\overline{\boldsymbol{u}}$, $\boldsymbol{u}'$. If in (5.4) space derivatives are taken into account up to the order $\varkappa$ and time derivatives up to the order ν, an error has to be expected of the order

$$O\left\{\left(\frac{\lambda_{\text{cor}}}{\bar{\lambda}}\right)^{\varkappa}, \left(\frac{\tau_{\text{cor}}}{\bar{\tau}}\right)^{\nu}\right\}.$$

If the back-reaction of the magnetic field is taken into account, the velocity fields $\overline{\boldsymbol{u}}$ and $\boldsymbol{u}'$ depend on $\overline{\boldsymbol{B}}$. It is worth noting that in this case the validity of (5.4) is not affected, whereas the tensors $g_{ij\ldots n}^{(\varkappa\nu)}$ are now functionals of $\overline{\boldsymbol{u}}$, $\boldsymbol{u}'$ and $\overline{\boldsymbol{B}}$.

5.2. Isotropic tensors

A property of the turbulence, e.g. isotropy, has to be reflected in the structure of the tensors $g_{i\ldots n}$. For an isotropic turbulence all mean quantities derived from it remain unchanged if undergoing a rotation. Tensors with this property are called isotropic tensors.

We already know isotropic tensors, e.g. the Kronecker tensor δ_{ij}. This tensor has the same components in all coordinate systems, this is clearly equivalent to the property of having unchanged components if undergoing an arbitrary rotation. The ε-tensor is a further example of an isotropic tensor of third rank and the product $\delta_{ij}\,\delta_{kl}$ is an isotropic tensor of fourth rank.

We shall now prove the following statements which are important in our deductions:

(i) If the tensor a_{ij} of second rank is isotropic, then

$$a_{ij} = \alpha\,\delta_{ij}. \tag{5.5}$$

(ii) If the tensor a_{ijk} of third rank is isotropic, then

$$a_{ijk} = \beta\varepsilon_{ijk}. \tag{5.6}$$

(iii) If the tensor a_{ijkl} of fourth rank is isotropic, then

$$a_{ijkl} = a\,\delta_{ij}\delta_{kl} + b\,\delta_{ik}\delta_{jl} + c\,\delta_{il}\delta_{jk}. \tag{5.7}$$

We denote the matrix of a rotation by D_{ij} where

$$D_{ij}D_{kj} = \delta_{ik}, \quad D_{ij}D_{il} = \delta_{jl}. \tag{5.8}$$

Hence a tensor undergoing this rotation is transformed by

$$a_{ij}^{(\text{rot})} = D_{ik}D_{jl}a_{kl}, \tag{5.9}$$

and the relation

$$a_{ij} = D_{ik}D_{jl}a_{kl} \tag{5.10}$$

holds for an isotropic tensor of rank two, where D_{ij} is an arbitrary rotation.

Let us first consider the 180°-rotation about the 3-axis given by

$$(D^{(3)}_{ij}) = \begin{pmatrix} -1 & 0 & 0 \\ 0 & -1 & 0 \\ 0 & 0 & 1 \end{pmatrix}. \tag{5.11}$$

Inserting (5.11) into (5.10) we find

$$a_{13} = -a_{13},\, a_{31} = -a_{31},\, a_{23} = -a_{23},\, a_{32} = -a_{32}\,, \tag{5.12}$$

and other relations which are fulfilled identically. From (5.12) we conclude

$$a_{13} = a_{31} = a_{23} = a_{32} = 0\,. \tag{5.31}$$

If we now consider the 180°-rotation about the 2-axis given by

$$(D^{(2)}_{ij}) = \begin{pmatrix} -1 & 0 & 0 \\ 0 & 1 & 0 \\ 0 & 0 & -1 \end{pmatrix}, \tag{5.14}$$

we find in addition

$$a_{12} = a_{21} = 0\,. \tag{5.15}$$

Let us now consider the 90°-rotation about the 3-axis given by

$$(\tilde{D}^{(3)}_{ij}) = \begin{pmatrix} 0 & -1 & 0 \\ 1 & 0 & 0 \\ 0 & 0 & 1 \end{pmatrix}. \tag{5.16}$$

Inserting it into (5.10) we obtain

$$a_{11} = a_{22}\,, \tag{5.17}$$

and in the same way we shall find

$$a_{11} = a_{33}\,, \tag{5.18}$$

if the 90°-rotation about the 2-axis is considered.

(5.12), (5.13), (5.15) and (5.17), (5.18) show that an isotropic tensor of second rank is indeed given by (5.5).

Statement (ii) may be proved analogously. For an isotropic tensor of third rank the relation

$$a_{ijk} = D_{il}D_{jm}D_{kn}a_{lmn} \tag{5.19}$$

must hold. Let us first consider those components with two or three equal indices. By inserting $D_{ij}^{(3)}$ or $D_{ij}^{(2)}$ into (5.19) relations are found which show that these components are equal to their opposite values and therefore vanish. Accordingly, only components with three different indices can be non-zero. Applying now $\tilde{D}_{ij}^{(3)}$ given by (5.16) to a_{123} we find

$$a_{123} = -a_{213}. \tag{5.20}$$

Also considering the 90°-rotation about the 1-axis and the 2-axis we see that by an exchange of any pair of indices the sign of the component will change. Thus, an isotropic tensor of third rank results which is given by (5.6).

For an isotropic tensor of fourth rank a_{ijkl} the relation

$$a_{ijkl} = D_{im} D_{jn} D_{kp} D_{lq} a_{mnpq} \tag{5.21}$$

must hold for any rotation D_{ij}. Considering the 180°-rotations about the coordinate axes we easily find that all components vanish apart from those which have an even number of equal indices. From an application of the 90°-rotations about the coordinate axes to (5.21) we conclude that any pair of indices can be exchanged without altering the value of the component. Thus we find four groups of equal components:

$$\begin{aligned}
a_{1111} &= a_{2222} = a_{3333} = A, \\
a_{1122} &= a_{2211} = a_{1133} = a_{3311} = a_{2233} = a_{3322} = B, \\
a_{1212} &= a_{2121} = a_{1313} = a_{3131} = a_{2323} = a_{3232} = C, \\
a_{1221} &= a_{2112} = a_{1331} = a_{3113} = a_{2332} = a_{3223} = D.
\end{aligned} \tag{5.22}$$

It is remarkable at this point that we have four free parameters in (5.22) whereas in (5.7) there are three free parameters only. For a final decision we shall now consider an arbitrary rotation about the 3-axis given by

$$D_{ij} = \begin{pmatrix} \cos\varphi & -\sin\varphi & 0 \\ \sin\varphi & \cos\varphi & 0 \\ 0 & 0 & 1 \end{pmatrix}, \tag{5.23}$$

where φ denotes the angle of rotation. For the component a_{1111} we obtain by inserting (5.23) into (5.20)

$$\begin{aligned}
a_{1111} = a_{1111} \cos^4\varphi + a_{2222} \sin^4\varphi + \cos^2\varphi \sin^2\varphi (a_{1122} \\
+ a_{2211} + a_{1212} + a_{2121} + a_{1221} + a_{2112}).
\end{aligned} \tag{5.24}$$

With the notation (5.22) and the identity $\cos^4\varphi + \sin^4\varphi = 1 - 2\cos^2\varphi \sin^2\varphi$ we have the relation

$$A = A + 2\cos^2\varphi \sin^2\varphi\,(-A + B + C + D), \tag{5.25}$$

which is fulfilled for any φ if

$$A = B + C + D. \tag{5.26}$$

In agreement with (5.7) the number of free parameters is reduced to three, and it is easily confirmed that (5.22) and (5.26) are identical with (5.7) if $B = a$, $C = b$, $D = c$.

As far as isotropic tensors of arbitrary rank are concerned we shall only mention that such tensors of even rank can be represented by a sum of products of δ-tensors, and those of odd rank by a sum of tensors, where any summand is a product of one ε-tensor with δ-tensors.

Finally, an application of statements (i) and (iii) may be presented. We intend to evaluate the integrals

$$\int f(\xi)\, \xi_i \xi_j \, d\xi \quad \text{and} \quad \int f(\xi)\, \xi_i \xi_j \xi_k \xi_l \, d\xi$$

where $f(\xi)$ is a function of $\xi = \sqrt{\xi^2}$, which vanishes sufficiently strongly if $\xi \to \infty$, thus providing for convergence. Both integrals are clearly isotropic tensors, therefore from (i) and (iii) follows

$$\int f(\xi)\, \xi_i \xi_j \, d\xi = A\, \delta_{ij}, \tag{5.27}$$

$$\int f(\xi)\, \xi_i \xi_j \xi_k \xi_l \, d\xi = B(\delta_{ij}\delta_{kl} + \delta_{ik}\delta_{jl} + \delta_{il}\delta_{jk}).$$

For the ansatz of the second integral we have already used its symmetry with respect to all four indices. Contracting (5.27) we find

$$\int f(\xi)\, \xi^2 \, d\xi = 3A, \quad \int f(\xi)\, \xi^4 \, d\xi = 15B, \tag{5.28}$$

and finally

$$\int f(\xi)\, \xi_i \xi_j \, d\xi = \delta_{ij} \frac{1}{3} \int f(\xi)\, \xi^2 \, d\xi, \tag{5.29}$$

$$\int f(\xi)\, \xi_i \xi_j \xi_k \xi_l \, d\xi = (\delta_{ij}\delta_{kl} + \delta_{ik}\delta_{jl} + \delta_{il}\delta_{jk}) \frac{1}{15} \int f(\xi)\, \xi^4 \, d\xi. \tag{5.30}$$

5.3. Structures of the tensors $g_{ij\ldots n}$

Let us consider a physical system and the processes within it, and another system derived from this first one by rotating it. According to our fundamental knowledge of nature the processes in the rotated system are just those which are obtained by rotating the corresponding processes of the original system. Tensors decribing physical quantities in one system are, therefore, to be derived from the corresponding tensors in the other system by the same rotation. Fundamental relations between physical quantities are independent of whether the one system or the other is considered. Consequently, relations between tensors can only be provided by tensors which are invariant with respect to rotation, hence those relations are provided by the isotropic tensors. In our special case we can conclude that the tensors $g_{ij\ldots n}$ introduced in (5.4) are to be related by isotropic tensors to the tensor quantities characterizing the turbulent motion under consideration.

It is fundamental knowledge too that we can derive a real physical system by reflecting a given one, e.g. as described in section 3.3. A fundamental relation connecting physical quantities has, therefore, to remain unchanged if a reflection is carried out. We have mentioned before that there are vector quantities of different behaviour with respect to reflections, the polar vectors and the axial vectors. In order to generalize this definition we consider a tensor of rank n, $a_{i_1 i_2 \ldots i_n}$, being constructed as a product of n vectors, where k of these vectors are axial and $n - k$ polar. If a reflection is carried out, according to (3.2), (3.3), this tensor will transform like

$$a^{(\mathrm{ref})}_{i_1 \ldots i_n}(\boldsymbol{x}, t) = (-1)^{n-k}\, a_{i_1 \ldots i_n}(-\boldsymbol{x}, t). \tag{5.31}$$

If k is odd we have, by definition, a pseudo-tensor, for even k a tensor. Hence a tensor of arbitrary rank will behave like

$$a^{(\mathrm{ref})}_{i_1 \ldots i_n}(\boldsymbol{x}, t) = (-1)^{n}\, a_{i_1 \ldots i_n}(-\boldsymbol{x}, t), \tag{5.32}$$

if reflected, and a pseudo-tensor like

$$a^{(\mathrm{ref})}_{i_1 \ldots i_n}(\boldsymbol{x}, t) = (-1)^{n+1}\, a_{i_1 \ldots i_n}(-\boldsymbol{x}, t). \tag{5.33}$$

Since δ_{ik} and ε_{ikl} are constant by definition, according to (5.32) and (5.33), δ_{ik} is a tensor but ε_{ikl} a pseudo-tensor.

An equality of two tensor quantities can only exist if both are either tensors or pseudo-tensors, because of the symmetry with respect to the reflections just mentioned. Therefore, since the turbulent electromotive force $\mathfrak{E}$ is a polar vector but the mean magnetic field an axial one, all the quantities $g_{ij\ldots n}$ introduced in (5.4) must be pseudo-tensors.

5.4. Examples for the turbulent electromotive force

With the considerations carried out in the foregoing section a rather definite structure of the functional becomes visible which describes the connection between $\mathfrak{E}$ and $\overline{\boldsymbol{B}}$. In the following we shall restrict ourselves to the case where a sufficiently accurate description is possible if in (5.4) only the spatial derivatives up to the first order are taken into account. Then an error of the order $O\left\{\left(\frac{\lambda_{\text{cor}}}{\bar{\lambda}}\right)^2, \frac{\tau_{\text{cor}}}{\bar{\tau}}\right\}$ must be expected. Simplifying the notation by writing

$$a_{ik} = g_{ik}^{(00)}, \quad b_{ikl} = g_{ikl}^{(10)}, \tag{5.34}$$

from (5.4) we arrive at the relation

$$\mathfrak{E}_i = a_{ik}\overline{B}_k + b_{ikl}\frac{\partial \overline{B}_k}{\partial x_l}, \tag{5.35}$$

where a_{ik} and b_{ikl} are pseudo-tensors depending on $\overline{\boldsymbol{u}}$ and $\boldsymbol{u}'$.

As a first example we choose a vanishing mean velocity field, $\overline{\boldsymbol{u}} = 0$, and a homogeneous isotropic steady turbulent velocity field $\boldsymbol{u}'$. The only tensors available for the construction of a_{ik} and b_{ikl} are the isotropic tensors, and therefore we have, according to section 5.2.

$$a_{ik} = \alpha\, \delta_{ik}, \quad b_{ikl} = \beta \varepsilon_{ikl}, \tag{5.36}$$

where α is a pseudo-scalar. Hence we find for the turbulent electromotive force

$$\boldsymbol{\mathfrak{E}} = \alpha \overline{\boldsymbol{B}} - \beta \operatorname{curl} \overline{\boldsymbol{B}}, \tag{5.37}$$

thus confirming relation (3.5). If the turbulent field $\boldsymbol{u}'$ is also assumed to be mirrorsymmetric, α must vanish, since it has to change its sign if $\boldsymbol{u}'$ is reflected, whereas $\boldsymbol{u}'$ does not change.

It is worth mentioning that a real turbulence undergoing no external influence is in good approximation a homogeneous isotropic mirrorsymmetric turbulence if it has grown old. For example, a turbulence excited in a homogeneous flow by a grid has these properties at a certain distance from the grid. There are no conditions known, where such a turbulence is non-mirrorsymmetric. Thus our foregoing example is rather academic. None the less we shall discuss it, since it is of great interest in dynamo theory. As we shall see in one of our following examples there are, indeed, real turbulences with non-vanishing α, but they are not isotropic.

As a second example we take again $\overline{\boldsymbol{u}} = 0$, with the turbulent motion being anisotropic now. Responsible for the anisotropy shall be a vector field $\boldsymbol{g}$, which can be the gradient of the turbulence intensity or the gradient of the density. Constructing the pseudo-tensors a_{ik}, b_{ikl} we shall restrict ourselves to linear expressions in the quantity $\boldsymbol{g}$ thus finding

$$a_{ik} = \gamma \varepsilon_{ikl} g_l, \quad b_{ikl} = \beta \varepsilon_{ikl}. \tag{5.38}$$

For the turbulent electromotive force we obtain

$$\boldsymbol{\mathfrak{E}} = -\beta \operatorname{curl} \overline{\boldsymbol{B}} - \gamma \boldsymbol{g} \times \overline{\boldsymbol{B}}. \tag{5.39}$$

The electromotive force perpendicular to $\boldsymbol{g}$ and $\overline{\boldsymbol{B}}$ shall later on be interpreted as a pumping effect.

In our third example the mean motion may be a rigid rotation and the turbulence $\boldsymbol{u}'$ shall be influenced by the motion $\overline{\boldsymbol{u}}$ and the angular velocity $\boldsymbol{\Omega}$. Again, with the restriction to linear expressions in $\overline{\boldsymbol{u}}$ and $\boldsymbol{\Omega}$ we find

$$a_{ik} = \gamma' \varepsilon_{ikl} \bar{u}_l, \quad b_{ikl} = \beta \varepsilon_{ikl} - \beta_1 \Omega_i \, \delta_{kl} - \beta_2 \Omega_k \, \delta_{il} - \beta_3 \Omega_l \, \delta_{ik}. \tag{5.40}$$

From this we obtain for the turbulent electromotive force

$$\mathfrak{E} = -\beta \operatorname{curl} \overline{\boldsymbol{B}} - \gamma' \overline{\boldsymbol{u}} \times \overline{\boldsymbol{B}} - \beta_3 (\boldsymbol{\Omega} \cdot \operatorname{grad}) \, \overline{\boldsymbol{B}} - \beta_2 \operatorname{grad} (\boldsymbol{\Omega} \cdot \overline{\boldsymbol{B}}); \tag{5.41}$$

the term with β_1 does not contribute since $\overline{\boldsymbol{B}}$ is source-free. Suppose we neglect the influence of the mean motion, i.e. $\gamma' = 0$. Then both examples deal with turbulences under an influence described by one vector quantity. However, since in the one case $\boldsymbol{g}$ is a polar vector whereas in the other case $\boldsymbol{\Omega}$ is an axial one, we get quite different results.

(5.41) describes the turbulent electromotive force $\mathfrak{E}$ for a turbulence undergoing the influence of Coriolis forces. It can be rewritten in the form

$$\mathfrak{E} = -\beta \operatorname{curl} \overline{\boldsymbol{B}} - \gamma' \overline{\boldsymbol{u}} \times \overline{\boldsymbol{B}} + \mu \beta_3 \boldsymbol{\Omega} \times \overline{\boldsymbol{j}} - (\beta_3 + \beta_2) \operatorname{grad} (\boldsymbol{\Omega} \cdot \overline{\boldsymbol{B}}). \tag{5.42}$$

As we shall see later, the electromotive force proportional to $\boldsymbol{\Omega} \times \overline{\boldsymbol{j}}$ can lead to dynamo action.

Our fourth example is closely related to the turbulent motions in a convective layer on a rotating body. There, apart from the angular velocity, we have to take into account that the radial direction is preferred. This is already the case for geometrical reasons, moreover, the gradient of density or the gradient of turbulence intensity or both will be parallel to this direction. Thus we have a turbulence under influences described by the polar vector quantities $\overline{\boldsymbol{u}}$ and $\boldsymbol{g}$ and the axial vector $\boldsymbol{\Omega}$. Assuming again linearity we find

$$a_{ik} = -\alpha_1' (\boldsymbol{g} \cdot \boldsymbol{\Omega}) \, \delta_{ik} - \alpha_2' g_i \Omega_k - \alpha_3' g_k \Omega_i + \gamma \varepsilon_{ikl} g_l + \gamma'' \varepsilon_{ikl} \bar{u}_l, \tag{5.43}$$

$$b_{ikl} = \beta \varepsilon_{ikl} - \beta_1 \Omega_i \, \delta_{kl} - \beta_2 \Omega_k \, \delta_{il} - \beta_3 \Omega_l \, \delta_{ik}. \tag{5.44}$$

Hence, for the turbulent electromotive force in a convective layer on a rotating body we obtain

$$\begin{aligned} \mathfrak{E} = & -\beta \operatorname{curl} \overline{\boldsymbol{B}} - \gamma \boldsymbol{g} \times \overline{\boldsymbol{B}} - \gamma'' \overline{\boldsymbol{u}} \times \overline{\boldsymbol{B}} - \beta_3 (\boldsymbol{\Omega} \cdot \operatorname{grad}) \, \overline{\boldsymbol{B}} - \beta_2 \operatorname{grad} (\boldsymbol{\Omega} \cdot \overline{\boldsymbol{u}}) \\ & - \alpha_1' (\boldsymbol{g} \cdot \boldsymbol{\Omega}) \, \overline{\boldsymbol{B}} - \alpha_2' (\boldsymbol{g} \cdot \overline{\boldsymbol{B}}) \, \boldsymbol{\Omega} - \alpha_3' (\boldsymbol{\Omega} \cdot \overline{\boldsymbol{B}}) \, \boldsymbol{g}. \end{aligned} \tag{5.45}$$

We emphasize the appearance of the electromotive force $-\alpha_1' (\boldsymbol{g} \cdot \boldsymbol{\Omega}) \, \overline{\boldsymbol{B}}$, which is parallel to the mean magnetic field. In contrast to the first example it is clearly to be seen here that this electromotive force may be a quite natural phenomenon in a convective layer on a rotating body.

Up to this point of our considerations on the structure of $\mathfrak{E}$ we have not gained any information whether the coefficients α, β, γ, ... are non-zero. This requires additional investigations.

5.5. Representation of the tensors $g_{ij...n}$

Before arriving at first results by combining the derivations of this and the foregoing chapter we want to assess what we have achieved with our deductions so far.

The relations (5.37), (5.39), (5.41), (5.42), (5.45) give rather explicit expressions for the turbulent electromotive force $\mathfrak{E}$. Only the scalars α_1', α_2', α_3', β, β_1, β_2, β_3, γ, γ'' appear as unknown quantities. It is essential that these expressions for $\mathfrak{E}$ are deduced with a rather weak restriction: the two-scale property of the turbulence.

A final determination of $\mathfrak{E}$ would, of course, require a general solution of the turbulence problem, which does not exist. However, since in our case uncertainty exists for a few scalars only, the insufficiency of the theory of turbulence is of minor importance here.

We shall now use the results of the foregoing chapter for the derivation of more explicit expressions for the pseudo-tensors $g_{ij...n}$. Representing $\overline{\boldsymbol{B}}$ in (4.54) by Taylor's formula and comparing the results with (5.4) we easily find

$$a_{ij} = g_{ij}^{(00)} = \iint K_{ij}(\boldsymbol{x}, \xi, t, \tau)\, \mathrm{d}\xi\, \mathrm{d}\tau, \tag{5.46}$$

$$b_{ijk} = g_{ijk}^{(10)} = -\iint K_{ij}(\boldsymbol{x}, \xi, t, \tau)\, \xi_k\, \mathrm{d}\xi\, \mathrm{d}\tau, \tag{5.47}$$

and, generally,

$$g_{ij}^{(0\nu)} = \frac{(-1)^\nu}{\nu!} \iint K_{ij}(\boldsymbol{x}, \xi, t, \tau)\, \tau^\nu\, \mathrm{d}\xi\, \mathrm{d}\tau, \tag{5.48}$$

$$g_{ijk}^{(1\nu)} = \frac{(-1)^{\nu+1}}{\nu!} \iint K_{ij}(\boldsymbol{x}, \xi, t, \tau)\, \xi_k \tau^\nu\, \mathrm{d}\xi\, \mathrm{d}\tau, \tag{5.49}$$

$$g_{ijk...n}^{(\varkappa\nu)} = \frac{(-1)^{\varkappa+\nu}}{\varkappa!\,\nu!} \iint K_{ij}(\boldsymbol{x}, \xi, t, \tau)\, \xi_k \ldots \xi_n \tau^\nu\, \mathrm{d}\xi\, \mathrm{d}\tau. \tag{5.50}$$

The tensor K_{ij} can be used as in (4.55) or in another form for a higher approximation level.

In order to give an impression of some of the following deductions we now deduce α and β for a homogeneous isotropic and steady turbulence and a vanishing mean velocity field. Using the second order correlation approximation we obtain from (5.46), (5.47) with (4.55) and (4.57)

$$a_{ij} = \varepsilon_{ilm}\varepsilon_{mnp}\varepsilon_{pqj} \iint \frac{\partial G(\xi. \tau)}{\partial \xi_n} Q_{lq}(-\xi, -\tau)\, d\xi\, d\tau, \tag{5.51}$$

$$b_{ijk} = -\varepsilon_{ilm}\varepsilon_{mnp}\varepsilon_{pqj} \iint \frac{\partial G(\xi, \tau)}{\partial \xi_n} \xi_k Q_{lq}(-\xi, -\tau)\, d\xi\, d\tau. \tag{5.52}$$

Since for homogeneous turbulence the correlation tensor Q_{lq} as defined by (4.3) does not depend on $\boldsymbol{x}, t$ we simply write

$$Q_{lq}(\xi, \tau) = \overline{u'_l(\boldsymbol{x}, t)\, u'_q(\boldsymbol{x} + \xi, t + \tau)}, \tag{5.53}$$

and by exchanging $\boldsymbol{x}, t$ for $\boldsymbol{x} - \xi, t - \tau$ we find the symmetry relation

$$Q_{lq}(-\xi, -\tau) = Q_{ql}(\xi, \tau). \tag{5.54}$$

We already know that $a_{ij} = \alpha\, \delta_{ij}$. Hence we get by contracting (5.51)

$$\alpha = \frac{1}{3} \varepsilon_{ilm}\varepsilon_{mnp}\varepsilon_{pqi} \iint \frac{\partial G(\xi, \tau)}{\partial \xi_n} Q_{ql}(\xi, \tau)\, d\xi\, d\tau$$

$$= \frac{1}{3} \varepsilon_{nlq} \iint \frac{\partial G(\xi, \tau)}{\partial \xi_n} Q_{ql}(\xi, \tau)\, d\xi\, d\tau = -\frac{1}{3} \iint G(\xi, \tau)\, \varepsilon_{nlq} \frac{\partial Q_{ql}(\xi, \tau)}{\partial \xi_n}\, d\xi\, d\tau.$$

Inserting now the definition of the correlation tensor (5.53) we finally obtain (KRAUSE [1])

$$\alpha = -\frac{1}{3} \iint G(\xi, \tau)\, \overline{\boldsymbol{u}'(\boldsymbol{x}, t) \cdot \operatorname{curl} \boldsymbol{u}'(\boldsymbol{x} + \xi, t + \tau)}\, d\xi\, d\tau. \tag{5.55}$$

This formula is a generalization of both the formulae (3.29) and (3.47). We find these earlier results as limiting cases of (5.55). Let us first consider the high-conductivity limit characterized by $\tau_{\text{cor}} \ll \mu\sigma\lambda_{\text{cor}}^2$. Concerning the spatial integration we note that $G(\xi, \tau)$ is significantly different from zero only if $\xi \lessapprox \sqrt{\tau/\mu\sigma}$. Since the scalar product vanishes if $\tau > \tau_{\text{cor}}$ we have $\xi \lessapprox \sqrt{\tau_{\text{cor}}/\mu\sigma} \ll \lambda_{\text{cor}}$ for the region where the integrand is significantly different from zero in the high-conductivity limit. On these conditions the scalar product does not vary and we obtain (3.29), since

$$\int G(\xi. \tau)\, d\xi = 1. \tag{5.56}$$

Alternatively, $\mu\sigma\lambda_{\text{cor}}^2 \ll \tau_{\text{cor}}$, we can carry out the τ-integration by assuming the scalar product $\overline{\boldsymbol{u}' \operatorname{curl} \boldsymbol{u}'}$ to be constant. Since

$$\int G(\xi, \tau)\, d\tau = \frac{\mu\sigma}{4\pi\xi}, \tag{5.57}$$

we obtain (3.47).

As for a_{ij} we know that the tensor b_{ijk} has the form $\beta\varepsilon_{ijk}$. Hence, by multiplying (5.52) by ε_{ijk} we find

$$\beta = \frac{1}{6}\,\varepsilon_{ijk}b_{ijk} = -\varepsilon_{ijk}\varepsilon_{ilm}\varepsilon_{mnp}\varepsilon_{pqj}\iint \frac{\partial G(\xi,\tau)}{\partial \xi_n}\,\xi_k Q_{ql}(\boldsymbol{\xi},\tau)\,\mathrm{d}\boldsymbol{\xi}\,\mathrm{d}\tau$$

$$= -\,(\delta_{kl}\delta_{nq} + \delta_{kq}\delta_{nl})\,\frac{1}{6}\iint \frac{\partial G(\xi,\tau)}{\partial \xi_n}\,\xi_k Q_{ql}(\boldsymbol{\xi},\tau)\,\mathrm{d}\boldsymbol{\xi}\,\mathrm{d}\tau. \tag{5.58}$$

We now take into account that G depends on the magnitude ξ only, hence we have

$$\frac{\partial G(\xi,\tau)}{\partial \xi_n} = \frac{\xi_n}{\xi}\,\frac{\partial G(\xi,\tau)}{\partial \xi}, \tag{5.59}$$

and find

$$\beta = -\,\frac{1}{3}\iint \frac{1}{\xi}\,\frac{\partial G(\xi,\tau)}{\partial \xi}\,\overline{(\boldsymbol{\xi}\cdot\boldsymbol{u}'(\boldsymbol{x},t))\,(\boldsymbol{\xi}\cdot\boldsymbol{u}'(\boldsymbol{x}+\boldsymbol{\xi},t+\tau)}\,\mathrm{d}\boldsymbol{\xi}\,\mathrm{d}\tau. \tag{5.60}$$

This formula confirms both the formulae (3.30) and (3.48), and generalizes them.

We can arrive at an alternative expression by introducing the longitudinal correlation function f defined by

$$\xi_k Q_{kl}(\boldsymbol{\xi},\tau) = f(\xi,\tau)\,\xi_l. \tag{5.61}$$

Inserting (5.59) and (5.61) into (5.58) we obtain

$$\beta = -\,\frac{1}{3}\iint \xi\,\frac{\partial G(\xi,\tau)}{\partial \xi}\,f(\xi,\tau)\,\mathrm{d}\boldsymbol{\xi}\,\mathrm{d}\tau, \tag{5.62}$$

a relation which is due to RÄDLER [2].

CHAPTER 6

HOMOGENEOUS TURBULENCE

6.1. Introductory remarks

We start our more detailed discussion with the treatment of homogeneous and steady turbulence. For this kind of turbulence all mean quantities are constant with respect to variations in space and time, and thus, in particular, the pseudo-tensors $g_{ij...h}$ introduced in the foregoing chapter are constant.

Moreover, since there is a close connection between the turbulent velocity field and the mean velocity field, we assume the mean velocity field to be constant, i.e. that we have $\overline{\boldsymbol{u}} = 0$ in a proper system of reference. Hence the Green's tensor G_{jp} is well known, given by (4.57) and (4.58).

We have already mentioned that the two-point two-time correlation tensor Q_{ij} is a function only of the relative coordinates $\boldsymbol{\xi}$, τ and that it fulfils the symmetry condition (5.54). There is still another condition: Let x_ν, t_ν, $\nu = 1, \ldots, N$, be space-time points. Then for any set of complex numbers $c_{i\nu}$, $i = 1, 2, 3$; $\nu = 1, \ldots, N$ we have

$$\overline{\left\{\sum_{\nu=1}^{N} c_{i\nu} u_i'(\boldsymbol{x}_\nu, t_\nu)\right\} \left\{\sum_{\mu=1}^{N} c_{k\mu}^* u_k'(\boldsymbol{x}_\mu, t_\mu)\right\}} \geqq 0. \tag{6.1}$$

Carrying out the multiplications and taking the average we obtain

$$\sum_{\nu,\mu=1}^{N} c_{i\nu} c_{k\mu} Q_{ik}(\boldsymbol{x}_\nu - \boldsymbol{x}_\mu, t_\nu - t_\mu) \geqq 0. \tag{6.2}$$

This inequality demonstrates Bochner's theorem (Bochner [1], cf. also Cramer [1]), which states that Q_{ik} can be the correlation tensor of a homogeneous stationary random field, $\boldsymbol{u}'$, if and only if it is positive semi-definite in the sense of (6.2).

If, in addition, the turbulence is isotropic, then for the construction of Q_{ik} there are only available the isotropic tensors δ_{ik} and ε_{ikl} and the vector $\boldsymbol{\xi}$. Therefore the general expression reads

$$Q_{ik}(\boldsymbol{\xi}, \tau) = A(\xi, \tau)\, \delta_{ik} + B(\xi, \tau)\, \xi_i \xi_k + C(\xi, \tau)\, \varepsilon_{ikl} \xi_l. \tag{6.3}$$

A, B, C depend only on the magnitude of $\boldsymbol{\xi}$, i.e. $\xi = \sqrt{\boldsymbol{\xi}^2}$, and therefore constitute even functions of $\boldsymbol{\xi}$. Then (5.54) implies that they are even also with respect to τ. Bochner's theorem (6.2) provides for additional restrictions on these functions. But we shall not record them here, since a more convenient form will be available by considering the Fourier transformed quantities.

6.2. Fourier transformation of homogeneous steady random fields

Let $F(\boldsymbol{x}, t)$ be a function of space and time. By $\hat{F}(\boldsymbol{k}, \omega)$ we denote its Fourier transform defined by

$$\hat{F}(\boldsymbol{k}, \omega) = \frac{1}{(2\pi)^4} \iint F(\boldsymbol{x}, t)\, \mathrm{e}^{-\mathrm{i}(\boldsymbol{kx}-\omega t)}\, \mathrm{d}\boldsymbol{x}\, \mathrm{d}t; \tag{6.4}$$

or the inversion formula

$$F(\boldsymbol{x}, t) = \iint \hat{F}(\boldsymbol{k}, \omega)\, \mathrm{e}^{\mathrm{i}(\boldsymbol{kx}-\omega t)}\, \mathrm{d}\boldsymbol{k}\, \mathrm{d}\omega. \tag{6.5}$$

Some basic relations may be noted:

(i) If $F(\boldsymbol{x}, t)$ is real it follows that

$$\hat{F}(-\boldsymbol{k}, -\omega) = \hat{F}^*(\boldsymbol{k}, \omega). \tag{6.6}$$

(ii) The Fourier transform of the product of the two functions $F(\boldsymbol{x}, t)$ and $G(\boldsymbol{x}, t)$ is given by the convolution integral

$$\widehat{(F \cdot G)}\, (\boldsymbol{k}, \omega) = (\hat{F} * \hat{G})\, (\boldsymbol{k}, \omega) = \iint \hat{F}(\boldsymbol{k}', \omega')\, \hat{G}(\boldsymbol{k} - \boldsymbol{k}', \omega - \omega')\, \mathrm{d}\boldsymbol{k}'\, \mathrm{d}\omega'. \tag{6.7}$$

(iii) For $\boldsymbol{k} = 0$, $\omega = 0$ in (ii) we obtain Parseval's theorem:

$$\iint F(\boldsymbol{x}, t)\, G(\boldsymbol{x}, t)\, \mathrm{d}\boldsymbol{x}\, \mathrm{d}t = (2\pi)^4 \iint \hat{F}(\boldsymbol{k}, \omega)\, \hat{G}^*(\boldsymbol{k}, \omega)\, \mathrm{d}\boldsymbol{k}\, \mathrm{d}\omega, \tag{6.8}$$

or, in scalar product notation,

$$(F, G) = (2\pi)^4\, (\hat{F}, \hat{G}). \tag{6.9}$$

For a homogeneous and steady random field like $\boldsymbol{u}'(\boldsymbol{x}, t)$ the Fourier transform in the classical sense is not defined, since $\boldsymbol{u}'$ generally does not vanish if $\boldsymbol{x}$ and t tend to infinity. They can, however, be defined as functionals. That is, we understand the symbol (F, φ) not as a scalar product of functions F and φ but as a functional $(F, \cdot)$ operating on the finite function φ, and we take Parseval's theorem as a definition of the Fourier transform $(\hat{F}, \cdot)$ of the functional $(F, \cdot)$:

$$(\hat{F}, \hat{\varphi}) = \left(\frac{1}{2\pi}\right)^4 (F, \varphi). \tag{6.10}$$

$\hat{\varphi}$ exists in the classical sense because φ is a finite function, i.e. a function which is zero outside a finite region. For the sake of convenience we shall write again F, $\hat{F}$ instead of $(F, \cdot)$, $(\hat{F}, \cdot)$.

We are now in a position to work with Fourier transforms of homogeneous steady random fields. It will be useful to know that in the definition of $\hat{F}$ the functions φ need not be finite but may be functions which are exponentially decreasing as $\boldsymbol{x}$, t tend to infinity. Thus if F is a homogeneous steady random field (taken as a functional) and G a function of that kind, then $\hat{F}$ is given by

$$(\hat{F}, \hat{G}) = \frac{1}{(2\pi)^4} \iint F(\boldsymbol{x}, t)\, G(\boldsymbol{x}, t)\, \mathrm{d}\boldsymbol{x}\, \mathrm{d}t, \tag{6.11}$$

As a useful example we choose $F(\boldsymbol{x}, t) \equiv 1$. From (6.11) and (6.4) we obtain

$$(\hat{F}, \hat{G}) = \frac{1}{(2\pi)^4} \iint G(\boldsymbol{x}, t)\, \mathrm{d}\boldsymbol{x}\, \mathrm{d}t = \hat{G}(0, 0), \tag{6.12}$$

and $\hat{F}$ turns out to be the Dirac δ-functional centered at $\boldsymbol{k} = 0$, $t = 0$.

As a second example we take $F(\boldsymbol{x}, t) = x_i$. We find

$$(\hat{F}, \hat{G}) = \frac{1}{(2\pi)^4} \iint x_i G(\boldsymbol{x}, t)\, \mathrm{d}\boldsymbol{x}\, \mathrm{d}t = \left(i\, \frac{\partial \hat{G}(\boldsymbol{k}, \omega)}{\partial k_i}\right)_{\boldsymbol{k}=0, \omega=0}.$$

i.e., that $\hat{F}$ is proportional to a derivative of the Dirac δ-functional.

Thus we have the pairs

$$F(\boldsymbol{x}, t) = 1, \quad \hat{F}(\boldsymbol{k}, \omega) = \delta(\boldsymbol{k}, \omega); \tag{6.13}$$

and

$$F(\boldsymbol{x}, t) = x_i, \quad \hat{F}(\boldsymbol{k}, \omega) = \mathrm{i}\, \frac{\partial \delta(\boldsymbol{k}, \omega)}{\partial k_i}. \tag{6.14}$$

As a third example we note the pair

$$F(\boldsymbol{x}, t) = x_i x_j, \quad \hat{F}(\boldsymbol{k}, \omega) = -\, \frac{\partial^2 \delta(\boldsymbol{k}, \omega)}{\partial k_i\, \partial k_j}. \tag{6.15}$$

6.3. A basic relation connecting the means of the Fourier transforms with the Fourier transform of the correlation tensor

For the functionals $\hat{u}_i$, $\hat{u}_j$ we have in the above sense

$$\begin{aligned} \overline{(\hat{u}_i, \hat{F})\, (\hat{u}_j, \hat{G})} &= \frac{1}{(2\pi)^8} \iiiint \overline{u_i'(\boldsymbol{x}, t)\, F(\boldsymbol{x}, t)\, u_j'(\boldsymbol{x}', t')\, G(\boldsymbol{x}', t')}\, \mathrm{d}\boldsymbol{x}'\, \mathrm{d}t'\, \mathrm{d}\boldsymbol{x}\, \mathrm{d}t \\ &= \frac{1}{(2\pi)^8} \iiiint Q_{ij}(\boldsymbol{x}' - \boldsymbol{x}, t' - t)\, F(\boldsymbol{x}, t)\, G(\boldsymbol{x}', t')\, \mathrm{d}\boldsymbol{x}\, \mathrm{d}t\, \mathrm{d}\boldsymbol{x}'\, \mathrm{d}t'. \end{aligned} \tag{6.16}$$

Taking into account that the Fourier transform of $Q_{ij}(\boldsymbol{x}' - \boldsymbol{x}, t' - t)$ with respect to $\boldsymbol{x}'$, t' is given by $\hat{Q}_{ij}(\boldsymbol{k}', \omega') \exp[-i(\boldsymbol{k}'\boldsymbol{x} - \omega' t)]$ we obtain by the definition of the Fourier transforms for real functions F and G

$$\overline{(\hat{u}_i, \hat{F})\,(\hat{u}_j, \hat{G})} = \int\int \hat{Q}_{ij}(\boldsymbol{k}', \omega')\, \hat{F}(\boldsymbol{k}', \omega')\, \hat{G}^*(\boldsymbol{k}', \omega')\, \mathrm{d}\boldsymbol{k}'\, \mathrm{d}\omega' \tag{6.17}$$
$$= \int\int\int\int \hat{Q}_{ij}(\boldsymbol{k}', \omega')\, \delta(\boldsymbol{k} + \boldsymbol{k}', \omega + \omega')\, \hat{F}^*(\boldsymbol{k}, \omega)\, \hat{G}^*(\boldsymbol{k}', \omega')\, \mathrm{d}\boldsymbol{k}'\, \mathrm{d}\omega'\, \mathrm{d}\boldsymbol{k}\, \mathrm{d}\omega\,.$$

Let S be the space of all real-valued admitted functions. Then (6.16) and (6.17) define functionals on the product space $S \times S$:

$$\left(\frac{1}{2\pi}\right)^8 (\overline{u_i' \times u_j'}, F \times G) = (\overline{\hat{u}_i \times \hat{u}_j}, \hat{F} \times \hat{G})$$

(with obvious symbolic notation), where $\overline{\hat{u}_i \times \hat{u}_j}$ is the Fourier transform of the functional $\overline{u_i' \times u_j'}$. These functionals are integral transformations with the kernels

$$\overline{u_i'(\boldsymbol{x}, t)\, u_j'(\boldsymbol{x}', t')} = Q_{ij}(\boldsymbol{x}' - \boldsymbol{x}, t' - t)$$

and

$$\overline{\hat{u}_i(\boldsymbol{k}, \omega)\, \hat{u}_j(\boldsymbol{k}', \omega')} = \hat{Q}_{ij}(\boldsymbol{k}', \omega')\, \delta(\boldsymbol{k} + \boldsymbol{k}', \omega + \omega') \tag{6.18}$$

respectively.

6.4. Bochner's theorem

$Q_{ij}(\xi, \tau)$ is a real quantity, therefore we have from (6.6)

$$\hat{Q}_{ij}(-\boldsymbol{k}, -\omega) = \hat{Q}_{ij}^*(\boldsymbol{k}, \omega)\,. \tag{6.19}$$

Moreover, from the symmetry relation (5.54) we obtain

$$\hat{Q}_{ij}(\boldsymbol{k}, \omega) = \hat{Q}_{ji}(-\boldsymbol{k}, -\omega)\,. \tag{6.20}$$

Combining the relations (6.19) and (6.20) we find

$$\hat{Q}_{ji}(\boldsymbol{k}, \omega) = \hat{Q}_{ij}^*(\boldsymbol{k}, \omega)\,, \tag{6.21}$$

i.e. the Fourier transform of the correlation tensor is Hermitian.

The inequality (6.2) can be cast in the form

$$\int\int \hat{Q}_{ij}(\boldsymbol{k}, \omega) \sum_{\nu,\mu=1}^{N} c_{i\nu} \mathrm{e}^{\mathrm{i}(\boldsymbol{k}\boldsymbol{x}_\nu - \omega t_\nu)}\, c_{j\mu}^* \mathrm{e}^{-\mathrm{i}(\boldsymbol{k}\boldsymbol{x}_\mu - \omega t_\mu)}\, \mathrm{d}\boldsymbol{k}\, \mathrm{d}\omega \geqq 0\,. \tag{6.22}$$

If we introduce the quantities

$$X_i = \sum_{\nu=1}^{N} c_{i\nu} e^{i(\boldsymbol{k}\boldsymbol{x}_\nu - \omega t_\nu)}, \tag{6.23}$$

we can write instead of (6.22)

$$\int\int \hat{Q}_{ij}(\boldsymbol{k}, \omega)\, X_i(\boldsymbol{k}, \omega)\, X_j^*(\boldsymbol{k}, \omega)\, d\boldsymbol{k}\, d\omega \geqq 0, \tag{6.24}$$

This inequality holds for arbitrary $c_{i\nu}$, $\boldsymbol{x}_\nu$, t_ν if and only if

$$\hat{Q}_{ij}(\boldsymbol{k}, \omega)\, X_i X_j^* \geqq 0, \tag{6.25}$$

i.e. if $\hat{Q}_{ij}(\boldsymbol{k}, \omega)$ is a positive semidefinite Hermitian tensor. This is a form of Bochner's theorem far more convenient than (6.2).

6.5. Isotropic turbulence

In (6.3) we already introduced the correlation tensor of the second rank of a homogeneous isotropic steady turbulence. Having in mind the determination of the Fourier transformed tensor we note the relations

$$\widehat{B\xi_i\xi_j} = -\frac{\partial^2 \hat{B}}{\partial k_i\, \partial k_j}, \quad \widehat{C\xi_i} = i\frac{\partial \hat{C}}{\partial k_i}, \tag{6.26}$$

which are easily derived from (6.7), (6.14) and (6.15) by regarding that

$$\left(\hat{F} * \frac{\partial \hat{G}}{\partial k_i}\right) = \frac{\partial}{\partial k_i}(\hat{F} * \hat{G}), \tag{6.27}$$

for arbitrary functions $\hat{F}$ and $\hat{G}$.

From (6.3) and (6.26) it follows for the Fourier transformed correlation tensor of the second rank of a homogeneous isotropic steady turbulence

$$\hat{Q}_{ij}(\boldsymbol{k}, \omega) = \hat{A}(k, \omega)\, \delta_{ij} - \frac{\partial^2 \hat{B}(k, \omega)}{\partial k_i\, \partial k_j} + i\varepsilon_{ijk} \frac{\partial \hat{C}(k, \omega)}{\partial k_k}. \tag{6.28}$$

The functions $\hat{A}$, $\hat{B}$, $\hat{C}$ depend only on the magnitude of $\boldsymbol{k}$. On carrying out the differentiations we therefore obtain

$$\hat{Q}_{ij}(\boldsymbol{k}, \omega) = \left(\hat{A} - \frac{1}{k}\frac{\partial \hat{B}}{\partial k}\right)\delta_{ij} - \left(\frac{\partial^2 \hat{B}}{\partial k^2} - \frac{1}{k}\frac{\partial \hat{B}}{\partial k}\right)\frac{k_i k_j}{k^2} + \frac{i}{k}\frac{\partial \hat{C}}{\partial k}\varepsilon_{ijk} k_k. \tag{6.29}$$

We now derive the conditions which Bochner's theorem imposes on the functions $\hat{A}$, $\hat{B}$, $\hat{C}$. For simplicity we write

$$\hat{Q}_{ij}(\boldsymbol{k}, \omega) = \hat{F}(k, \omega)\, \delta_{ij} + \hat{G}(k, \omega)\, k_i k_j + i\hat{H}(k, \omega)\, \varepsilon_{ijk} k_k, \tag{6.30}$$

where

$$\hat{F} = \hat{A} - \frac{1}{k}\frac{\partial \hat{B}}{\partial k}, \quad \hat{G} = -\frac{1}{k^2}\left(\frac{\partial^2 \hat{B}}{\partial k^2} - \frac{1}{k}\frac{\partial \hat{B}}{\partial k}\right), \quad \hat{H} = \frac{1}{k}\frac{\partial \hat{C}}{\partial k}. \tag{6.31}$$

$\hat{Q}_{ij}$ is positive semidefinite if and only if

(i) the trace of $\hat{Q}_{ij}$ is non-negative, i.e. if

$$3\hat{F} + k^2\hat{G} = 0; \tag{6.32}$$

(ii) the trace of the tensor of the algebraic complements, $\hat{q}_{ij}$,is non-negative, where $\hat{q}_{ij}$ is defined by

$$\hat{q}_{ij}(\boldsymbol{k}, \omega) = \varepsilon_{imn}\varepsilon_{jpq}\hat{Q}_{mp}\hat{Q}_{nq}. \tag{6.33}$$

Thus we have

$$\varepsilon_{imn}\varepsilon_{ipq}\hat{Q}_{mp}\hat{Q}_{nq} = 4\hat{F}(\hat{F} + k^2\hat{G}) + 2(\hat{F}^2 - k^2\hat{H}^2) \geqq 0; \tag{6.34}$$

(iii) the determinant of the tensor $\hat{Q}_{ij}$ is non-negative, i.e.

$$\varepsilon_{ijk}\varepsilon_{pqr}\hat{Q}_{ip}\hat{Q}_{jq}\hat{Q}_{kr} = 6(\hat{F} + k^2\hat{G})\,(\hat{F}^2 - k^2\hat{H}^2) \geqq 0. \tag{6.35}$$

(6.32), (6.34) and (6.35) can be replaced by the more convenient conditions

$$\hat{F} + k^2\hat{G} \geqq 0, \quad \hat{F} \geqq |k\hat{H}|. \tag{6.36}$$

(6.32), (6.34) and (6.35) are clearly a consequence of (6.36). In order to prove the converse let us assume (6.32), (6.34) and (6.35) are valid and one of the conditions (6.36) is violated. Because of (6.35) the other relation (6.36) must be violated too. Then (6.34) requires $\hat{F} < 0$, thus giving $2\hat{F} + (\hat{F} + k^2\hat{G}) < 0$, in contradiction to (6.32). Consequently, according to Bochner's theorem the tensor $\hat{Q}_{ij}$ given by (6.3) is the correlation tensor of the second rank of a homogeneous isotropic steady random field, if and only if

$$\hat{A} - \frac{\partial^2 \hat{B}}{\partial k^2} \geqq 0, \quad \hat{A} - \frac{1}{k}\frac{\partial \hat{B}}{\partial k} \geqq \left|\frac{\partial \hat{C}}{\partial k}\right|. \tag{6.37}$$

(Krause and Roberts [2]). This follows from (6.36) in combination with (6.31).

6.6. Two special cases: Incompressible turbulence and random sound waves

For incompressible turbulence we have

$$\frac{\partial u_i'(\boldsymbol{x}, t)}{\partial x_i} = 0, \tag{6.38}$$

and, therefore, for the correlation tensor

$$\frac{\partial Q_{ij}(\boldsymbol{\xi}, \tau)}{\partial \xi_i} = 0, \quad \frac{\partial Q_{ij}(\boldsymbol{\xi}, \tau)}{\partial \xi_j} = 0. \tag{6.39}$$

In Fourier space these conditions read

$$k_i \hat{u}_i(\boldsymbol{k}, \omega) = 0, \tag{6.40}$$

and for the correlation tensor

$$k_i \hat{Q}_{ij}(\boldsymbol{k}, \omega) = 0, \quad k_j \hat{Q}_{ij}(\boldsymbol{k}, \omega) = 0. \tag{6.41}$$

If, in addition, the turbulence is isotropic we find from (6.29)

$$\hat{A} - \frac{\partial^2 \hat{B}}{\partial k^2} = 0, \tag{6.42}$$

and the correlation tensor simplifies to

$$\hat{Q}_{ij}(\boldsymbol{k}, \omega) = \hat{Q}(k, \omega)\,(k^2\,\delta_{ij} - k_i k_j) + \frac{i}{k}\,\frac{\partial \hat{C}(k, \omega)}{\partial k}\,\varepsilon_{ijk} k_k, \tag{6.43}$$

where

$$k^2 \hat{Q}(k, \omega) = \hat{A}(k, \omega) - \frac{1}{k}\,\frac{\partial \hat{B}(k, \omega)}{\partial k}, \tag{6.44}$$

and, due to Bochner's theorem (6.37),

$$\hat{Q}(k, \omega) \geq \left| \frac{1}{k^2}\,\frac{\partial \hat{C}(k, \omega)}{\partial k} \right|. \tag{6.45}$$

Hence, in particular, $\hat{Q}$ is a non-negative function.

We note here the incompressibility condition in $\boldsymbol{\xi}$-space

$$\frac{1}{\xi}\,\frac{\partial A}{\partial \xi} + \xi\,\frac{\partial B}{\partial \xi} + 4B = 0, \tag{6.46}$$

which can easily be derived by inserting (6.3) into (6.39). In terms of the longitudinal correlation function f introduced in (5.61) this condition can be rewritten

$$A = f + \frac{\xi}{2}\,\frac{\partial f}{\partial \xi}, \quad B = -\frac{1}{2\xi}\,\frac{\partial f}{\partial \xi}. \tag{6.47}$$

A different case we have assuming the turbulence is given by random sound waves thus having a potential:

$$u_i'(\boldsymbol{x}, t) = -\frac{\partial \Phi(\boldsymbol{x}, t)}{\partial x_i}. \tag{6.48}$$

Φ is a homogeneous random function. For the correlation tensor we find

$$Q_{ij}(\xi, \tau) = \overline{\frac{\partial \Phi(x, t)}{\partial x_i} \frac{\partial \Phi(x + \xi, t + \tau)}{\partial \xi_j}}$$

$$= \frac{\partial}{\partial x_i} \overline{\left[\Phi(x, t) \frac{\partial \Phi(x + \xi, t + \tau)}{\partial \xi_j}\right]} - \overline{\Phi(x, t) \frac{\partial^2 \Phi(x + \xi, t + \tau)}{\partial \xi_i \, \partial \xi_j}}$$

$$= -\frac{\partial^2 R(\xi. \tau)}{\partial \xi_i \, \partial \xi_j}, \tag{6.49}$$

where R is the correlation function

$$R(\xi, \tau) = \overline{\Phi(x, t) \, \Phi(x + \xi, t + \tau)}. \tag{6.50}$$

In Fourier space we have

$$\hat{Q}_{ij}(k, \omega) = k_i k_j \hat{R}(k, \omega), \tag{6.51}$$

and Bochner's theorem (6.36) simply reads

$$\hat{R}(k, \omega) \geqq 0. \tag{6.52}$$

6.7. Fourier transform of the Green's function tensor. Evaluation of integrals in the limiting cases

As we said, this chapter shall be restricted to cases of vanishing mean flow. For these the Green's function tensor G_{jp} is known, given by (4.57) and (4.58). The special dependence on the arguments allows us to write

$$G_{jp}(x, \xi, t, \tau) = G_{jp}(x - \xi, t - \tau) \tag{6.53}$$

and to introduce the Fourier transform by

$$G_{jp}(x - \xi, t - \tau) = \int\int \hat{G}_{jp}(k, \omega) \, e^{i[k(x - \xi) - \omega(t - \tau)]} \, dk \, d\omega, \tag{6.54}$$

where

$$\hat{G}_{jp}(k, \omega) = \delta_{jp} \hat{G}(k, \omega). \tag{6.55}$$

We can easily derive the function $\hat{G}(k, \omega)$ from the Fourier transformed condition (iii) in section 4.5., which because of (6.4) reads

$$(-i\omega + \eta k^2) \, \hat{G}_{jp} = \frac{\delta_{jp}}{(2\pi)^4}. \tag{6.56}$$

Hence we obtain

$$\hat{G}(\boldsymbol{k}, \omega) = \frac{1}{(2\pi)^4} \frac{1}{\eta k^2 - \mathrm{i}\omega}. \tag{6.57}$$

As was already seen in section 5.5. in our investigations there are certain integrals of interest in which the integrand is a product of the Green's function with another function of space and time. Let A be such a quantity. In virtue of Parseval's theorem (6.8) we can write

$$A = \iint G(\xi, \tau)\, F(\boldsymbol{\xi}, \tau)\, \mathrm{d}\boldsymbol{\xi}\, \mathrm{d}\tau = (2\pi)^4 \iint \hat{G}(k, \omega)\, \hat{F}^*(\boldsymbol{k}, \omega)\, \mathrm{d}\boldsymbol{k}\, \mathrm{d}\omega, \tag{6.58}$$

or explicitily,

$$A = \int_0^\infty \iint \left(\frac{\mu\sigma}{4\pi\tau}\right)^{3/2} \exp\left(-\frac{\mu\sigma\xi^2}{4\tau}\right) F(\boldsymbol{\xi}, \tau)\, \mathrm{d}\boldsymbol{\xi}\, \mathrm{d}\boldsymbol{\tau} = \iint \frac{\hat{F}^*(\boldsymbol{k}, \omega)}{\eta k^2 - \mathrm{i}\omega}\, \mathrm{d}\boldsymbol{k}\, \mathrm{d}\omega. \tag{6.59}$$

Often we wish to know this quantity A in one of the limiting cases, the high-conductivity limit, i.e. $\tau_{\mathrm{cor}} \ll \mu\sigma\lambda_{\mathrm{cor}}^2$ or $\eta = \dfrac{1}{\mu\sigma} \to 0$, or the low-conductivity limit, i.e. $\tau_{\mathrm{cor}} \gg \mu\sigma\lambda_{\mathrm{cor}}^2$ or $\eta = \dfrac{1}{\mu\sigma} \to \infty$.

Inserting $\boldsymbol{\zeta} = \sqrt{\mu\sigma}\, \boldsymbol{\xi}$ into (6.59) we get

$$A = \int_0^\infty \iint \frac{1}{(4\pi\tau)^{3/2}} \mathrm{e}^{-\zeta^2/4\tau} F\left(\frac{\boldsymbol{\zeta}}{\sqrt{\mu\sigma}}, \tau\right) \mathrm{d}\boldsymbol{\zeta}\, \mathrm{d}\tau$$

$$\to \int_0^\infty \iint \frac{1}{(4\pi\tau)^{3/2}} \mathrm{e}^{-\zeta^2/4\tau} F(0, \tau)\, \mathrm{d}\boldsymbol{\zeta}\, \mathrm{d}\tau = \int_0^\infty F(0, \tau)\, \mathrm{d}\tau, \text{ if } \sigma \to \infty. \tag{6.60}$$

Hence we obtain in the high-conductivity limit

$$A = \int_0^\infty F(0, \tau)\, \mathrm{d}\tau = \iint \frac{\hat{F}^*(\boldsymbol{k}, \omega)}{-\mathrm{i}\omega}\, \mathrm{d}\boldsymbol{k}\, \mathrm{d}\omega. \tag{6.61}$$

In case the integral in Fourier space does not exist, a special interpretation of relation (6.61) is needed: Let $\psi(t)$ be the integral of the δ-functional defined by

$$\psi(t) = \int_{-\infty}^{t} \delta(t')\, \mathrm{d}t'. \tag{6.62}$$

Then

$$\hat{\psi}(\omega) = \frac{1}{2\pi} \frac{1}{-\mathrm{i}\omega}. \tag{6.63}$$

Now we can understand the second integral in (6.61) as the value at $\hat{F}^*(\boldsymbol{k}, \omega)$ of the Fourier transform of the functional $(2\pi)^4\, \psi(t)\, \delta(\boldsymbol{x})$. Now by the definition of Fourier transforms of functionals the first integral in (6.61) is equal to the second.

Substituting into (6.59) $\varrho = \dfrac{\mu\sigma}{4\tau}$, we find

$$A = \frac{\mu\sigma}{4\sqrt{\pi}^3} \int_0^\infty \int \frac{1}{\sqrt{\varrho}}\, e^{-\varrho\xi^2}\, F\left(\xi, \frac{\mu\sigma}{4\varrho}\right) d\xi\, d\varrho \to \frac{\mu\sigma}{4\sqrt{\pi}^3} \int_0^\infty \int \frac{e^{-\varrho\xi^2}}{\sqrt{\varrho}}\, F(\xi, 0)\, d\xi\, d\varrho$$

if $\sigma \to 0$. Carrying out the ϱ-integration we obtain in the low-conductivity limit

$$A = \frac{\mu\sigma}{4\pi} \int \frac{F(\xi, 0)}{\xi}\, d\xi = \mu\sigma \iint \frac{\hat{F}^*(\boldsymbol{k}, \omega)}{k^2}\, d\boldsymbol{k}\, d\omega\,. \tag{6.64}$$

In this case the equality of the two integrals is obvious because a function A is the regular solution of the equation

$$\Delta A(\xi) = -\mu\sigma F(\xi, 0)\,, \tag{6.65}$$

there is the well-known representation

$$A(0) = \frac{\mu\sigma}{4\pi} \int \frac{F(\xi, 0)}{\xi}\, d\xi\,. \tag{6.66}$$

Let $A(\xi)$ be represented by the Fourier integral

$$A(\xi) = \int \tilde{A}(\boldsymbol{k})\, e^{i\boldsymbol{k}\xi}\, d\boldsymbol{k}\,, \tag{6.67}$$

then from (6.65) follows

$$\tilde{A}(\boldsymbol{k}) = \frac{\mu\sigma}{k^2} \int \hat{F}(\boldsymbol{k}, \omega)\, d\omega\,, \tag{6.68}$$

and therefore

$$A(0) = \int \tilde{A}(\boldsymbol{k})\, d\boldsymbol{k} = \mu\sigma \iint \frac{\hat{F}(\boldsymbol{k}, \omega)}{k^2}\, d\boldsymbol{k}\, d\omega\,. \tag{6.69}$$

As $F(\xi, 0)$ is real, the equality of the integrals in (6.64) is evident.

CHAPTER 7

MEAN-FIELD ELECTRODYNAMICS FOR HOMOGENEOUS TURBULENCE IN THE CASE OF VANISHING MEAN FLOW

7.1. Determination of the tensor a_{ij}

In section 5.5. we already derived the representation of the tensor a_{ij} in the frame of the second order correlation approximation. Taking into account Parseval's theorem (6.8) and the symmetry relation (5.54) we have according to (5.51)

$$\begin{aligned} a_{ij} &= \varepsilon_{ilm}\varepsilon_{mnp}\varepsilon_{pqj} \iint \frac{\partial G(\xi,\tau)}{\partial \xi_n} Q_{ql}(\xi,\tau)\,\mathrm{d}\xi\,\mathrm{d}\tau \\ &= -\varepsilon_{ilm}\varepsilon_{mnp}\varepsilon_{pqj} \iint G(\xi,\tau)\frac{\partial Q_{ql}(\xi,\tau)}{\partial \xi_n}\,\mathrm{d}\xi\,\mathrm{d}\tau \\ &= +\mathrm{i}\varepsilon_{ilm}\varepsilon_{mnp}\varepsilon_{pqj} \iint \frac{k_n \hat{Q}_{lq}(\boldsymbol{k},\omega)}{\eta k^2 - \mathrm{i}\omega}\,\mathrm{d}\boldsymbol{k}\,\mathrm{d}\omega\,. \end{aligned} \tag{7.1}$$

Of special interest is the quantity α which we define by

$$\alpha = \frac{1}{3} a_{ii}, \tag{7.2}$$

thus including non-isotropic conditions too. By contracting (7.1) we find

$$\begin{aligned} \alpha &= -\frac{1}{3}\varepsilon_{nlq} \iint G(\xi,\tau)\frac{\partial Q_{ql}(\xi,\tau)}{\partial \xi_n}\,\mathrm{d}\xi\,\mathrm{d}\tau \\ &= -\frac{1}{3}\iint G(\xi,\tau)\,\overline{\boldsymbol{u}'(\boldsymbol{x},t)\cdot \operatorname{curl} \boldsymbol{u}'(\boldsymbol{x}+\xi, t+\tau)}\,\mathrm{d}\xi\,\mathrm{d}\tau \\ &= +\frac{\mathrm{i}}{3}\varepsilon_{nlq} \iint \frac{k_n \hat{Q}_{lq}(\boldsymbol{k},\omega)}{\eta k^2 - \mathrm{i}\omega}\,\mathrm{d}\boldsymbol{k}\,\mathrm{d}\omega\,. \end{aligned} \tag{7.3}$$

The first two lines show that the definition (7.2) of α leads back to the same formula as for the isotropic case (c.f. (5.55)), the last line is due to Parseval's theorem.

From the analysis in section 5.4. we can take that it is of interest to know whether the tensor a_{ij} is symmetric or not. For this reason we consider the vector $\gamma \boldsymbol{g}$ defined by

$$\gamma g_k = \frac{1}{2} \varepsilon_{ijk} a_{ij}. \tag{7.4}$$

In order to fix the value of γ let $\boldsymbol{g}$ be a unit vector. From (7.1) we obtain

$$\gamma g_k = -\frac{1}{2} \iint G(\xi, \tau) \left[\frac{\partial Q_{kl}(\xi, \tau)}{\partial \xi_l} + \frac{\partial Q_{lk}(\xi, \tau)}{\partial \xi_l}\right] \mathrm{d}\xi \, \mathrm{d}\tau. \tag{7.5}$$

Using the symmetry condition (5.54) and exchanging ξ for $-\xi$ in the second summand we can rewrite (7.5) in the form

$$\gamma g_k = -\frac{1}{2} \iint G(\xi, \tau) \left[\frac{\partial Q_{kl}(\xi, \tau)}{\partial \xi_l} - \frac{\partial Q_{kl}(\xi, -\tau)}{\partial \xi_l}\right] \mathrm{d}\xi \, \mathrm{d}\tau. \tag{7.6}$$

This relation enables us to conclude that a_{ij} is symmetric, i.e. $\gamma = 0$, for a number of interesting cases:

(i) The tensor a_{ij} is symmetric for an incompressible turbulence.

(ii) The tensor a_{ij} is symmetric for a turbulence where the correlation tensor is an even function of the relative time τ.

This is obvious from (6.75). An interesting example illustrating (ii) is isotropic turbulence. By taking into account (6.58) and (6.64) it can easily be confirmed that:

(iii) The tensor a_{ij} is symmetric in the low-conductivity limit.

Combining statements (i) and (iii) we confirm Moffatt's result [2] that the tensor a_{ij} is symmetric for an incompressible turbulence in the low-conductivity limit.

The equation of continuity,

$$\frac{\partial \varrho}{\partial t} + \operatorname{div} \varrho \boldsymbol{u} = 0, \tag{7.7}$$

furnishes by way of second order correlation approximation the relation

$$\frac{\partial}{\partial \tau} \overline{u_k'(\boldsymbol{x}, t)\, \varrho'(\boldsymbol{x} + \xi, t + \tau)} + \bar{\varrho} \frac{\partial}{\partial \xi_l} Q_{kl}(\xi, \tau) = 0, \tag{7.8}$$

if we put for the mass density $\varrho = \bar{\varrho} + \varrho'$. This enables us to rewrite (7.6) in the form

$$\gamma \boldsymbol{g} = \frac{1}{2\bar{\varrho}} \iint G(\xi, \tau) \frac{\partial}{\partial \tau} \overline{\boldsymbol{u}'(\boldsymbol{x}, t)\, [\varrho'(\boldsymbol{x} + \xi, t + \tau) + \varrho'(\boldsymbol{x} + \xi, t - \tau)]}\, \mathrm{d}\xi \, \mathrm{d}\tau, \tag{7.9}$$

from which, with the aid of (6.61), we obtain in the high-conductivity limit

$$\gamma \boldsymbol{g} = -\frac{1}{\bar{\varrho}} \overline{\boldsymbol{u}'(\boldsymbol{x}, t)\, \varrho'(\boldsymbol{x}, t)} = \overline{\boldsymbol{u}} = 0, \tag{7.10}$$

since we assume, for physical reasons, that there is no mean mass transport, i.e. $\overline{\varrho \boldsymbol{u}} = \bar{\varrho}\bar{\boldsymbol{u}} + \overline{\varrho' \boldsymbol{u}'} = 0$.

Thus we can add a fourth statement:

(iv) The tensor a_{ij} is symmetric in the high-conductivity limit.

7.2. The pumping effect

If we take into account only the turbulent electromotive force $\mathfrak{E}$ provided by the antisymmetric part of the tensor a_{ij} then, by the definition (7.4) of $\boldsymbol{g}$, the behaviour of the mean magnetic field is governed by the equation

$$\frac{\partial \overline{\boldsymbol{B}}}{\partial t} + \operatorname{curl}\,(\gamma \boldsymbol{g} \times \overline{\boldsymbol{B}}) - \frac{1}{\mu\sigma} \Delta\, \overline{\boldsymbol{B}} = 0. \tag{7.11}$$

From this equation it is clear that the action of $\mathfrak{E}$ on the mean magnetic field is the same as that of a constant velocity field $\boldsymbol{u} = -\gamma \boldsymbol{g}$, i.e. the mean magnetic field can be transported through the turbulent medium with the velocity $-\gamma \boldsymbol{g}$, although there is no mean motion. Following Drobyshevskij and Yuferev [1] we call the action of such a turbulence "pumping effect".

Drobyshevskij and Yuferev proved the pumping effect to exist for an incompressible stationary cellular motion by numerical calculations. Since such a motion resembles a random motion in the low-conductivity limit one must suspect that statements (i) and (iii) derived in the foregoing paragraph are due to the approximation level only. This is, indeed, the case as shown by Moffatt [6], who considered the electromotive force in a third order approximation. Let us, therefore, derive the vector $\gamma \boldsymbol{g}$ on the basis of the third order correlation approximation.

According to (4.56), (4.57) and (5.46) we have for a homogeneous steady turbulence

$$\begin{aligned} a_{ij} = &-\varepsilon_{ikl}\varepsilon_{lmn}\varepsilon_{npj} \iint G(\xi, \tau)\, \frac{\partial Q_{pk}(\xi, \tau)}{\partial \xi_m}\, \mathrm{d}\xi\, \mathrm{d}\tau \\ &+ \varepsilon_{ikl}\varepsilon_{lmn}\varepsilon_{npq}\varepsilon_{qrs}\varepsilon_{stj} \iiiint G(\xi, \tau)\, G(\xi', \tau') \\ &\quad \frac{\partial^2 Q_{kpt}(-\xi, -\xi', -\tau, -\tau')}{\partial \xi_m\, \partial \xi'_r}\, \mathrm{d}\xi\, \mathrm{d}\tau\, \mathrm{d}\xi'\, \mathrm{d}\tau', \end{aligned} \tag{7.12}$$

where the correlation tensor of the third rank, Q_{kpt}, is defined by (4.12). We clearly have the relation

$$Q_{kpt}(-\xi, -\xi', -\tau, -\tau') = Q_{tpk}(\xi', \xi, \tau', \tau). \tag{7.13}$$

Hence we obtain from (7.4) and (7.12), (7.13)

$$\gamma g_j = -\frac{1}{2}\iint G(\xi,\tau)\left[\frac{\partial Q_{jl}(\xi,\tau)}{\partial \xi_l} - \frac{Q_{jl}(\xi,-\tau)}{\partial \xi_l}\right] \mathrm{d}\xi\,\mathrm{d}\tau$$

$$+ \frac{1}{2}\iiiint G(\xi,\tau)\,G(\xi',\tau')\frac{\partial^2}{\partial \xi_k\,\partial \xi_l'}[Q_{ljk}(\xi,\xi',\tau,\tau') - Q_{kjl}(\xi,\xi',\tau,\tau')$$

$$- Q_{jlk}(\xi,\xi',\tau,\tau') - Q_{lkj}(\xi,\xi',\tau,\tau')]\,\mathrm{d}\xi\,\mathrm{d}\xi'\,\mathrm{d}\tau\,\mathrm{d}\tau'. \quad (7.14)$$

For a homogeneous incompressible turbulence the correlation tensor of third rank fulfils the conditions

$$\frac{\partial Q_{pqr}(\xi,\xi',\tau,\tau')}{\partial \xi_p} = 0, \quad (7.15)$$

$$\frac{\partial Q_{pqr}(\xi,\xi',\tau,\tau')}{\partial \xi_q} = \frac{\partial Q_{pqr}(\xi,\xi',\tau,\tau')}{\partial \xi_q'}, \quad (7.16)$$

$$\frac{\partial Q_{pqr}(\xi,\xi',\tau,\tau')}{\partial \xi_r'} = 0, \quad (7.17)$$

which are easily obtained from the definition (4.12) if $\operatorname{div} \boldsymbol{u}' = 0$ is taken into account. Consequently we have for an incompressible turbulence

$$\gamma g_j = \frac{1}{2}\iiiint G(\xi,\tau)\,G(\xi',\tau')\frac{\partial^2}{\partial \xi_k\,\partial \xi_l'}[Q_{ljk}(\xi,\xi',\tau,\tau')$$

$$- Q_{jlk}(\xi,\xi',\tau,\tau') - Q_{lkj}(\xi,\xi',\tau,\tau')]\,\mathrm{d}\xi\,\mathrm{d}\xi'\,\mathrm{d}\tau\,\mathrm{d}\tau'. \quad (7.18)$$

The model studied by Drobyshevskij and Yuferev and by Moffatt is one example in which the value of γ given by (7.18) does not vanish.

7.3. Dynamo action of homogeneous turbulence

The symmetric part of the tensor a_{ij} describes qualitatively new features of the behaviour of the magnetic field: Dynamo action of the turbulent motion will overcome the Ohmic dissipation thus providing for growing or—in case of equilibrium—non-decaying magnetic fields.

We get insight into this situation most easily by considering homogeneous isotropic turbulence having helicity, i.e. the turbulent electromotive force $\mathfrak{E}$ is given by (5.37) with $\alpha \neq 0$ and the mean magnetic field is governed by the equation

$$\frac{\partial \overline{\boldsymbol{B}}}{\partial t} = \alpha \operatorname{curl} \overline{\boldsymbol{B}} + \eta_T \Delta \overline{\boldsymbol{B}}, \quad \operatorname{div} \overline{\boldsymbol{B}} = 0, \quad (7.19)$$

where we introduced the total magnetic diffusivity,

$$\eta_T = \eta + \beta . \tag{7.20}$$

We have already demonstrated by physical arguments (chapter 3) that dynamo excitation may originate from a turbulence having helicity. Here we shall start from the ansatz

$$\boldsymbol{B} = \hat{\overline{\boldsymbol{B}}} e^{i(\boldsymbol{k}\cdot\boldsymbol{x} - \omega t)}, \quad \boldsymbol{k} \cdot \hat{\overline{\boldsymbol{B}}} = 0, \tag{7.21}$$

and look for non-decaying solutions of equation (7.19), i.e. for solutions with Re $(-i\omega) = 0$. Inserting (7.21) into (7.19) we obtain the algebraic system

$$(p\,\delta_{ij} - i\alpha\varepsilon_{ipj}k_p)\,\hat{\overline{B}}_j = 0, \quad p = -i\omega + \eta_T k^2 . \tag{7.22}$$

The condition for having non-zero solutions leads us to the dispersion relation

$$\begin{aligned} &\det\,(p\,\delta_{ij} - i\alpha\varepsilon_{ipj}k_p) \\ &= \frac{1}{6}\varepsilon_{ijk}\varepsilon_{lmn}(p\,\delta_{il} - i\alpha\varepsilon_{ipl}k_p)\,(p\,\delta_{jm} - i\alpha\varepsilon_{jqm}k_q)\,(p\,\delta_{kn} - i\alpha\varepsilon_{krn}k_r) \\ &= p^3 - \alpha^2k^2p = p(p + \alpha k)\,(p - \alpha k) = 0 , \end{aligned} \tag{7.23}$$

with the two solutions

$$-i\omega = -\eta_T k^2 + |\alpha|\,k, \quad -i\omega = -\eta_T k^2 - |\alpha|\,k . \tag{7.24}$$

The third root, $p = 0$, does not correspond to a solenoidal field. The first of the solutions (7.24) is clearly positive for sufficiently small k, hence we find indeed, growing fields. In figure 7.1 the two branches of the solutions of the dispersion relation are depicted.

In this way we just deduced a fundamental statement: A non-mirrorsymmetric homogeneous isotropic turbulence provides for dynamo action.

Let us now consider the case of a general homogeneous turbulence with the electromotive force $\mathfrak{E}$ given by (5.35). We split the tensor a_{ij} into its symmetric and its antisymmetric part by writing

$$a_{ij} = \alpha_{ij} + \gamma\varepsilon_{ijk}g_k . \tag{7.25}$$

The induction equation for the mean field then reads

$$\frac{\partial \overline{B}_i}{\partial t} = \varepsilon_{ijk}\frac{\partial}{\partial x_j}\left[(\alpha_{kl} + \gamma\varepsilon_{klm}g_m)\,\overline{B}_l + b_{klm}\frac{\partial \overline{B}_l}{\partial x_m}\right] + \eta\,\Delta\overline{B}_i , \tag{7.26}$$

and with (7.21) we find the algebraic system

$$(q\,\delta_{il} - i\varepsilon_{ijk}k_j\alpha_{kl} + \varepsilon_{ijk}k_jb_{klm}k_m)\,\hat{\overline{B}}_l = 0, \tag{7.27}$$

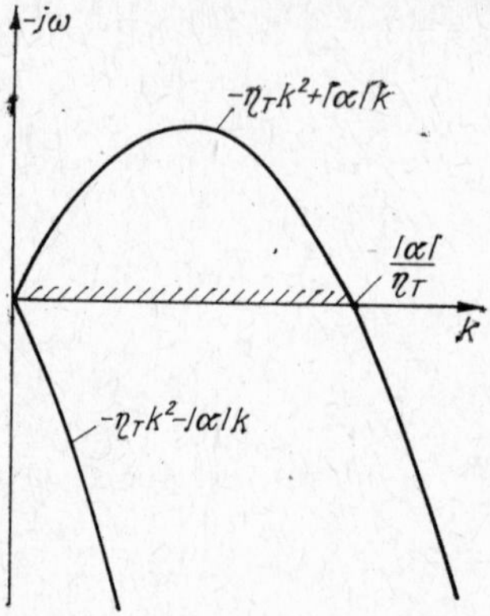

Fig. 7.1. The two branches of the solution of the dispersion relation for homogeneous isotropic non-mirrorsymmetric turbulence. Growing modes exist for wave vectors with a magnitude smaller than $\dfrac{|\alpha|}{\eta_T}$

where

$$q = p - \mathrm{i}(\gamma \boldsymbol{k} \cdot \boldsymbol{g}) = -\mathrm{i}\omega + \eta k^2 - \mathrm{i}(\gamma \boldsymbol{k} \cdot \boldsymbol{g}). \tag{7.28}$$

Investigating dynamo excitation provided by homogeneous turbulence we neglect the tensor b_{klm} since its contribution is of second order in $\boldsymbol{k}$. Therefore, if growing solutions are found, the understanding in the following will always be that there are growing solutions for sufficiently small wave vectors $\boldsymbol{k}$.

Neglecting the tensor b_{klm} in (7.27) we have to consider the system

$$(q\ \delta_{il} - \mathrm{i}k_j \varepsilon_{ijk} \alpha_{kl})\ \hat{\bar{B}}_l = 0, \tag{7.29}$$

where non-decaying solutions require Re $(q) > 0$. This problem was first treated by Moffatt [2], and later by Krause [6].

For the determinant of (7.29) we find

$$\det\ (q\ \delta_{il} - \mathrm{i}k_j \varepsilon_{ijk} \alpha_{kl}) = q(q^2 - A_{pq} k_p k_q), \tag{7.30}$$

where A_{pq} is the tensor of the algebraic complements of α_{kl}:

$$A_{pq} = \frac{1}{2}\, \varepsilon_{pkl} \varepsilon_{qmn} \alpha_{km} \alpha_{ln}. \tag{7.31}$$

The root $q = 0$, is of no interest here, since it corresponds to a decaying field, if it exists at all. Thus we have to consider the dispersion relation

$$q^2 - A_{pq} k_p k_q = 0. \tag{7.32}$$

From this we can see that for $A_{pq} k_p k_q > 0$ growing solutions, i.e. those with Re $(-\mathrm{i}\omega) > 0$, exist for sufficiently small magnitude of $\boldsymbol{k}$.

Now we can easily state some theorems concerning dynamo excitation:

(i) A necessary and sufficient condition for the existence of dynamo excitation for some wave vectors $\boldsymbol{k}$ is that the tensor A_{pq} is not negative semi-definite.

(i′) There is no dynamo excitation if, and only if, the tensor A_{pq} is negative semi-definite.

Since

$$\det (A_{pq}) = (\det (\alpha_{pq}))^2 , \tag{7.33}$$

the tensor A_{pq} cannot be negative semi-definite if α_{pq} is non-singular. Therefore we can state

(ii) Dynamo excitation exists for some wave vectors $\boldsymbol{k}$ in case the tensor α_{pq} is non-singular.

This statement clearly shows that dynamo excitation does not require $\alpha \neq 0$, although this condition is of some importance as will be revealed in the following.

(iii) $\alpha \neq 0$ is a necessary condition for having dynamo excitation for all wave vectors $\boldsymbol{k}$ of sufficiently small magnitude.

In order to prove this, we remark that dynamo excitation for all $\boldsymbol{k}$ of sufficiently small magnitude requires a positive definite quadratic form $A_{pq}k_pk_q$. This is the case if, and only if, α_{pq} is either positive or negative definite, consequently $\alpha \neq 0$.

We define the tensor $\tilde{\alpha}_{ij}$ by

$$\alpha_{ij} = \alpha\, \delta_{ij} + \tilde{\alpha}_{ji} , \quad \tilde{\alpha}_{ii} = 0 . \tag{7.34}$$

With this definition we can formulate a sufficient condition:

(iv) $\alpha^2 > \tilde{\alpha}_{kl}\tilde{\alpha}_{kl}$ is a sufficient condition for having dynamo excitation for all wave vectors $\boldsymbol{k}$ of sufficiently small magnitude.

In order to prove this statement we derive from (7.31) and (7.34)

$$A_{pq}k_pk_q = \frac{1}{2} [(2\alpha^2 - \tilde{\alpha}_{kl}\tilde{\alpha}_{kl})\, k^2 - 2\alpha\tilde{\alpha}_{kl}k_kk_l + 2\tilde{\alpha}_{jk}\tilde{\alpha}_{jl}k_kk_l] , \tag{7.35}$$

an expression which can be rewritten as

$$A_{pq}k_pk_q = \frac{1}{2} [(\alpha^2 - \tilde{\alpha}_{kl}\tilde{\alpha}_{kl})\, k^2 + (\alpha\, \delta_{jk} - \tilde{\alpha}_{jk})\, (\alpha\, \delta_{jl} - \tilde{\alpha}_{jl})\, k_kk_l \\ + \tilde{\alpha}_{jk}\tilde{\alpha}_{jl}k_kk_l] . \tag{7.36}$$

The second and the third summands are clearly non-negative and the first is positive according to our assumption in (iv), hence A_{pq} is positive definite.

We apply the statement (i′) to two-dimensional turbulence. For such turbulence the velocity field vectors lie in planes orthogonal to a given direction, say $\boldsymbol{e}$, and do not vary in the direction of $\boldsymbol{e}$. Accordingly, the correlation tensor in the Fourier space can easily be written down, namely

$$\hat{Q}_{jn}(\boldsymbol{k}, \omega) = \delta(\boldsymbol{k} \cdot \boldsymbol{e})\, \hat{Q}^{(2)}_{jn}(\tilde{\boldsymbol{k}}, \omega), \tag{7.37}$$

where

$$\tilde{\boldsymbol{k}} = \boldsymbol{k} - (\boldsymbol{e} \cdot \boldsymbol{k})\, \boldsymbol{e}, \tag{7.38}$$

and

$$e_j \tilde{Q}^{(2)}_{jn} = 0, \quad e_n \hat{Q}^{(2)}_{jn} = 0. \tag{7.39}$$

Inserting (7.37) into (7.1) we find

$$a_{ij} = \mathrm{i}(\varepsilon_{ilq}\, \delta_{nj} - \varepsilon_{ilj}\, \delta_{nq}) \iint \frac{\tilde{k}_n \hat{Q}^{(2)}_{lq}(\tilde{\boldsymbol{k}}, \omega)}{\eta \tilde{k}^2 - \mathrm{i}\omega}\, \mathrm{d}\tilde{\boldsymbol{k}}\, \mathrm{d}\omega. \tag{7.40}$$

Let us now assume that $\boldsymbol{e} = (0, 0, 1)$ and, consequently, $\tilde{\boldsymbol{k}} = (k_1, k_2, 0)$. Then it is easily seen that

$$a_{11} = a_{22} = a_{12} = a_{21} = a_{33} = 0.$$

Hence α_{ij} takes the form

$$\alpha_{ij} = \begin{pmatrix} 0 & 0 & a \\ 0 & 0 & b \\ a & b & 0 \end{pmatrix},$$

which implies

$$A_{pq} = \begin{pmatrix} -b^2 & ab & 0 \\ ab & -a^2 & 0 \\ 0 & 0 & 0 \end{pmatrix}.$$

It is clearly to be seen that A_{pq} is negative semi-definite. Hence we find within the frame of the second order correlation approximation the statement (Krause [16]):

(v) Two-dimensional turbulence cannot provide for dynamo excitation.

This result is in agreement with a statement derived by Zel'dovich [1] (1956) that a two-dimensional velocity field cannot provide for dynamo excitation.

In order to quote yet another interesting result we consider an anisotropic turbulence on a rotating system. If restricted to linear expressions with respect to the direction of anisotropy, $\boldsymbol{g}$, and the angular velocity, $\boldsymbol{\Omega}$, we have according to (5.43)

$$\alpha_{ij} = -\alpha_1(\boldsymbol{g} \cdot \boldsymbol{\Omega})\, \delta_{ij} - \alpha_2(g_i \Omega_j + g_j \Omega_i), \tag{7.41}$$

with certain scalars α_1, α_2. From (7.31) we easily obtain

$$A_{pq} = \alpha_1^2(\boldsymbol{g}\cdot\boldsymbol{\Omega})^2\,\delta_{pq} + \alpha_1\alpha_2(\boldsymbol{g}\cdot\boldsymbol{\Omega})\,[2(\boldsymbol{g}\cdot\boldsymbol{\Omega})\,\delta_{pq} - (g_p\Omega_q + g_q\Omega_p)] - \alpha_2^2(\boldsymbol{g}\times\boldsymbol{\Omega})_p\,(\boldsymbol{g}\times\boldsymbol{\Omega})_q. \tag{7.42}$$

Now we can prove the statement:

(vi) $\alpha_1 \neq 0$ is a necessary and sufficient condition for dynamo excitation.

A_{pq} is clearly negative semi-definite if $\alpha_1 = 0$. Consequently, $\alpha_1 \neq 0$ is necessary for dynamo excitation according to (i').

We further have $A_{pq}g_pg_q = \alpha_1^2(\boldsymbol{g}\cdot\boldsymbol{\Omega})^2g^2$ and, therefore, A_{pq} is not negative semi-definite if $\alpha_1 \neq 0$. Consequently, this condition is sufficient for dynamo excitation according to (i).

7.4. Determination of the tensor b_{ijk}: The turbulent magnetic diffusivity

In section 5.5. we already derived the representation of the tensor b_{ijk} within the frame of the second order correlation approximation. Taking into account Parseval's theorem (6.8) and the symmetry relations (6.19), (6.20) and (6.21) we obtain from (5.52)

$$\begin{aligned} b_{ijk} &= -\varepsilon_{ilm}\varepsilon_{mnp}\varepsilon_{pqj}\iint \frac{\partial G(\xi,\tau)}{\partial\xi_n}\,\xi_kQ_{ql}(\xi,\tau)\,\mathrm{d}\xi\,\mathrm{d}\tau \\ &= \varepsilon_{ilm}\varepsilon_{mnp}\varepsilon_{pqj}\iint G(\xi,\tau)\,\frac{\partial}{\partial\xi_n}\,[\xi_kQ_{ql}(\xi,\tau)]\,\mathrm{d}\xi\,\mathrm{d}\tau \\ &= -\varepsilon_{ilm}\varepsilon_{mnp}\varepsilon_{pqj}\iint \frac{k_n}{\eta k^2 - \mathrm{i}\omega}\,\frac{\partial\hat{Q}_{lq}(\boldsymbol{k},\omega)}{\partial k_k}\,\mathrm{d}\boldsymbol{k}\,\mathrm{d}\omega. \end{aligned} \tag{7.43}$$

Our main interest is directed to the quantity β, the turbulent magnetic diffusivity, which we define by

$$\beta = \frac{1}{6}\,\varepsilon_{ijk}b_{ijk}. \tag{7.44}$$

From (7.43) we obtain

$$\beta = -\frac{1}{3}\iint \frac{1}{\xi}\,\frac{\partial G(\xi,\tau)}{\partial\xi}\,\xi_k\xi_lQ_{kl}(\xi,\tau)\,\mathrm{d}\xi\,\mathrm{d}\tau, \tag{7.45}$$

an expression which is identical with that given in section 5.5. or, in Fourier space

$$\begin{aligned} \beta &= -\frac{1}{6}\iint \frac{k_n}{\eta k^2 - \mathrm{i}\omega}\left[\frac{\partial\hat{Q}_{nq}(\boldsymbol{k},\omega)}{\partial k_q} + \frac{\partial\hat{Q}_{qn}(\boldsymbol{k},\omega)}{\partial k_q}\right]\mathrm{d}\boldsymbol{k}\,\mathrm{d}\omega \\ &= \frac{1}{3}\iint \frac{1}{\eta k^2 - \mathrm{i}\omega}\left\{\hat{Q}_{qq}(\boldsymbol{k},\omega) - \frac{1}{2}\,\frac{\partial}{\partial k_q}\,[k_n(\hat{Q}_{nq}(\boldsymbol{k},\omega) + \hat{Q}_{qn}(\boldsymbol{k},\omega))]\right\}\mathrm{d}\boldsymbol{k}\,\mathrm{d}\omega. \end{aligned} \tag{7.46}$$

This opens possibilities of gaining information about the sign of β. For incompressible turbulence, for example, we have

$$\beta = \frac{1}{3} \iint \frac{\hat{Q}_{qq}(\boldsymbol{k}, \omega)}{\eta k^2 - \mathrm{i}\omega} \,\mathrm{d}\boldsymbol{k}\,\mathrm{d}\omega$$
$$= \frac{1}{3} \iint \frac{\eta k^2 \hat{Q}_{qq}(\boldsymbol{k}, \omega)}{\eta^2 k^4 + \omega^2} \,\mathrm{d}\boldsymbol{k}\,\mathrm{d}\omega + \frac{1}{3} \iint \frac{\omega \hat{Q}_{qq}(\boldsymbol{k}, \omega)}{\eta^2 k^4 + \omega^2} \,\mathrm{d}\boldsymbol{k}\,\mathrm{d}\omega . \tag{7.47}$$

From the symmetry relations (6.19), (6.20) and (6.21) we can derive

$$\hat{Q}_{qq}(\boldsymbol{k}, \omega) = \hat{Q}^*_{qq}(\boldsymbol{k}, \omega) = \hat{Q}_{qq}(-\boldsymbol{k}, -\omega), \tag{7.48}$$

hence the trace of the Fourier-transformed correlation tensor is an even function of its arguments, and, therefore, the second integral in (7.47) vanishes. Consequently, for incompressible turbulence we have

$$\beta = \frac{\eta}{3} \iint \frac{k^2 \hat{Q}_{qq}(\boldsymbol{k}, \omega)}{\eta^2 k^4 + \omega^2} \,\mathrm{d}\boldsymbol{k}\,\mathrm{d}\omega \geqq 0, \tag{7.49}$$

since the trace, $\hat{Q}_{qq}$, by Bochner's theorem is a non-negative function (KRAUSE and ROBERTS [1, 2]).

As already mentioned, a different result is obtained, if the random velocity field has a potential. Inserting (6.51) into (7.46) we find

$$\beta = \frac{1}{3} \iint \frac{1}{\eta k^2 - \mathrm{i}\omega} \left[k^2 \hat{\Phi}(\boldsymbol{k}, \omega) - \frac{\partial}{\partial k_q} (k_q k^2 \hat{\Phi}(\boldsymbol{k}, \omega) \right] \mathrm{d}\boldsymbol{k}\,\mathrm{d}\omega$$
$$= -\frac{1}{3} \iint \frac{k^2}{\eta k^2 - \mathrm{i}\omega} \left[\hat{\Phi}(\boldsymbol{k}, \omega) + \frac{\partial}{\partial k_q} (k_q \hat{\Phi}(\boldsymbol{k}, \omega)) \right] \mathrm{d}\boldsymbol{k}\,\mathrm{d}\omega . \tag{7.50}$$

In the low-conductivity limit we easily derive with (6.64)

$$\beta = -\frac{1}{3\eta} \iint \hat{\Phi}(\boldsymbol{k}, \omega) \,\mathrm{d}\boldsymbol{k}\,\mathrm{d}\omega = -\frac{\overline{\varphi^2}}{3\eta}, \tag{7.51}$$

in agreement with (3.60).

For investigating the high-conductivity limit we start from relation (7.45), which we write in the form

$$\beta = \frac{1}{3} \iint G(\xi, \tau) \left[Q_{kk}(\xi, \tau) + \xi_l \frac{\partial Q_{kl}(\xi, \tau)}{\partial \xi_k} \right] \mathrm{d}\xi\,\mathrm{d}\tau . \tag{7.52}$$

In this limit we find with (6.61)

$$\beta = \frac{1}{3} \int_0^\infty Q_{kk}(0, \tau) \,\mathrm{d}\tau = \frac{1}{6} \int Q_{kk}(0, \tau) \,\mathrm{d}\tau , \tag{7.53}$$

where we also take into account that, because of (5.54), the trace $Q_{kk}(\xi, \tau)$ is an even function of its arguments. Representing now Q_{kk} by its Fourier transform we find

$$\beta = \frac{\pi}{3} \int \hat{Q}_{kk}(\boldsymbol{k}, 0)\, \mathrm{d}\boldsymbol{k} \geqq 0, \tag{7.54}$$

thus clearly showing that in virtue of Bochner's theorem (6.25) β is non-negative in the high-conductivity limit.

With the defnition (6.89) of the quantity β we can write

$$b_{ijk} = \beta\varepsilon_{ijk} + \beta_{ijk}, \tag{7.55}$$

where

$$\varepsilon_{ijk}\beta_{ijk} = 0. \tag{7.56}$$

If we consider now only the electromotive force $\mathfrak{E}$ due to the term $\beta\varepsilon_{ijk}$, the behaviour of the mean magnetic field is governed by the equation

$$\frac{\partial \overline{\boldsymbol{B}}}{\partial t} = (\eta + \beta)\, \Delta \overline{\boldsymbol{B}}; \tag{7.57}$$

thus β proves to be the turbulent magnetic diffusivity.

For physical reasons one expects β to be positive and therefore should be quite surprised by its negative sign for a potential flow (equ. (7.51)). However the total magnetic diffusivity, $\eta + \beta$, is positive, as the estimation in section 3.8 shows.

For investigating the possibility of a negative value of the total magnetic diffusivity we can restrict ourselves to a potential random field, since according to (4.49), the contribution of the solenoidal part of an arbitrary field is positive in any case.

From (7.50) we can take

$$\beta = -\frac{1}{3\eta} \iint \frac{\eta k^2(\eta k^2 + \mathrm{i}\omega)}{(\eta k^2 - \mathrm{i}\omega)^2} \hat{\Phi}(\boldsymbol{k}, \omega)\, \mathrm{d}\boldsymbol{k}\, \mathrm{d}\omega$$

$$= \frac{1}{3\eta} \iint \frac{\eta^2 k^4(3\omega^2 - \eta^2 k^4)}{(\eta^2 k^4 + \omega^2)^2} \hat{\Phi}(\boldsymbol{k}, \omega)\, \mathrm{d}\boldsymbol{k}\, \mathrm{d}\omega. \tag{7.58}$$

Since $\hat{\Phi} \geqq 0$ and

$$-1 \leqq \frac{\eta^2 k^4(3\omega^2 - \eta^2 k^4)}{(\eta^2 k^4 + \omega^2)^2} \leqq \frac{9}{16}, \tag{7.59}$$

we can evaluate the integral of the right-hand side of (7.58) and obtain

$$\beta = \frac{1}{3\eta} \vartheta \iint \hat{\Phi}(\boldsymbol{k}, \omega)\, \mathrm{d}\boldsymbol{k}\, \mathrm{d}\omega = \frac{\vartheta}{3} \mathrm{Rm}^2 \eta, \quad -1 \leqq \vartheta \leqq \frac{9}{16}. \tag{7.60}$$

Consequently, $\left|\frac{\vartheta}{3} \mathrm{Rm}^2\right| > 1$ is a necessary condition for having a negative value of the total magnetic diffusivity, i.e. $\mathrm{Rm}^2 > 3$. The condition for the validity of the second order correlation approximation, (4.30), then requires $\mathrm{S} \ll 1$, and we have all the more $\mathrm{S}/\mathrm{Rm} = \tau_{\mathrm{cor}}/\mu\sigma\lambda_{\mathrm{cor}}^2 \ll 1$. Hence we have the high-conductivity limit where $\beta > 0$, as we have just shown (cf. (7.54)).

In this way we have the result that the total magnetic diffusivity, $\eta + \beta$, within the frame of the second order correlation approximation is non-negative (Krause and Roberts [2]).

Finally we note that an evaluation of formulae (7.45) or (7.46) for β in the limiting cases leads to the results already given in (3.30) and (3.48).

7.5. Turbulence undergoing the influence of Coriolis forces

In section 7.2. we discussed the effect of an anisotropic turbulence on the turbulent electromotive force $\mathfrak{E}$ in the case that the anisotropy is due to a certain vector field $\boldsymbol{g}$. There is, furthermore, a situation of interest, where the anisotropy is due to a field represented by a pseudo-vector. A realization is simply given by a turbulence on a rotating system, where the only influence is that of the Coriolis forces. For simplicity we suppose this influence to be weak enough in order to neglect all quantities of higher than first order with regard to the angular velocity $\boldsymbol{\Omega}$. Since a_{ij} and b_{ijk} are pseudo-tensors we then have

$$a_{ij} = 0, \quad b_{ijk} = \beta\varepsilon_{ijk} - \beta_1\Omega_i\, \delta_{jk} - \beta_2\Omega_j\, \delta_{ki} - \beta_3\Omega_k\, \delta_{ij}, \tag{7.61}$$

or using the notation introduced in (7.55)

$$\beta_{ijk} = -\beta_1\Omega_i\, \delta_{jk} - \beta_2\Omega_j\, \delta_{ki} - \beta_3\Omega_k\, \delta_{ij}. \tag{7.62}$$

It is easily confirmed that condition (7.56) holds.

With the aid of (7.43) the constants $\beta_1, \beta_2, \beta_3$ can be related to the correlation tensor of the velocity field. For this purpose we take from (7.61) the relations

$$b_{ijj} = (-3\beta_1 - \beta_2 - \beta_3)\, \Omega_i, \tag{7.63}$$

$$b_{jij} = (-\beta_1 - 3\beta_2 - \beta_3)\, \Omega_i, \tag{7.64}$$

$$b_{jji} = (-\beta_1 - \beta_2 - 3\beta_3)\, \Omega_i. \tag{7.65}$$

For the sum of them we obtain

$$5\Omega_i(\beta_1 + \beta_2 + \beta_3) = -(b_{ijj} + b_{jij} + b_{jji}). \tag{7.66}$$

For β_3 we thus have

$$\beta_3\Omega_i = -\frac{1}{10}(4b_{jji} - b_{jij} - b_{ijj}), \tag{7.67}$$

and inserting (7.43) we get (RÄDLER [14])

$$\beta_3\Omega_i = -\frac{\varepsilon_{jkl}}{10}\iint \frac{1}{\xi}\frac{\partial G(\xi, \tau)}{\partial \xi}(3\xi_i\xi_j - \xi^2\,\delta_{ij})\,Q_{kl}(\xi, \tau)\,\mathrm{d}\xi\,\mathrm{d}\tau. \tag{7.68}$$

We do not give the corresponding expression for β_1 and β_2. These quantities are of less interest, because for the turbulent electromotive force $\mathfrak{E}$ with b_{ijk} given by (7.61) we have

$$\mathfrak{E} = -\beta\mu\bar{\boldsymbol{j}} - (\beta_2 + \beta_3)\,\mathrm{grad}\,(\boldsymbol{\Omega}\cdot\overline{\boldsymbol{B}}) + \mu\beta_3\boldsymbol{\Omega}\times\bar{\boldsymbol{j}}. \tag{7.69}$$

One notices that β_1 does not appear since the magnetic field is source-free, and β_2 is only involved in the summand which is a gradient. This part of the electromotive force is always compensated by space charges and does not influthe magnetic field.

The most interesting feature of the result (7.69) is the occurrence of an electromotive force, $\mu\beta_3\boldsymbol{\Omega}\times\boldsymbol{j}$, often called "$\Omega\times j$-effect". It reminds us of the Hall effect. Just as the α-effect occurring with homogeneous isotropic nonmirrorsymmetric turbulence, the $\Omega\times j$-effect is also caused by helical motions, but it does not require a predominance of right- or left-handed helical motions. The anisotropy in combination with the spatial variation of the mean magnetic field produces a mean current despite the equipartition of both types of helical motions.

We shall now treat the interesting question whether dynamo excitation is possible for the turbulence considered here. From (7.27) we can see that we have to ask for roots of the equation

$$\det\,(p\,\delta_{il} + \varepsilon_{ijk}k_jk_m\beta_{klm}) = 0. \tag{7.70}$$

with $\mathrm{Re}\,(-\mathrm{i}\omega) > 0$, $p = -\mathrm{i}\omega + \eta_T k^2$.

Inserting (7.62) into (7.70) one readily finds

$$\det\,(p\,\delta_{il} + \varepsilon_{ijk}k_jk_m\beta_{klm}) = p(p^2 + k^2(\boldsymbol{\Omega}\cdot\boldsymbol{k})^2\,\beta_3^2) = 0, \tag{7.71}$$

and sees that no root giving a positive real part of $-\mathrm{i}\omega$ exists. Hence we can conclude that a turbulence of the type considered here cannot provide for dynamo excitation.

7.6. Two-dimensional turbulence

A type of turbulence which deserves some interest is two-dimensional turbulence. This turbulence is defined by the properties that (i) the velocity vectors lie completely in planes orthogonal to a given direction, $\boldsymbol{e}$, and (ii) do not vary along this direction, i.e.,

$$\boldsymbol{e} \cdot \boldsymbol{u}' = 0, \quad (\boldsymbol{e} \cdot \operatorname{grad})\, \boldsymbol{u}' = 0, \tag{7.72}$$

or the equivalent in Fourier space

$$\boldsymbol{e} \cdot \hat{\boldsymbol{u}} = 0, \quad \hat{\boldsymbol{u}}(\boldsymbol{k}, \omega) = \delta(\boldsymbol{e} \cdot \boldsymbol{k})\, \hat{\boldsymbol{u}}^{(2)}(\tilde{\boldsymbol{k}}, \omega), \tag{7.73}$$

where

$$\tilde{\boldsymbol{k}} = \boldsymbol{k} - (\boldsymbol{e} \cdot \boldsymbol{k})\, \boldsymbol{e}. \tag{7.74}$$

For the correlation tensor we consequently have

$$\begin{aligned} &Q_{ik}(\boldsymbol{\xi}, \tau) = Q_{ik}^{(2)}(\tilde{\boldsymbol{\xi}}, \tau), \quad e_i Q_{ik}^{(2)}(\tilde{\boldsymbol{\xi}}, \tau) = 0, \quad e_k Q_{ik}^{(2)}(\tilde{\boldsymbol{\xi}}, \tau) = 0, \\ &\tilde{\boldsymbol{\xi}} = \boldsymbol{\xi} - (\boldsymbol{e} \cdot \boldsymbol{\xi})\, \boldsymbol{e}, \end{aligned} \tag{7.75}$$

or in Fourier space

$$\hat{Q}_{ik}(\boldsymbol{k}, \tau) = \delta(\boldsymbol{e} \cdot \boldsymbol{k})\, \hat{Q}_{ik}^{(2)}(\tilde{\boldsymbol{k}}, \omega), \quad e_i \hat{Q}_{ik}^{(2)}(\tilde{\boldsymbol{k}}, \omega) = 0, \quad e_k Q_{ik}^{(2)}(\tilde{\boldsymbol{k}}, \omega) = 0. \tag{7.76}$$

We already found in 7.3. that no dynamo action will arise from two-dimensional turbulence. Here we shall concentrate our interest on the tensor b_{ijk} and the two-dimensional magnetic diffusivity. We start from (7.43) using the Fourier representation. Defining the differentiation within the planes orthogonal to $\boldsymbol{e}$ by

$$\frac{\partial}{\partial \tilde{k}_i} = (\delta_{ij} - e_i e_j) \frac{\partial}{\partial k_j}, \tag{7.77}$$

after carrying out a partial integration we can write

$$\begin{aligned} b_{ijk} = \varepsilon_{ilm} \varepsilon_{mnp} \varepsilon_{pqj} \iint \Bigg[&\frac{\partial}{\partial \tilde{k}_k} \left(\frac{k_n}{\eta k^2 - \mathrm{i}\omega} \right) + \frac{e_k e_n}{\eta k^2 - \mathrm{i}\omega} \\ &- e_k k_n \frac{2\eta (\boldsymbol{e} \cdot \boldsymbol{k})}{(\eta k^2 - \mathrm{i}\omega^2)} \Bigg] \hat{Q}_{lq}(\boldsymbol{k}, \omega)\, \mathrm{d}\boldsymbol{k}\, \mathrm{d}\omega . \end{aligned} \tag{7.78}$$

Inserting now the specified tensor (7.76) we easily obtain

$$\begin{aligned} b_{ijk} = \varepsilon_{ilm} \varepsilon_{mnp} \varepsilon_{pqj} &\iint \frac{\partial}{\partial \tilde{k}_k} \left(\frac{\tilde{k}_n}{\eta \tilde{k}^2 - \mathrm{i}\omega} \right) \hat{Q}_{lq}^{(2)}(\tilde{\boldsymbol{k}}, \omega)\, \mathrm{d}\tilde{\boldsymbol{k}}\, \mathrm{d}\omega \\ &+ e_j e_k \varepsilon_{ilq} \iint \frac{\hat{Q}_{lq}^{(2)}(\tilde{\boldsymbol{k}}, \omega)}{\eta \tilde{k}^2 - \mathrm{i}\omega}\, \mathrm{d}\tilde{\boldsymbol{k}}\, \mathrm{d}\omega , \end{aligned} \tag{7.79}$$

or, using Parseval's theorem (6.11),

$$b_{ijk} = -\varepsilon_{ilm}\varepsilon_{mnp}\varepsilon_{pqj} \iint \frac{1}{\tilde{\xi}} \frac{\partial G^{(2)}(\tilde{\xi}, \tau)}{\partial \tilde{\xi}} \tilde{\xi}_k \tilde{\xi}_n Q_{ql}^{(2)}(\tilde{\xi}, \tau) \, \mathrm{d}\tilde{\xi} \, \mathrm{d}\tau - e_j e_k \varepsilon_{ilq} \iint G^{(2)}(\tilde{\xi}, \tau) \, Q_{lq}^{(2)}(\tilde{\xi}, \tau) \, \mathrm{d}\tilde{\xi} \, \mathrm{d}\tau. \tag{7.80}$$

$G^{(2)}(\tilde{\xi}, \tau)$ denotes the two-dimensional Green's function given by

$$G_{(2)}(\tilde{\xi}, \tau) = \frac{\mu\sigma}{4\pi\tau} \exp\left(-\frac{\mu\sigma\tilde{\xi}^2}{4\tau}\right). \tag{7.81}$$

The integrations in (7.79) and (7.80) are to be extended over planes orthogonal to $\boldsymbol{e}$, i.e. over a two-dimensional manifold.

For a more detailed discussion we restrict ourselves to the two-dimensional isotropic mirrorsymmetric case. Then the second integral in (7.79) or (7.80) clearly does not contribute, since it depends on $\boldsymbol{u}'(\boldsymbol{x}, t) \cdot \operatorname{curl} \boldsymbol{u}'(\boldsymbol{x} + \boldsymbol{\xi}, t + \tau)$, a quantity which is different from zero only in the case of an imbalance of clockwise and counter-clockwise motions. In addition, the tensor b_{ijk} can only be composed of the tensors e_i, $\delta_{ij} - e_i e_j$, $\varepsilon_{ijl} e_l$, and since it is skew we have the only possible ansatz

$$b_{ijk} = b_1 e_i \varepsilon_{jkl} e_l + b_2 e_j \varepsilon_{kil} e_l + b_3 e_k \varepsilon_{ijl} e_l, \tag{7.82}$$

with certain constants b_1, b_2, b_3. In the non-mirrorsymmetric case additional summands composed of the tensor $e_i(\delta_{jk} - e_j e_k), \ldots, e_i e_j e_k$ are possible.

Multiplying (7.82) and (7.80) by e_k one finds $b_3 = 0$. Multiplications by $\varepsilon_{jkr} e_r$ and $\varepsilon_{kir} e_r$ show that

$$b_1 = b_2 = \beta^{(2)} = -\frac{1}{2} \iint \tilde{\xi} \frac{\partial G^{(2)}(\tilde{\xi}, \tau)}{\partial \tilde{\xi}} f^{(2)}(\tilde{\xi}, \tau) \, \mathrm{d}\tilde{\xi} \, \mathrm{d}\tau, \tag{7.83}$$

where $f^{(2)}$ is the two-dimensional longitudinal correlation function defined by

$$\tilde{\xi}^2 f^{(2)}(\tilde{\xi}, \tau) = \tilde{\xi}_k \tilde{\xi}_n Q_{kn}^{(2)}(\tilde{\xi}, \tau), \tag{7.84}$$

in analogy to the three-dimensional case (Krause and Rüdiger [1]). Consequently, we find for the turbulent electromotive force

$$\mathfrak{E}_i = \beta^{(2)}(e_i \varepsilon_{jkl} + e_j \varepsilon_{kil}) \, e_l \frac{\partial \overline{B}_j}{\partial x_k} = \beta^{(2)}(\delta_{kl} - e_k e_l) \, \varepsilon_{ijl} \frac{\partial \overline{B}_j}{\partial x_k}. \tag{7.85}$$

If we insert the latter expression given here into the induction equation for the mean field, i.e. into

$$\frac{\partial \overline{B}_i}{\partial t} - \eta \, \Delta \overline{B}_i = \varepsilon_{ijk} \frac{\partial \mathfrak{E}_k}{\partial x_j}, \tag{7.86}$$

we obtain the equation

$$\frac{\partial \overline{B}_i}{\partial t} - \eta\,\Delta \overline{B}_i = \beta^{(2)}(\delta_{jk} - e_j e_k)\,\frac{\partial^2 \overline{B}_i}{\partial x_j\,\partial x_k}. \tag{7.87}$$

In the general case this equation describes an anisotropic diffusion. There are, however, two configurations where the mean magnetic field itself has two-dimensional geometry. The first one is provided by a magnetic field parallel to $\boldsymbol{e}$, i.e.

$$\overline{\boldsymbol{B}} = (\boldsymbol{e}\cdot\overline{\boldsymbol{B}})\,\boldsymbol{e} = \overline{B}\boldsymbol{e}, \quad \boldsymbol{e}\cdot \operatorname{grad} \overline{B} = 0. \tag{7.88}$$

In this case the one component of the mean magnetic field, $\overline{B}$, satisfies the two-dimensional diffusion equation

$$\frac{\partial \overline{B}}{\partial t} - \eta_T\,\Delta^{(2)}\overline{B} = 0, \tag{7.89}$$

where η_T is the total magnetic diffusivity

$$\eta_T = \eta + \beta^{(2)} \tag{7.90}$$

analogously defined as in the three-dimensional case, and $\Delta^{(2)}$ the two-dimensional Δ-operator

$$\Delta^{(2)} = (\delta_{ij} - e_i e_j)\,\frac{\partial^2}{\partial x_i\,\partial x_j}. \tag{7.91}$$

The second case is given by a magnetic field having the structure of the turbulent motion, i.e.

$$\boldsymbol{e}\cdot\overline{\boldsymbol{B}} = 0,\ (\boldsymbol{e}\cdot\nabla)\,\overline{\boldsymbol{B}} = 0, \tag{7.92}$$

therefore, in particular,

$$\overline{B}_i = \tilde{\overline{B}}_i = (\delta_{ij} - e_i e_j)\,\overline{B}_j.$$

Then, again, the mean magnetic field $\overline{\boldsymbol{B}}$ is governed by the two-dimensional equation

$$\frac{\partial \tilde{\overline{B}}_i}{\partial t} - \eta_T\,\Delta^{(2)}\tilde{\overline{B}}_i = 0. \tag{7.93}$$

Finally, we remark that there is no fundamental difference between the three-dimensional turbulent magnetic diffusivity, β, and the two-dimensional one, $\beta^{(2)}$. This is easily seen if (7.83) is compared with (5.62). In addition, evaluating (7.83) in the high-conductivity limit analogously to the three-dimensional case in section 7.4. we find

$$\beta^{(2)} = \frac{1}{2} \int_0^\infty Q_{kk}(0, \tau)\, \mathrm{d}\tau = \frac{1}{2} \overline{u'^2}\, \tau_{\mathrm{cor}}. \tag{7.94}$$

7.7. Higher order correlation approximation: Vainshtein's recurrence formula

We denote by $\tilde{\boldsymbol{u}}(\boldsymbol{k}, t)$ and $\tilde{\boldsymbol{B}}(\boldsymbol{k}, t)$ the Fourier transforms of the velocity field and the magnetic field with respect to the space coordinates only. As before, the mean velocity field should be zero. The induction equation then reads

$$\frac{\partial \tilde{B}_i}{\partial t} + \eta k_k \tilde{B}_i = i\, \varepsilon_{ijk} \varepsilon_{klm} k_j (\tilde{u}_l * \tilde{B}_m), \tag{7.95}$$

where the asterisk denotes the convolution operation introduced in (6.7).

We represent the magnetic field $\tilde{\boldsymbol{B}}$ by the series

$$\tilde{\boldsymbol{B}}(\boldsymbol{k}, t) = \sum_{n=0}^{\infty} \tilde{\boldsymbol{B}}^{(n)}(\boldsymbol{k}, t), \tag{7.96}$$

where

$$\tilde{\boldsymbol{B}}^{(0)}(\boldsymbol{k}, t) = \tilde{\boldsymbol{B}}(\boldsymbol{k}) \exp(-\eta k^2 t), \tag{7.97}$$

and

$$\frac{\partial \tilde{B}_i^{(n)}}{\partial t} + \eta k^2 \tilde{B}_i^{(n)} = i \varepsilon_{ijk} \varepsilon_{klm} k_j (\tilde{u}_l * \tilde{B}_m^{(n-1)}), \quad \tilde{B}_i^{(n)}(\boldsymbol{k}, 0) = 0, \quad n \geqq 1. \tag{7.98}$$

From (7.96), (7.97) and (7.98) it is clearly to be seen that the field $\tilde{\boldsymbol{B}}$ defined by (7.96) solves the induction equation (7.95) in case of convergence.

Equation (7.98) can be integrated by

$$\tilde{B}_i^{(n)} = i \varepsilon_{ijk} \varepsilon_{klm} k_j \int_0^t \mathrm{e}^{-\eta k^2 (t-t')} (\tilde{u}_l * \tilde{B}_m{}^{(n-1)})\, \mathrm{d}t', \quad n \geqq 1, \tag{7.99}$$

and, applying this formula twice, we obtain

$$\tilde{B}_i^{(n)}(\boldsymbol{k}, t) = -\varepsilon_{ijk} \varepsilon_{klm} \varepsilon_{mnp} \varepsilon_{pqr} k_j \int_0^t \int_0^{t'} \iint \tilde{u}_l (\boldsymbol{k} - \boldsymbol{k}', t')\, k_n' \times \tilde{u}_q(\boldsymbol{k}' - \boldsymbol{k}'', t'')\, \tilde{B}_r^{(n-2)}(\boldsymbol{k}'', t'')\, \mathrm{e}^{-\eta[k^2(t-t') + k'^2(t'-t'')]}\, \mathrm{d}\boldsymbol{k}'\, \mathrm{d}\boldsymbol{k}''\, \mathrm{d}t'\, \mathrm{d}t''. \tag{7.100}$$

We can successively substitute $\tilde{\boldsymbol{B}}^{(n-3)}$, $\tilde{\boldsymbol{B}}^{(n-4)}$, and so on, in this way finally arriving at a representation of $\tilde{\boldsymbol{B}}^{(n)}$ by $\tilde{\boldsymbol{B}}^{(0)}$.

We are interested in the mean value $\overline{\tilde{\boldsymbol{B}}^{(n)}}$. Restricting ourselves to the main features we find an expression

$$\overline{\tilde{B}_i^{(n)}} = \int_0^t \int_0^{t'} \dots \int_0^{t^{(n-1)}} \dots \overline{\tilde{u}(t')\, \tilde{u}(t'') \dots \tilde{u}(t^{(n)}} \dots \tilde{B}^{(0)}(t^{(n)}) \dots \mathrm{d}t'\, \mathrm{d}t'' \dots \mathrm{d}t^{(n)}. \tag{7.101}$$

Following VAINSHTEIN [1] (c.f. also KAZANTSEV [1]) we now consider a Gaussian distribution for the velocity field, i.e. all correlation tensors of odd rank vanish and those of even rank are given by the sum of products of all possible pairs,

$$\overline{u_{i_1}(t')\, u_{i_2}(t'') \dots u_{i_{2n}}(t^{(2n)})} = \sum \prod \overline{\tilde{u}_{i_\alpha}(t^{(\alpha)})\, \tilde{u}_{i_\beta}(t^{(\beta)})}. \tag{7.102}$$

As a consequence the mean values of the magnetic field constituents with an odd number vanish:

$$\overline{\tilde{\boldsymbol{B}}^{(2n+1)}} = 0. \tag{7.103}$$

For those with an even number we obtain from (7.101) and (7.102)

$$\overline{\tilde{B}_i^{(2n)}} = \int_0^t \int_0^{t'} \dots \int_0^{t^{(2n-1)}} \sum \prod \overline{\tilde{u}_{i_\alpha}(\boldsymbol{k}^{(\alpha)}, t^{(\alpha)})\, \tilde{u}_{i_\beta}(\boldsymbol{k}^{(\beta)}, t^{(\beta)})}\, \tilde{B}^{(0)}(\, \mathrm{d}t' \dots, \tag{7.104}$$

where it is important to note that

$$t \geqq t' \geqq t'' \geqq \dots \geqq t^{(2n-1)} = t^{(2n)}. \tag{7.105}$$

In addition to the requirement that the probability distribution of the velocity field is Gaussian, VAINSHTEIN assumes that the correlation time is very short as compared with all other characteristic time scales and takes for the correlation tensor of second rank

$$Q_{kl}(\boldsymbol{\xi}, \tau) = 2\tau_{\mathrm{cor}}\, \delta(\tau)\, Q_{kl}(\boldsymbol{\xi}), \tag{7.106}$$

where $Q_{kk}(0) = \overline{u'^2}$.

From (7.104) we now obtain

$$\overline{\tilde{B}_i^{(2n)}} = \int_0^t \int_0^t \dots \int_0^{t^{(2n-1)}} \sum \prod \tilde{Q}_{i_\alpha i_\beta}(\boldsymbol{k}^{(\alpha)} - \boldsymbol{k}^{(\beta)})\, \delta(t^{(\alpha)} - t^{(\beta)}) \dots \tilde{B}^{(0)}\, \mathrm{d}t' \dots, \tag{7.107}$$

where $\tilde{Q}_{kl}$ denotes the Fourier-transform of Q_{kl} introduced in (7.106) with respect to the space coordinates.

An evaluation of (7.107) can easily be carried out: There is one, and only one, summand in the integrand, where the arguments $t^{(\alpha)}$ have their natural order given in (7.105). In all the other summands there is at least one pair $t^{(\alpha)}$, $t^{(\beta)}$ where α and β are not neighbours. In this case we have

$$\int_0^{t^{(\beta-1)}} \delta(t^{(\alpha)} - t^{(\beta)})\, \mathrm{d}t^{(\beta)} = \begin{cases} 0, & \text{if } t^{(\beta-1)} < t^{(\alpha)} \\ \dfrac{1}{2}, & \text{if } t^{(\beta-1)} = t^{(\alpha)} \end{cases}; \tag{7.108}$$

and this expression vanishes if the integration over $t^{(\beta-1)}$ is carrried out. Consequently, the sum under the integral reduces to one term only, thus leading to

$$\overline{B_i^{(2n)}} = \int_0^t \int_0^{t'} \dots \int_0^{t^{(2n-1)}} \dots \tilde{Q}_{i_1 i_2}(\boldsymbol{k}^{(1)} - \boldsymbol{k}^{(2)}\, \delta(t' - t'')$$

$$\tilde{Q}_{i_3 i_4}(\boldsymbol{k}^{(3)} - \boldsymbol{k}^{(4)})\, \delta\,(t^{(3)} - t^{(4)}) \dots \tilde{B}^{(0)} \mathrm{d}t' \dots . \tag{7.109}$$

Moreover, it becomes clear that $\tilde{Q}_{i_1 i_2}$ is completely disengaged from all that follows, and all that follows is the mean of $\tilde{\boldsymbol{B}}^{(2n-2)}$. In this way we obtain the recurrence formula

$$\overline{\tilde{B}_i^{(2n)}}(\boldsymbol{k}, t) = -\varepsilon_{ijk}\varepsilon_{klm}\varepsilon_{mnp}\varepsilon_{pqr} k_j \int_0^t \int_0^{t'} \iint \overline{k'_n \tilde{u}_l(\boldsymbol{k} - \boldsymbol{k}', t')\, \tilde{u}_q(\boldsymbol{k}' - \boldsymbol{k}'', t'')}$$

$$\overline{\tilde{B}_r^{(2n-2)}(\boldsymbol{k}'', t'')}\, \mathrm{e}^{-\eta[k^2(t-t') + k'^2(t-t'')]}\, \mathrm{d}\boldsymbol{k}'\, \mathrm{d}\boldsymbol{k}''\, \mathrm{d}t'\, \mathrm{d}t'' . \tag{7.110}$$

Taking into account that according to (6.18) and (7.106) we have

$$\overline{\tilde{u}_l(\boldsymbol{k} - \boldsymbol{k}', t')\, \tilde{u}_q(\boldsymbol{k}' - \boldsymbol{k}'', t'')} = \delta(\boldsymbol{k} - \boldsymbol{k}'')\, \delta(t' - t'')\, \tilde{Q}_{lq}(\boldsymbol{k}' - \boldsymbol{k}'')\, 2\tau_{\mathrm{cor}}, \tag{7.111}$$

and by carrying out the integrations,

$$\overline{\tilde{B}_i^{(2n+2)}}(\boldsymbol{k}, t) = -\int_0^t A_{ir} \mathrm{e}^{-\eta k^2 (t-t')}\, \overline{\tilde{B}_r^{(2n)}}(\boldsymbol{k}, t')\, \mathrm{d}t' , \tag{7.112}$$

where

$$A_{ir} = \tau_{\mathrm{cor}} \varepsilon_{ijk}\varepsilon_{klm}\varepsilon_{mnp}\varepsilon_{pqr} k_j \int k'_n \tilde{Q}_{lq}(\boldsymbol{k}' - \boldsymbol{k})\, \mathrm{d}\boldsymbol{k}' . \tag{7.113}$$

For isotropic conditions we have

$$\tau_{\mathrm{cor}} \int k'_n Q_{lq}(\boldsymbol{k}' - \boldsymbol{k})\, \mathrm{d}\boldsymbol{k}' = \tau_{\mathrm{cor}}[k_n \int \tilde{Q}_{lq}(\boldsymbol{k}')\, \mathrm{d}\boldsymbol{k}' + \int k'_n \tilde{Q}_{lq}(\boldsymbol{k}')\, \mathrm{d}\boldsymbol{k}']$$

$$= \beta k_n \delta_{lq} - \frac{\mathrm{i}}{2} \alpha \varepsilon_{nlq} . \tag{7.114}$$

Multiplying this equation successively by δ_{lq} and ε_{nlq} we easily obtain

$$\beta = \frac{\tau_{\text{cor}}}{3} \int \tilde{Q}_{qq}(\boldsymbol{k}')\,\mathrm{d}\boldsymbol{k}' = \frac{\tau_{\text{cor}}}{3} Q_{qq}(0) = \frac{1}{3}\overline{u'^2}\,\tau_{\text{cor}}, \tag{7.115}$$

and

$$\alpha = \tau_{\text{cor}} \frac{\mathrm{i}}{3} \int \varepsilon_{nlq} k'_n \tilde{Q}_{lq}(\boldsymbol{k}')\,\mathrm{d}\boldsymbol{k}' = \frac{\tau_{\text{cor}}}{6} \varepsilon_{nlq} \left.\frac{\partial Q_{lq}}{\partial \xi_n}\right|_{\xi=0}$$

$$= -\frac{\tau_{\text{cor}}}{3} \varepsilon_{lnq} \left.\overline{u'_l(\boldsymbol{x}, t) \frac{\partial u'_q(\boldsymbol{x} + \boldsymbol{\xi}, t)}{\partial \xi_n}}\right|_{\xi=0} = -\frac{1}{3}\overline{\boldsymbol{u}' \cdot \operatorname{curl} \boldsymbol{u}'}\,\tau_{\text{cor}}. \tag{7.116}$$

Inserting (7.114) into (7.113) we get

$$A_{ir} = -\mathrm{i}\alpha\varepsilon_{ijr}k_j + \beta(k^2\,\delta_{ir} - k_i k_r). \tag{7.117}$$

In order to obtain a final interpretation of our result we differentiate the recurrence formula (7.112) with respect to the time and find

$$\frac{\partial \overline{\tilde{B}_i^{(2n+2)}}}{\partial t} = -A_{ir}\overline{\tilde{B}_r^{(2n)}} - k^2\,\overline{\tilde{B}_i^{(2n+2)}}. \tag{7.118}$$

Summing up these equations by taking into account that

$$\frac{\partial \tilde{\boldsymbol{B}}^{(0)}}{\partial t} = -\eta k^2 \tilde{\boldsymbol{B}}^{(0)}$$

we find for the mean magnetic field

$$\frac{\partial \bar{\tilde{B}}_i}{\partial t} = -A_{ir}\bar{\tilde{B}}_r - \eta k^2 \bar{\tilde{\boldsymbol{B}}}_i, \tag{7.119}$$

which, in the isotropic case, reduces to

$$\frac{\partial \overline{\tilde{\boldsymbol{B}}(\boldsymbol{k}, t)}}{\partial t} = -(\eta + \beta)\,k^2\,\overline{\tilde{\boldsymbol{B}}(\boldsymbol{k}, t)} + \mathrm{i}\alpha\boldsymbol{k}\times\overline{\tilde{\boldsymbol{B}}(\boldsymbol{k}, t)}, \tag{7.120}$$

or if written in the $(\boldsymbol{x}, t)$-space

$$\frac{\partial \overline{\boldsymbol{B}}(\boldsymbol{x}, t)}{\partial t} = \operatorname{curl} \alpha\overline{\boldsymbol{B}}(\boldsymbol{x}, t) + (\eta + \beta)\,\Delta\overline{\boldsymbol{B}}(\boldsymbol{x}, t). \tag{7.121}$$

This shows that the action of the turbulence is again that of an electromotive force $\mathfrak{E}$ given by (3.5), or (5.37), however, the main point is that the values of α and β are just the same as derived within the frame of the second order correlation approximation in the high-conductivity limit ((3.31), (3.32)).

7.8. The dispersion relation

As already mentioned in section 5.1., for applications of the mean-field theory it is appropriate to assume the turbulence to be two-scaled. Otherwise the use of the ergodic theorem is questionable. In addition, we saw that the two-scale property resulted in mathematical simplifications (possibility of expansion of the mean field), which allowed a more detailed discussion of the subject. In the case of homogeneous turbulence, however, there is a mathematical procedure, first proposed by LERCHE [1, 6], which provides an alternative to the expansion technique.

Let us consider the integro-differential equation (4.50) deduced within the frame of the second order correlation approximation. For homogeneous turbulence and vanishing mean flow it reads

$$\left(\frac{\partial}{\partial t} - \eta \Delta\right) \overline{B}_i(\boldsymbol{x}, t) = \varepsilon_{ijk}\varepsilon_{klm}\varepsilon_{mnp}\varepsilon_{pqr} \frac{\partial}{\partial x_j} \iint \frac{\partial G(\xi, \tau)}{\partial \xi_n} Q_{ql}(\xi, \tau)\, \overline{B}_r(\boldsymbol{x} - \xi, t - \tau)\, \mathrm{d}\xi\, \mathrm{d}\tau . \tag{7.122}$$

Since the integral on the right is of convolution type the equation takes a particularly simple form in Fourier-transformed quantities, namely,

$$[p\, \delta_{ir} - A_{ir}(\boldsymbol{k}, \omega)]\, \hat{\overline{B}}_r(\boldsymbol{k}, \omega) = 0 , \tag{7.123}$$

where p is given by

$$p = -\mathrm{i}\omega + \eta k^2 \tag{7.124}$$

and

$$A_{ir}(\boldsymbol{k}, \omega) = \mathrm{i}\varepsilon_{ijk}\varepsilon_{klm}\varepsilon_{mnp}\varepsilon_{pqr} k_j \iint \frac{\partial G(\xi, \tau)}{\partial \xi_n} Q_{ql}(\xi, \tau)\, \mathrm{e}^{-\mathrm{i}(\boldsymbol{k}\cdot\xi - \omega\tau)}\, \mathrm{d}\xi\, \mathrm{d}\tau . \tag{7.125}$$

Non-trivial solutions $\hat{\overline{B}}_i$ of (7.123) only exist if

$$\det\, [p\, \delta_{ir} - A_{ir}(\boldsymbol{k}, \omega)] = 0 , \tag{7.126}$$

an equation which represents a dispersion relation for determining the frequencies ω for given $\boldsymbol{k}$.

At first we remark that for small $\boldsymbol{k}$ this dispersion relation is identical with that for two-scale turbulence discussed in sections 7.3. and 7.5. It is easily confirmed that in second order in $\boldsymbol{k}$

$$A_{ir} = \mathrm{i}\varepsilon_{ijk} a_{kr} k_j - \varepsilon_{ijk} k_j k_s b_{krs} , \tag{7.127}$$

where the tensors a_{kr} and b_{krs} are given by (7.1) and (7.43).

Secondly, we find by inspection of (7.125) that the integral is the Fourier transform of $\frac{\partial G}{\partial \xi_n} Q_{ql}$, up to a factor $(2\pi)^{-4}$. Therefore we can rewrite this formula using (6.7) and (6.57)

$$A_{ir}(\boldsymbol{k}, \omega) = -\varepsilon_{ijk}\varepsilon_{klm}\varepsilon_{mnp}\varepsilon_{pqr}k_j \iint \frac{(k_n - k_n')\,\hat{Q}_{ql}(\boldsymbol{k}', \omega')}{\eta(\boldsymbol{k} - \boldsymbol{k}')^2 - \mathrm{i}(\omega - \omega')}\,\mathrm{d}\boldsymbol{k}'\,\mathrm{d}\omega'. \tag{7.128}$$

This shows that $A_{ir} = \mathrm{O}(1)$ if $k \to \infty$, i.e. the dispersion relation (7.126) yields $p = 0$, or $-\mathrm{i}\omega = -\eta k^2$, if $k \to \infty$. This result is clear for physical reasons: The small-wavelength Fourier modes decay uninfluenced by the turbulent motion.

Statements for arbitrary $\boldsymbol{k}$ can hardly be gained in the general case. Therefore, we restrict ourselves to isotropic mirrorsymmetric turbulences. Since $k_i A_{ij} = 0$ we then have

$$A_{ij}(\boldsymbol{k}, \omega) = a(k, \omega)\left(\delta_{ij} - \frac{k_i k_j}{k^2}\right), \tag{7.129}$$

and the dispersion relation simply reads $p = a$, if the uninteresting solution $p = 0$ is excluded. Thus, contracting (7.125), we have

$$p = \mathrm{i}k_j \iint \frac{1}{\xi}\frac{\partial G(\xi, \tau)}{\partial \xi}\,\xi_k Q_{jk}(\boldsymbol{\xi}, \tau)\,\mathrm{e}^{-\mathrm{i}(\boldsymbol{k}\cdot\boldsymbol{\xi} - \omega\tau)}\,\mathrm{d}\boldsymbol{\xi}\,\mathrm{d}\tau, \tag{7.130}$$

or, if definition (5.61) of the longitudinal correlation function is taken into account,

$$\begin{aligned} p &= \mathrm{i}\iint \frac{(\boldsymbol{k}\cdot\boldsymbol{\xi})}{\xi}\frac{\partial G(\xi, \tau)}{\partial \xi}\,f(\xi, \tau)\,\mathrm{e}^{-\mathrm{i}(\boldsymbol{k}\cdot\boldsymbol{\xi} - \omega\tau)}\,\mathrm{d}\boldsymbol{\xi}\,\mathrm{d}\tau \\ &= -k\frac{\partial}{\partial k}\iint \frac{1}{\xi}\frac{\partial G(\xi, \tau)}{\partial \xi}\,f(\xi, \tau)\,\mathrm{e}^{-\mathrm{i}(\boldsymbol{k}\cdot\boldsymbol{\xi} - \omega\tau)}\,\mathrm{d}\boldsymbol{\xi}\,\mathrm{d}\tau. \end{aligned} \tag{7.131}$$

Equivalent representations of the dispersion relation are found if a is determined by contracting (7.128). We then obtain

$$\begin{aligned} p &= -\iint \frac{k_j(k_k - k_k')\,\hat{Q}_{jk}(\boldsymbol{k}', \omega')}{\eta(\boldsymbol{k} - \boldsymbol{k}')^2 - \mathrm{i}(\omega - \omega')}\,\mathrm{d}\boldsymbol{k}'\,\mathrm{d}\omega' \\ &= -\iint \frac{k_j(k_j - k_j')}{\eta(\boldsymbol{k} - \boldsymbol{k}')^2 - \mathrm{i}(\omega - \omega')}\,\hat{f}(k', \omega')\,\mathrm{d}\boldsymbol{k}'\,\mathrm{d}\omega'. \end{aligned} \tag{7.132}$$

The latter form of the dispersion relation is most useful for scrutinizing the interesting question whether growing solutions will exist, i.e. Re $(-\mathrm{i}\omega)$ positive or not. Let us write $\omega = \omega_r + \mathrm{i}\omega_i$, thus Re $(-\mathrm{i}\omega) = \omega_i$.

First, suppose the fluid is incompressible, so that equation (7.132) becomes

$$-\mathrm{i}\omega = -\eta k^2 - \iint \frac{k_j k_k \hat{Q}_{jk}(\boldsymbol{k}', \omega')}{\eta(\boldsymbol{k} - \boldsymbol{k}')^2 - \mathrm{i}(\omega - \omega')} \mathrm{d}\boldsymbol{k}' \, \mathrm{d}\omega' . \tag{7.133}$$

Suppose a root exists for which Re $(-\mathrm{i}\omega) = \omega_i \geqq 0$. It is then permissible to take the ω'-contour in (7.133) to be the real axis. Take the real part of both sides of (7.133) to obtain

$$\mathrm{Re}\,(-\mathrm{i}\omega) = \omega_i = -\eta k^2 - \iint \frac{(\boldsymbol{k} - \boldsymbol{k}')^2 + \omega_i}{\eta^2(\boldsymbol{k} - \boldsymbol{k}')^4 + (\omega - \omega')^2} k_j k_k \hat{Q}_{jk}(\boldsymbol{k}', \omega') \, \mathrm{d}\boldsymbol{k}' \, \mathrm{d}\omega' . \tag{7.134}$$

In this connection it should be noted that in the mirrorsymmetric case, per definition, we have

$$\hat{Q}_{jk}(-\boldsymbol{k}, \omega) = \hat{Q}_{jk}(\boldsymbol{k}, \omega) , \tag{7.135}$$

thus being able to derive from (6.19) and (6.20) that $\hat{Q}_{jk}$ is a real quantity.

The left-hand side of (7.134) is positive semi-definite by assumption, but the right-hand side is negative definite because of Bochner's theorem (6.25). This contradiction reduces the initial postulate Re $(-\mathrm{i}\omega) = \omega_i \geqq 0$ to an absurdity (Krause and Roberts [1, 2]).

Considering the high-conductivity limit, $\mu\sigma\lambda_{\mathrm{cor}}^2 \gg \tau_{\mathrm{cor}}$, we realize that the summand $\eta(\boldsymbol{k} - \boldsymbol{k}')^2$ in the denominator remarkably contributes only if $k^2 > \mu\sigma/\tau_{\mathrm{cor}}$. Furthermore, since $\hat{Q}_{jk}(\boldsymbol{k}', \omega')$ is significantly different from zero only if $k'^2 \leqq 1/\lambda_{\mathrm{cor}}^2 \ll \mu\sigma/\tau_{\mathrm{cor}}$, we can simplify the dispersion relation (7.132) to

$$-\mathrm{i}\omega = -\eta k^2 - \iint \frac{k_j k_k \hat{Q}_{jk}(\boldsymbol{k}', \omega')}{\eta k^2 - \mathrm{i}(\omega - \omega')} \mathrm{d}\boldsymbol{k}' \, \mathrm{d}\omega' . \tag{7.136}$$

Now the same arguments will apply as did to (7.134): Assuming $\omega_i \geqq 0$ the real part of the left-hand side of (7.136) is positive semi-definite whereas the right-hand side is negative because of Bochner's theorem (Krause and Roberts [1, 2]).

The following estimations will show that no growing solutions can be found within the frame of the second order correlation approximation at all.

For any homogeneous isotropic mirrorsymmetric turbulence the Fourier-transformed correlation tensor can be written as

$$\hat{Q}_{ij}(\boldsymbol{k}, \omega) = \hat{Q}(k, \omega) \, (k^2 \, \delta_{ij} - k_i k_j) + \hat{Q}'(k, \omega) \, k_i k_j , \tag{7.137}$$

where

$$\hat{Q}(k, \omega) \geqq 0, \quad \hat{Q}'(k, \omega) \geqq 0, \tag{7.138}$$

because of Bochner's theorem (6.36). From the above investigations of incompressible turbulence it is clear that the contribution of the first summand in (7.137) on the right-hand side of the dispersion relation is always in the direction of negative ω_i. Therefore, the most favourable situation for having growing modes is provided by a potential random field, i.e. $\hat{Q} = 0$ and $\hat{Q}' = \hat{\Phi} \geqq 0$, with the denotation introduced in section 6.6. The dispersion relation (7.132) then simplifies to

$$p = \iint \frac{(\boldsymbol{k}\cdot\boldsymbol{k}')\,(k'^2 - (\boldsymbol{k}\cdot\boldsymbol{k}'))}{\eta(\boldsymbol{k}\cdot - \boldsymbol{k}')^2 - \mathrm{i}(\omega - \omega')}\,\hat{\Phi}(k', \omega')\,\mathrm{d}\boldsymbol{k}'\,\mathrm{d}\omega', \tag{7.139}$$

and assuming $\omega_i > 0$ we can write

$$\omega_i = \eta k^2 \Bigg[-1 + \frac{1}{\eta k^2}\iint \frac{(\boldsymbol{k}\cdot\boldsymbol{k}')\,(k'^2 - (\boldsymbol{k}\cdot\boldsymbol{k}'))\,(\eta(\boldsymbol{k} - \boldsymbol{k}')^2 + \omega_i)}{(\eta(\boldsymbol{k} - \boldsymbol{k}')^2 + \omega_i)^2 + (\omega_r - \omega')^2}\,\hat{\Phi}(k', \omega')\,\mathrm{d}\boldsymbol{k}'\,\mathrm{d}\omega'\Bigg], \tag{7.140}$$

where the ω'-contour is the real axis. Therefore $\omega_i > 0$ requires

$$\frac{1}{\eta k^2}\iint \frac{(\boldsymbol{k}\cdot\boldsymbol{k}')\,(k'^2 - (\boldsymbol{k}\cdot\boldsymbol{k}'))\,(\eta(\boldsymbol{k} - \boldsymbol{k}')^2 + \omega_i)}{(\eta(\boldsymbol{k} - \boldsymbol{k}')^2 + \omega_i)^2 + (\omega_r - \omega')^2}\,\hat{\Phi}(k', \omega')\,\mathrm{d}\boldsymbol{k}'\,\mathrm{d}\omega' > 1. \tag{7.141}$$

An inspection of the integrand shows that the region $|\boldsymbol{k}'| \ll |\boldsymbol{k}|$ always gives a negative contribution. The same holds for $|\boldsymbol{k}'| \approx |\boldsymbol{k}|$: The integrand is negative for $\boldsymbol{k}' \approx -\boldsymbol{k}$, positive for $\boldsymbol{k}' \approx \boldsymbol{k}$ but small, therefore altogether we have a negative contribution. Consequently, it is best to assume that $\Phi(\boldsymbol{k}', \omega') \approx 0$ for $|\boldsymbol{k}'| \lessapprox |\boldsymbol{k}|$, and $\Phi(\boldsymbol{k}', \omega') > 0$ for $|\boldsymbol{k}'| \gg |\boldsymbol{k}|$. The region where $\hat{\Phi} > 0$ is also characterized by the condition $|\boldsymbol{k}'| \approx 1/\lambda_{\mathrm{cor}}$, thus we have the most favourable situation for having growing modes in case $1/\lambda_{\mathrm{cor}} \gg |\boldsymbol{k}|$, i.e. for two-scale turbulence. However, in section 7.4. it was already shown that even for two-scale turbulence no growing modes exist. Consequently, we can state: Within the framework of the second order correlation approximation there is no evidence that homogeneous isotropic mirrorsymmetric turbulence can provide for dynamo excitation.

There are, however, suggestions that growing modes may exist in higher order approximations. Kasantzev [1] deduced a result of this kind using diagram techniques. The considerations of Gailitis [4] and of Kraichnan [5], however, give a much clearer idea by considering the helicity fluctuations in a homogeneous isotropic mirrorsymmetric turbulence to provide for dynamo excitation. The lowest level to prove this is the fourth order correlation approximation.

GAILITIS [4] considered a double-scaled turbulence. Two kinds of turbulence elements are assumed to exist with the scales $\lambda_{\text{cor},1}, \tau_{\text{cor},1}$ and $\lambda_{\text{cor},2}, \tau_{\text{cor},2}$ related by $\lambda_{\text{cor},1} \gg \lambda_{\text{cor},2}$, and $\tau_{\text{cor},1} \gg \tau_{\text{cor},2}$. According to a result to be deduced later (chapter 9) the small-scale turbulence imbedded in one whirl of the large scale turbulence is expected to show helicity. In this way the large whirl can act as a dynamo exciting a magnetic field with the characteristic length scale $\lambda_{\text{cor},1}$.

KRAICHNAN [2, 4] followed a more deductive course. He derived in fourth order a negative contribution to the turbulent magnetic diffusivity, which is due to the helicity fluctuations. It is likely that for a proper choice of the characteristic scales this negative contribution will predominate, thus providing for growing modes.

Finally we remark that an extended investigation of the dispersion relation was carried out by GILLILAND [1].

7.9. The mean square of the fluctuating magnetic field

We assume the magnetic field to be constant. Then, in the frame of the second order correlation approximation the magnetic field fluctuations are described by the induction equation

$$\left(\frac{\partial}{\partial t} - \eta \Delta\right) B_i' = \varepsilon_{ijk}\varepsilon_{klm}\overline{B}_m \frac{\partial u_l'}{\partial x_j}, \tag{7.142}$$

or in Fourier space

$$(-\mathrm{i}\omega + \eta k^2)\, \hat{B}_i = \varepsilon_{ijk}\varepsilon_{klm}\overline{B}_m \mathrm{i} k_j \hat{u}_l. \tag{7.143}$$

This gives for the correlation tensor B_{in} introduced in (4.5)

$$\begin{aligned} B_{in}(\xi, \tau) &= \iiiint \overline{\hat{B}_i(\boldsymbol{k}, \omega)\, \hat{B}_n(\boldsymbol{k}', \omega')}\, \mathrm{e}^{\mathrm{i}(\boldsymbol{x}(\boldsymbol{k}+\boldsymbol{k}') - t(\omega+\omega'))}\, \mathrm{e}^{\mathrm{i}(\boldsymbol{k}'\xi - \omega' t)}\, \mathrm{d}\boldsymbol{k}\, \mathrm{d}\boldsymbol{k}'\, \mathrm{d}\omega\, \mathrm{d}\omega' \\ &= -\varepsilon_{ijk}\varepsilon_{klm}\varepsilon_{npq}\varepsilon_{qrs} \iiiint \frac{k_j k_p' \overline{\hat{u}_l(\boldsymbol{k}, \omega)\, \hat{u}_r(\boldsymbol{k}', \omega')}}{(\eta k^2 - \mathrm{i}\omega)\,(\eta k'^2 - \mathrm{i}\omega')} \\ &\qquad \mathrm{e}^{\mathrm{i}(\boldsymbol{x}(\boldsymbol{k}+\boldsymbol{k}') - t(\omega+\omega'))}\, \mathrm{e}^{\mathrm{i}(\boldsymbol{k}'\xi - \omega' t)}\, \mathrm{d}\boldsymbol{k}\, \mathrm{d}\omega\, \mathrm{d}\boldsymbol{k}'\, \mathrm{d}\omega' \overline{B}_m \overline{B}_s. \end{aligned} \tag{7.144}$$

Inserting (6.18) we obtain (BRÄUER and KRAUSE [1])

$$B_{in}(\xi, \tau) = \varepsilon_{ijk}\varepsilon_{klm}\varepsilon_{npq}\varepsilon_{qrs} \iint \frac{k_j k_p}{\eta^2 k^4 + \omega^2} \hat{Q}_{lr}(\boldsymbol{k}, \omega)\, \mathrm{e}^{\mathrm{i}(\boldsymbol{k}\xi - \omega\tau)}\, \mathrm{d}\boldsymbol{k}\, \mathrm{d}\omega\, \overline{B}_m \overline{B}_s. \tag{7.145}$$

From this formula we find for the mean square of the magnetic field fluctuations

$$\overline{B'^2} = B_{ii}(0, 0) = q_{ms} \overline{B}_m \overline{B}_s, \tag{7.146}$$

with

$$q_{ms} = \iint [\hat{Q}_{ll}(\boldsymbol{k}, \omega)\, k_m k_s - \hat{Q}_{sq}(\boldsymbol{k}, \omega)\, k_m k_q - \hat{Q}_{lm}(\boldsymbol{k}, \omega)\, k_l k_s + \delta_{ms} \hat{Q}_{lq}(\boldsymbol{k}, \omega)\, k_l k_q] \frac{\mathrm{d}\boldsymbol{k}\, \mathrm{d}\omega}{\eta^2 k^4 + \omega^2}. \tag{7.147}$$

For isotropic turbulence we clearly have $q_{ms} = q\, \delta_{ms}$ and obtain from (7.147) by contraction

$$q = \frac{1}{3} \iint \frac{k^2 \hat{Q}_{ll}(\boldsymbol{k}, \omega) + k_l k_q \hat{Q}_{lq}(\boldsymbol{k}, \omega)}{\eta^2 k^4 + \omega^2}\, \mathrm{d}\boldsymbol{k}\, \mathrm{d}\omega. \tag{7.148}$$

Bochner's theorem guarantees the factor q to be positive.

We can now rigorously confirm relations (3.81) and (3.87). For simplicity we assume div $\boldsymbol{u}' = 0$. Then we find from (7.148) for $\eta \to \infty$

$$q \approx \frac{1}{3\eta^2} \iint \frac{\hat{Q}_{ll}(\boldsymbol{k}, \omega)}{k^2}\, \mathrm{d}\boldsymbol{k}\, \mathrm{d}\omega \approx \frac{1}{3\eta^2}\, u'^2\, \lambda_{\mathrm{cor}}^2 = \frac{1}{3} \mathrm{Rm}^2, \tag{7.149}$$

in agreement with (3.88). If inserting $\omega = \eta w$ in (7.148) for the high-conductivity limit, $\eta \to 0$, we obtain

$$q \approx \frac{1}{3\eta} \iint \frac{k^2 \hat{Q}_{ll}(k, 0)}{k^4 + w^2}\, \mathrm{d}\boldsymbol{k}\, \mathrm{d}w = \frac{\pi}{3\eta} \int \hat{Q}_{ll}(k, 0)\, \mathrm{d}\boldsymbol{k}$$

$$= \frac{1}{3\eta} \int \overline{\boldsymbol{u}'(\boldsymbol{x}, t) \cdot \boldsymbol{u}'\,(\boldsymbol{x}, t + \tau)}\, \mathrm{d}\tau \approx \frac{\beta}{\eta} \approx \frac{\eta_T}{\eta}. \tag{7.150}$$

A fairly similar result was deduced by Low [1]. (7.150) allows an interesting interpretation which is indicated by writing q as the ratio of the magnetic diffusivities. From (7.146) and (7.150) we see that

$$\overline{B'^2} : \overline{B}^2 = \eta_T : \eta. \tag{7.151}$$

The left-hand side of this relation represents the ratio of the magnetic energy stored in the fluctuations to that stored in the mean magnetic field. It becomes large if the ratio of the turbulent magnetic diffusivity to the molecular diffusivity is large. With the data given in section 3.11. we obtain for the Sun $\eta_T \approx 10^4 \eta$, and see that we have to expect that the fluctuations store ten thousand times the energy that is stored in the mean field.

For the r.m.s. of the magnetic field fluctuations at the Sun's surface we find

$$B' : \overline{B} \approx 100. \tag{7.152}$$

This relation makes it clear that it might be difficult to derive reliable data of the mean magnetic field at the Sun's surface from the values observed there.

CHAPTER 8

THE TURBULENT ELECTROMOTIVE FORCE IN THE CASE OF NON-VANISHING MEAN FLOW

8.1. Introductory remarks

The considerations in the foregoing chapter already led to a number of important results and in this way quite an extensive insight in the subject was provided. However, just with regard to applications in cosmical physics neither homogeneity nor a non-vanishing mean flow can be assumed for those problems we intend to attack. The most obvious example in this respect is the convection in the upper layer of the Sun. In the convection zone of the Sun all physical quantities like pressure, temperature, mass density, etc., show a strong dependence on the radial coordinate. Therefore, the convection has to be looked upon as anisotropic and inhomogeneous, depending on the radial coordinate. Moreover, the turbulent convective motion undergoes the influence of the rotational motion, i.e. there is a non-vanishing mean motion. Consequently, an extension of the theory in two directions is desirable: Firstly, towards non-vanishing mean flow and, secondly, towards more general, especially inhomogeneous models of turbulence.

Our considerations in chapter 4 show that the first extension needs a determination of the Green's tensor for non-vanishing mean flow, a problem which can hardly be solved in a general way. However, for flows with constant rates of strain the Green's tensor can be represented in an almost explicit form (Krause [1, 2, 6, 9]). On the basis of this representation essential results can already be deduced, since, firstly, the rotational motion is a motion with constant rates of strain and, secondly, in the case of two-scale turbulence a motion of that kind can be considered to be a good approximation of a general motion.

As far as the extension to more general turbulence models is concerned we shall present here some examples of inhomogeneous turbulence undergoing the influence of Coriolis forces. It should be noted that we have arrived at a point that still lacks sufficient exploration. The results presented here, especially those of quantitative character, are therefore of limited applicability.

8.2. The Green's tensor for velocity fields with constant rates of strain

We represent the velocity field by

$$u_i = u_{ij}x_j + u_{0i}, \quad u_{ij}, u_{0i} \text{ constant}, \tag{8.1}$$

and introduce the tensor $\gamma_{ij}(t)$ by

$$\dot{\gamma}_{ij} = u_{ik}\gamma_{kj}, \quad \gamma_{ij}(0) = \delta_{ij}. \tag{8.2}$$

From the basic statements of the theory of ordinary differential equations it is clear that γ_{ij} exists and is uniquely defined. We have, e.g.,

$$\gamma_{ij}(t) = \begin{pmatrix} 1 & \omega^* t & 0 \\ 0 & 1 & 0 \\ 0 & 0 & 1 \end{pmatrix}, \tag{8.3}$$

if the shear flow

$$u_1 = \omega^* x_2, \quad u_2 = u_3 = 0, \tag{8.4}$$

is considered. For the rotational motion

$$\boldsymbol{u} = \boldsymbol{\Omega} \times \boldsymbol{x}, \quad \boldsymbol{\Omega} = \text{const}, \tag{8.5}$$

we obtain

$$\gamma_{ij}(t) = \frac{1}{\Omega^2}[\Omega_i\Omega_j + (\Omega^2\,\delta_{ij} - \Omega_i\Omega_j)\cos\Omega t] - \varepsilon_{ijp}\frac{\Omega_p}{\Omega}\sin\Omega t. \tag{8.6}$$

The first order result

$$\gamma_{ij}(t) = \delta_{ij} + u_{ij}t, \quad \text{if} \quad t \to 0, \tag{8.7}$$

holds for a general motion of type (8.1).

We note the following relations:

$$\gamma_{ij}(t_1 + t_2) = \gamma_{ik}(t_1)\,\gamma_{kj}(t_2), \tag{8.8}$$

$$\gamma_{ij}(-t) = \gamma_{ij}^{-1}(t), \tag{8.9}$$

$$u_{ij}\gamma_{jk}(t) = \gamma_{ij}(t)\,u_{jk}. \tag{8.10}$$

Relation (8.8) is clear since both the tensors, $\tilde{\gamma}_{ij}(t) = \gamma_{ij}(t + t_2)$ and $\tilde{\tilde{\gamma}}_{ij}(t) = \gamma_{ik}(t)\,\gamma_{kj}(t_2)$ solve equation (8.2) with the initial condition $\tilde{\gamma}_{ij}(0) = \tilde{\tilde{\gamma}}_{ij}(0) = \gamma_{ij}(t_2)$, hence both are identical. (8.9) follows immediately from (8.8) putting $t_1 = -t_2 = -t$. Differentiating the relation

$$\gamma_{ij}(t)\,\gamma_{jk}(-t) = \delta_{ik}$$

we obtain

$$\dot{\gamma}_{ij}(t)\,\gamma_{jk}(-t) - \gamma_{ij}(t)\,\dot{\gamma}_{jk}(-t) = 0$$

and with (8.2)

$$\gamma_{ik}(-t)\,[u_{il}\gamma_{lj}(t) - \gamma_{il}(t)\,u_{lj}] = 0 .$$

From this follows relation (8.10).

For the determinant, $\gamma = \det(\gamma_{ij})$, we have $\dot{\gamma} = \dot{\gamma}_{ij}(t)\,\gamma_{ji}(-t)\,\gamma(t)$, if (8.9) is taken into account. (8.2) yields

$$\dot{\gamma}(t) = u_{ik}\gamma_{kj}(t)\,\gamma_{ji}(-t)\,\gamma(t) = u_{ii}\gamma(t), \tag{8.11}$$

and we obtain finally

$$\gamma(t) = \mathrm{e}^{\operatorname{div} \boldsymbol{u}\cdot t}. \tag{8.12}$$

The equation of motion

$$\dot{\boldsymbol{x}} = \boldsymbol{u}, \tag{8.13}$$

with $\boldsymbol{u}$ given by (8.1) can now be integrated by

$$x_i(t) = \gamma_{ij}(t)\,x_{0j} + \int_0^t \gamma_{ij}(\tau)\,\mathrm{d}\tau\,u_{0j}. \tag{8.14}$$

With (8.2) and the inverse of (8.14) we find

$$\begin{aligned}\dot{x}_i = \dot{\gamma}_{ij}x_{0j} + \gamma_{ij}u_{0j} &= u_{ik}\left[x_k - \int_0^t \gamma_{kj}(\tau)\,\mathrm{d}\tau\,u_{0j}\right] + \gamma_{ij}u_{0j}\\ &= u_{ik}x_k + u_{0j}\left[\gamma_{ij} - \int_0^t \dot{\gamma}_{ij}(\tau)\,\mathrm{d}\tau\right] = u_{ik}x_k + u_{0i},\end{aligned}$$

thus confirming the above statement.

(8.14) describes the path of a particle moving with the flow which was found at $\boldsymbol{x}_0$ for $t = 0$. This allows us to introduce co-moving coordinates $\boldsymbol{x}'$ by

$$x_i' = \gamma_{ij}(-t)\,x_j + \int_0^{-t} \gamma_{ij}(\tau)\,\mathrm{d}\tau\,u_{0j}, \tag{8.15}$$

and the equations for the magnetic field read

$$\frac{\partial B_i}{\partial t} = (-\operatorname{div} \boldsymbol{u}\,\delta_{ij} + u_{ij})\,B_j + \frac{1}{\mu\sigma}\gamma_{jr}(-t)\,\gamma_{kr}(-t)\,\frac{\partial^2 B_i}{\partial x_j'\partial x_k'}, \tag{8.16}$$

$$\operatorname{div} \boldsymbol{B} = \gamma_{ri}(-t)\,\frac{\partial B_i}{\partial x_r'} = 0 . \tag{8.17}$$

Introducing Fourier transforms with respect to the space coordinates $\boldsymbol{x}'$ we obtain

$$\frac{\partial \tilde{B}_i(\boldsymbol{k}, t)}{\partial t} = \left\{ - \left[\operatorname{div} \boldsymbol{u} + \frac{1}{\mu\sigma} \gamma_{pr}(-t)\, \gamma_{qr}(-t)\, k_p k_q \right] \delta_{ij} + u_{ij} \right\} \tilde{B}_j(\boldsymbol{k}, t), \tag{8.18}$$

$$\gamma_{ri}(-t)\, k_r \tilde{B}_i(\boldsymbol{k}, t) = 0. \tag{8.19}$$

Using (8.2) and (8.12) we easily find the solution

$$\tilde{B}_i(\boldsymbol{k}, t) = \frac{\gamma_{ij}(t)}{\gamma(t)} \exp \left[- \frac{t}{\mu\sigma} c_{pq}^{-1}(t)\, k_p k_q \right] \tilde{B}_{0j}(\boldsymbol{k}), \tag{8.20}$$

with

$$k_j \tilde{B}_{0j}(\boldsymbol{k}) = 0. \tag{8.21}$$

We introduced in (8.20) the symmetric positive definite tensor c_{pq}^{-1} defined by

$$c_{pq}^{-1}(t) = \frac{1}{t} \int_0^t \gamma_{pr}(-\tau)\, \gamma_{qr}(-\tau)\, \mathrm{d}\tau, \tag{8.22}$$

where the notation indicates that later on we shall use the inverse tensor c_{pq} defined by $c_{pq} c_{qr}^{-1} = \delta_{pr}$. The field $\tilde{\boldsymbol{B}}_0(\boldsymbol{k})$ introduced in (8.20) is clearly the Fourier transform of the initial field $B_i(\boldsymbol{x}, 0)$.

Let $\boldsymbol{x}_t$ denote the coordinates of a point that moves with the velocity $\boldsymbol{u}$ and at the time $t = 0$ was found at $\boldsymbol{x}$. Explicitly we have according to (8.14)

$$(\boldsymbol{x}_t)_i = \gamma_{ij}(t)\, x_j + \int_0^t \gamma_{ij}(\tau)\, \mathrm{d}\tau\, u_{0j}. \tag{8.23}$$

Then the coordinate $\boldsymbol{x}'$ introduced in (8.15) is clearly given by

$$\boldsymbol{x}' = \boldsymbol{x}_{-t}. \tag{8.24}$$

According to the above analysis we have

$$B_i(\boldsymbol{x}, t) = \int \tilde{B}_i(\boldsymbol{k}, t)\, \mathrm{e}^{\mathrm{i}\boldsymbol{k}\cdot\boldsymbol{x}_{-t}}\, \mathrm{d}\boldsymbol{k}, \quad B_i(\boldsymbol{x}, 0) = \int \tilde{B}_{0i}(\boldsymbol{k})\, \mathrm{e}^{\mathrm{i}\boldsymbol{k}\cdot\boldsymbol{x}}\, \mathrm{d}\boldsymbol{k}. \tag{8.25}$$

Now inserting (8.20) we obtain the representation of $B_i(\boldsymbol{x}, t)$ in the form

$$B_i(\boldsymbol{x}, t) = \int G_{ij}(\boldsymbol{x}, \xi, t, 0)\, B_j(\xi, 0)\, \mathrm{d}\xi, \tag{8.26}$$

where the tensor G_{ij} is given by

$$G_{ij}(\boldsymbol{x}, \xi, t, \tau) = \gamma_{ij}(t - \tau)\, G'(\boldsymbol{x}_{-(t-\tau)} - \xi, t - \tau \mid \overline{\boldsymbol{u}}), \tag{8.27}$$

and

$$G'(\boldsymbol{x}, t \mid \overline{\boldsymbol{u}}) = \begin{cases} \dfrac{1}{\gamma(t)} \dfrac{1}{(2\pi)^3} \displaystyle\int \exp \left\{ - \frac{t}{\mu\sigma} c_{pq}^{-1}(t) k_p k_q + \mathrm{i}\boldsymbol{k}\boldsymbol{x} \right\} \mathrm{d}\boldsymbol{k}, & t \geqq 0, \\ 0\,, & t < 0. \end{cases} \tag{8.28}$$

(8.26) is the solution of the initial value problem of the induction equation for velocity fields with constant rates of strain. Consequently, the tensor G_{ij} introduced by (8.27) and (8.28) is, indeed, the Green's tensor as defined in section 4.5.

Let C_{kl} be a positive definite tensor. Then we have

$$\frac{1}{(2\pi)^3}\int \exp\left\{-C_{kl}k_k k_l + \mathrm{i}\boldsymbol{k}\cdot\xi\right\}\mathrm{d}\boldsymbol{k} = \frac{1}{\sqrt{4\pi^3}}\,\frac{1}{\sqrt{C}}\exp\left\{-\frac{1}{4}C_{pq}^{-1}\xi_p\,\xi_q\right\}, \tag{8.29}$$

a relation which is easily proved by using the principal-axis transformation of the tensor C_{kl}. This result enables us to evaluate the integral (8.28) and we find

$$G'(\boldsymbol{x},t\mid\boldsymbol{u}) = \begin{cases} \left(\dfrac{\mu\sigma}{4\pi t}\right)^{3/2}\dfrac{\sqrt{c}}{\gamma}\exp\left\{-\dfrac{\mu\sigma}{4t}c_{pq}x_p x_q\right\}, & \text{if } t \geqq 0, \\ 0\quad , & \text{if } t < 0. \end{cases} \tag{8.30}$$

There is an alternative representation of the Green's tensor G_{ij}, which proves to be useful later on. The tensor d_{pq} shall be defined by

$$d_{pq}(t) = c_{rs}(t)\,\gamma_{rp}(-t)\,\gamma_{sq}(-t). \tag{8.31}$$

Then we obtain from (8.27) and (8.30) the slightly different representation of the Green's tensor

$$G_{ij}(\boldsymbol{x},\boldsymbol{\xi},t,\tau) = \gamma_{ij}(t-\tau)\,G(\boldsymbol{x}-\boldsymbol{\xi}_{t-\tau},t-\tau\mid\boldsymbol{u}), \tag{8.32}$$

with the function G given by

$$G(\boldsymbol{x},t\mid\boldsymbol{u}) = \begin{cases} \left(\dfrac{\mu\sigma}{4\pi t}\right)^{3/2}\sqrt{d}\exp\left\{-\dfrac{\mu\sigma}{4t}d_{pq}x_p x_q\right\}, & \text{if } t \geqq 0, \\ 0\quad , & \text{if } t < 0. \end{cases} \tag{8.33}$$

We note the relation

$$d_{pq}^{-1}(t) = \frac{1}{t}\int_0^t \gamma_{pr}(\tau)\,\gamma_{qr}(\tau)\,\mathrm{d}\tau = c_{pq}^{-1}(-t), \tag{8.34}$$

which follows from (8.8), (8.22) and (8.31). As a consequence we have

$$d_{pq}(t) = c_{pq}(-t). \tag{8.35}$$

Some remarks shall conclude this section:

The tensors c_{pq} and d_{pq} are equal to the Kronecker tensor not only in the case $u_{ij} = 0$, but also for a rotational motion since $\gamma_{kr}(t)\,\gamma_{lr}(t) = \delta_{kl}$.

For the shear flow (8.3) we find from (8.34)

$$d_{pq}^{-1}(t) = c_{pq}^{-1}(-t) = \begin{pmatrix} 1 + \frac{1}{3}(\omega^* t)^2 & \frac{1}{2}\omega^* t & 0 \\ \frac{1}{2}\omega^* t & 1 & 0 \\ 0 & 0 & 1 \end{pmatrix}, \tag{8.36}$$

and

$$d_{pq}(t) = c_{pq}(-t) = \frac{1}{1 + \frac{1}{12}(\omega^* t)^2} \begin{pmatrix} 1 & -\frac{1}{2}\omega^* t & 0 \\ -\frac{1}{2}\omega^* t & 1 + \frac{1}{3}(\omega^* t)^2 & 0 \\ 0 & 0 & 1 \end{pmatrix}, \tag{8.37}$$

(RÄDLER [1], KRAUSE [2]).

For a general flow of type (8.1) we note the first order result

$$d_{pq}(t) = c_{pq}(-t) = \delta_{pq} - \frac{1}{2}(u_{pq} + u_{qp})\, t, \tag{8.38}$$

which shows the tensors c_{pq} and d_{pq} to be closely connected with the deformations which a fluid element undergoes following the motion.

8.3. Representation of the turbulent electromotive force

We start from the representation of $\mathfrak{E}$ in the frame of second order correlation approximation

$$\mathfrak{E}_i(\boldsymbol{x}, t) = \varepsilon_{ijk}\varepsilon_{lmn}\varepsilon_{npq} \iint \frac{\partial G_{kl}(\boldsymbol{x}, \boldsymbol{x} - \boldsymbol{\xi}, t, t - \tau)}{\partial \xi_m} Q_{jp}(\boldsymbol{x}, -\boldsymbol{\xi}, t, -\tau)\, \overline{B}_q(\boldsymbol{x} - \boldsymbol{\xi}, t - \tau)\, \mathrm{d}\boldsymbol{\xi}\, \mathrm{d}\tau, \tag{8.39}$$

which we obtain from (4.54) and (4.55). Inserting (8.32) we find

$$\mathfrak{E}_i(\boldsymbol{x}, t) = \varepsilon_{ijk}\varepsilon_{lmn}\varepsilon_{npq} \iint \gamma_{kl}(\tau) \frac{\partial G(\boldsymbol{x} - (\boldsymbol{x} - \boldsymbol{\xi})_\tau, \tau \mid \overline{\boldsymbol{u}})}{\partial \xi_m} Q_{jp}(\boldsymbol{x}, -\boldsymbol{\xi}, t, -\tau)\, \overline{B}_q(\boldsymbol{x} - \boldsymbol{\xi}, t - \tau)\, \mathrm{d}\boldsymbol{\xi}\, \mathrm{d}\tau. \tag{8.40}$$

We now introduce the new variable $\boldsymbol{\xi}'$ by

$$\boldsymbol{x} - \boldsymbol{\xi}' = (\boldsymbol{x} - \boldsymbol{\xi})_\tau, \tag{8.41}$$

and can thus rewrite the integral by

$$\mathfrak{E}_i(\boldsymbol{x}, t) = \varepsilon_{ijk}\varepsilon_{lmn}\varepsilon_{npq} \iint \frac{\gamma_{kl}(\tau)\,\gamma_{rm}(\tau)}{\gamma(\tau)} \frac{\partial G(\boldsymbol{\xi}, \tau \mid \overline{\boldsymbol{u}})}{\partial \xi_r}$$
$$Q_{jp}(\boldsymbol{x}, (\boldsymbol{x} - \boldsymbol{\xi})_{-\tau} - \boldsymbol{x}, t, -\tau)\, \overline{B}_q((\boldsymbol{x} - \boldsymbol{\xi})_{-\tau}, t - \tau)\, \mathrm{d}\boldsymbol{\xi}\, \mathrm{d}\tau; \tag{8.42}$$

deleting the dash at $\boldsymbol{\xi}$ when writing down this formula. If we insert the relations

$$\varepsilon_{lmn} = \gamma(\tau)\, \varepsilon_{stu}\gamma_{ls}(-\tau)\, \gamma_{mt}(-\tau)\, \gamma_{nu}(-\tau), \tag{8.43}$$
$$\varepsilon_{npq} = \gamma(-\tau)\, \varepsilon_{vwx}\gamma_{vn}(\tau)\, \gamma_{wp}(\tau)\, \gamma_{xq}(\tau), \tag{8.44}$$

(8.42) yields for a proper choice of the dummy indices

$$\mathfrak{E}_i(\boldsymbol{x}, t) = \varepsilon_{ijk}\varepsilon_{klm}\varepsilon_{mnp} \iint \frac{\partial G(\boldsymbol{\xi}, \tau \mid \overline{\boldsymbol{u}})}{\partial \xi_l} \gamma_{nq}(\tau)\, Q_{jq}(\boldsymbol{x}, (\boldsymbol{x} - \boldsymbol{\xi})_{-\tau} - \boldsymbol{x}, t, -\tau)$$
$$\frac{\gamma_{pr}(\tau)}{\gamma(\tau)}\, \overline{B}_r((\boldsymbol{x} - \boldsymbol{\xi})_{-\tau}, t - \tau)\, \mathrm{d}\boldsymbol{\xi}\, \mathrm{d}\tau. \tag{8.45}$$

For an interpretation of this formula let us first introduce a coordinate system $\boldsymbol{y}$ co-moving with the mean flow $\overline{\boldsymbol{u}}$ by

$$x_k = \gamma_{kl}(t - t_0)\, y_l + \int_0^{t-t_0} \gamma_{kl}(\tau)\, \mathrm{d}\tau\, \overline{u}_{0l}; \tag{8.46}$$

at the time t_0 both systems are identical. By taking the time derivative of (8.46) we obtain from (8.13), (8.14)

$$\frac{\mathrm{d}x_k}{\mathrm{d}t} = \overline{u}_{kl}x_l = \overline{u}_{0k} + \gamma_{kl}(t - t_0)\frac{\mathrm{d}y_l}{\mathrm{d}t}. \tag{8.47}$$

The velocity with respect to the co-moving frame shall be denoted by $\boldsymbol{v}$, i.e.

$$\boldsymbol{v} = \frac{\mathrm{d}\boldsymbol{y}}{\mathrm{d}t}.$$

Taking into account that

$$\frac{\mathrm{d}\boldsymbol{x}}{\mathrm{d}t} = \boldsymbol{u} = \overline{\boldsymbol{u}} + \boldsymbol{u}', \tag{8.48}$$

we have

$$v_k(\boldsymbol{y}, t) = \gamma_{kl}(t_0 - t)\, u_l'(\boldsymbol{x}, t), \tag{8.49}$$
$$u_k'(\boldsymbol{x}, t) = \gamma_{kl}(t - t_0)\, v_l(\boldsymbol{y}, t), \tag{8.50}$$

where the arguments are related by

$$\boldsymbol{x} = \boldsymbol{y}_{t-t_0}, \quad \boldsymbol{y} = \boldsymbol{x}_{t_0-t}. \tag{8.51}$$

For our applications to the integral (8.45) we identify t_0 with t and t with $t - \tau$.

From (8.49), (8.50) and (8.51) we get

$$\mathfrak{E}_i(\boldsymbol{x}, t) = \varepsilon_{ijk}\varepsilon_{klm}\varepsilon_{mnp} \iint \frac{\partial G(\xi, \tau \mid \overline{\boldsymbol{u}})}{\partial \xi_l} Q_{jn}(\boldsymbol{x}, -\xi, t, -\tau \mid \overline{\boldsymbol{u}}) \frac{\gamma_{pr}(\tau)}{\gamma(\tau)} \overline{B_r}((\boldsymbol{x} - \xi)_{-\tau}, t - \tau) \, \mathrm{d}\xi \, \mathrm{d}\tau, \tag{8.52}$$

where we introduced the correlation tensor

$$Q_{jn}(\boldsymbol{x}, \xi, t, \tau \mid \overline{\boldsymbol{u}}) = \overline{v_j(\boldsymbol{x}, t) \, v_n(\boldsymbol{x} + \xi, t + \tau)}. \tag{8.53}$$

which is formed from the velocity components with respect to the frame co-moving with the mean motion.

Considering the latter part in the integrand of (8.52) we draw attention to the well-known representation of the solution of the induction equation in the case of an ideally conducting medium, which can be written in the form

$$\frac{\boldsymbol{B}(\boldsymbol{x}, t)}{\varrho(\boldsymbol{x}, t)} = \left(\frac{\boldsymbol{B}(\boldsymbol{x}_0, t_0)}{\varrho(\boldsymbol{x}_0, t_0)} \cdot \operatorname{grad}_0\right) \boldsymbol{x}(\boldsymbol{x}_0, t_0), \tag{8.54}$$

where ϱ is the mass density and $\boldsymbol{x}(\boldsymbol{x}_0, t_0)$ the integral of the equation of motion $\dot{\boldsymbol{x}} = \boldsymbol{u}$. From the equation of continuity

$$\frac{\partial \varrho}{\partial t} + \operatorname{div} \varrho \boldsymbol{u} = 0, \tag{8.55}$$

and the definition of γ in (8.12) we can take

$$\frac{\varrho(\boldsymbol{x}, t)}{\varrho(\boldsymbol{x}_0, t_0)} = \frac{1}{\gamma(t - t_0)}. \tag{8.56}$$

Consequently, we see from (8.14) that the field $\dfrac{\gamma_{pr}(\tau)}{\gamma(\tau)} \overline{B_r}((\boldsymbol{x} - \xi)_{-\tau}, t - \tau)$ would be identical with $\overline{B_p}(\boldsymbol{x} - \xi, t)$ when the mean magnetic field is frozen-in in the mean motion. Or, in other words, we have

$$\overline{B_p}(\boldsymbol{x}, t) = \frac{\gamma_{pr}(\tau)}{\gamma(\tau)} \overline{B_r}((\boldsymbol{x} - \xi)_{-\tau}, t - \tau) \tag{8.57}$$

as long as the mean magnetic field can be looked upon as frozen-in in the mean motion. Hence, relation (8.57) can be used for an evaluation of (8.52) when the mean magnetic field does not greatly diffuse over a time interval of length τ_{cor}, an assumption which is already familiar to us from the considerations of two-scale turbulence in section 5.4. Under this condition (8.52) takes the form

$$\mathfrak{E}_i(\boldsymbol{x}, t) = \varepsilon_{ijk}\varepsilon_{klm}\varepsilon_{mnp} \iint \frac{\partial G(\xi, \tau \mid \overline{\boldsymbol{u}})}{\partial \xi_l} Q_{jn}(\boldsymbol{x}, -\xi, -\tau \mid \overline{\boldsymbol{u}}) \overline{B_p}(\boldsymbol{x} - \xi, t) \, \mathrm{d}\xi \, \mathrm{d}\tau. \tag{8.58}$$

This representation of the electromotive force $\mathfrak{E}$ differs only slightly from that of the case $\overline{\boldsymbol{u}} = 0$.

Assuming that the mean magnetic field has the two-scale property also with respect to the space coordinates we obtain from (8.5) for the tensors a_{ip}, b_{ipq}

$$a_{ip} = \varepsilon_{ijk}\varepsilon_{klm}\varepsilon_{mnp} \iint \frac{\partial G(\boldsymbol{\xi}, \tau \mid \overline{\boldsymbol{u}})}{\partial \xi_l} Q_{jn}(\boldsymbol{x}, -\boldsymbol{\xi}, t, -\tau \mid \overline{\boldsymbol{u}})\, d\boldsymbol{\xi}\, d\tau, \tag{8.59}$$

$$b_{ipq} = -\varepsilon_{ijk}\varepsilon_{klm}\varepsilon_{mnp} \iint \frac{\partial G(\boldsymbol{\xi}, \tau \mid \overline{\boldsymbol{u}})}{\partial \xi_l} \xi_q Q_{jn}(\boldsymbol{x}, -\boldsymbol{\xi}, t, -\tau \mid \overline{\boldsymbol{u}})\, d\boldsymbol{\xi}\, d\tau. \tag{8.60}$$

8.4. On the influence of a mean motion on the correlation tensor

The results (8.58), (8.59) and (8.60) of the foregoing investigations show that we still have to determine the correlation tensor undergoing the influence of a mean motion. Attempting this we have to start from an ensemble of equations including Navier-Stokes equation, the equation of continuity, equations of state, etc. We see already that there is less hope of finding a sufficiently general model which can be treated. Therefore we restrict ourselves to the incompressible case, i.e. we consider the Navier-Stokes equation in the form

$$\frac{\partial \boldsymbol{u}}{\partial t} + (\boldsymbol{u} \cdot \nabla\, p)\, \boldsymbol{u} = -\frac{1}{\varrho} \nabla\, p + \boldsymbol{F} + \nu \Delta \boldsymbol{u}, \tag{8.61}$$

and the equation of continuity

$$\operatorname{div} \boldsymbol{u} = 0. \tag{8.62}$$

p denotes the pressure, $\boldsymbol{F}$ the forces and ν the viscosity. By taking the average we obtain the Reynolds equations

$$\frac{\partial \bar{u}_i}{\partial t} + \bar{u}_k \frac{\partial \bar{u}_i}{\partial x_k} - \nu \Delta \bar{u}_i = -\frac{1}{\varrho} \frac{\partial \bar{p}}{\partial x_i} + \bar{F}_i - \frac{\partial Q_{ik}(\boldsymbol{x}, 0, t, 0)}{\partial x_k}, \tag{8.63}$$

$$\operatorname{div} \overline{\boldsymbol{u}} = 0. \tag{8.64}$$

The equation governing $\boldsymbol{u}'$ is obtained by subtracing (8.63) from (8.61). It is

$$\frac{\partial u_i'}{\partial t} - \nu \Delta u_i' = -\frac{1}{\varrho} \frac{\partial p'}{\partial x_i} - \frac{\partial}{\partial x_k} (\bar{u}_k u_i' + u_k' \bar{u}_i) + F_i'$$
$$+ \frac{\partial}{\partial x_k} (u_i' u_k' - Q_{ik}(\boldsymbol{x}, 0, t, 0)). \tag{8.65}$$

We assume that the turbulent motion is due to the force $\boldsymbol{F}'$, i.e. $\boldsymbol{F}'$ excites $\boldsymbol{u}'$ in some way, which we do not discuss explicitly.

We now confine ourselves to the second order correlation approximation and therefore neglect in (8.65) the term which is of second order in the fluctuations. We then have to deal with the equations

$$\frac{\partial u_i'}{\partial t} - \nu \Delta u_i' = -\frac{1}{\varrho} \frac{\partial p'}{\partial x_i} - \frac{\partial}{\partial x_k} (\bar{u}_k u_i' + u_k' \bar{u}_i) + F_i', \tag{8.66}$$

$$\operatorname{div} \boldsymbol{u}' = \frac{\partial u_i'}{\partial x_i} = 0. \tag{8.67}$$

This procedure is permitted whenever

$$\min(\mathrm{Re}, \mathrm{S}) \ll 1, \tag{8.68}$$

where S was already defined by (3.14), and Re is the Reynolds number

$$\mathrm{Re} = \frac{u'\lambda_{\mathrm{cor}}}{\nu}. \tag{8.69}$$

Having in mind to determine the correlation tensor in the coordinate system co-moving with the mean motion, we restrict ourselves to those motions which have constant rates of strain, and use the coordinates introduced in (8.46). From (8.49) with (8.2) and (8.10) it is found that

$$\frac{\partial v_i}{\partial t} = \gamma_{ij}(t_0 - t)\left[\frac{\partial u_j'}{\partial t} + \bar{u}_l \frac{\partial u_j'}{\partial x_l} - u_l' \frac{\partial \bar{u}_j}{\partial x_l}\right], \tag{8.70}$$

thus leading to a representation of equations (8.66), (8.67) in a co-moving frame,

$$\frac{\partial v_i}{\partial t} + 2\bar{u}_{ij}v_j - \nu\gamma_{jr}(t_0 - t)\,\gamma_{kr}(t_0 - t)\,\frac{\partial^2 v_i}{\partial y_j \partial y_k} =$$

$$- \gamma_{ij}(t_0 - t)\,\gamma_{kj}(t_0 - t)\,\frac{\partial p'}{\partial y_k} + f_i', \tag{8.71}$$

$$\frac{\partial v_i}{\partial y_i} = 0. \tag{8.72}$$

f' denotes the force in the co-moving system,

$$f_i'(\boldsymbol{y}, t) = \gamma_{ij}(t_0 - t)\, F_j'(\boldsymbol{x}, t). \tag{8.73}$$

One realizes the generalized Coriolis forces as the second term on the left of equation (8.71).

Introduce now Fourier transforms with respect to the space coordinates. From (8.71), (8.72) follows

$$\frac{\partial \tilde{v}_i}{\partial t} + 2\bar{u}_{ij}\,\tilde{v}_j + \nu\gamma_{jr}(t_0 - t)\,\gamma_{kr}(t_0 - t)\,k_j k_k \tilde{v}_i$$

$$= - \gamma_{ij}(t_0 - t)\,\gamma_{kj}(t_0 - t)\, i k_k \tilde{p}' + \tilde{f}_i', \tag{8.74}$$

$$k_i \tilde{v}_i = 0. \tag{8.75}$$

Elimination of the pressure yields

$$\frac{\partial \tilde{v}_i}{\partial t} + 2\left[\delta_{ip} - \frac{\gamma_{ij}(t_0 - t)\,\gamma_{lj}(t_0 - t)\,k_l k_p}{\gamma_{mr}(t_0 - t)\,\gamma_{nr}(t_0 - t)\,k_m k_n}\right]\bar{u}_{pq}\tilde{v}_q \tag{8.76}$$

$$+ \nu\gamma_{jr}(t_0 - t)\,\gamma_{kr}(t_0 - t)\,k_j k_k \tilde{v}_i = \left[\delta_{ip} - \frac{\gamma_{ij}(t_0 - t)\,\gamma_{lj}(t_0 - t)\,k_l k_p}{\gamma_{mr}(t_0 - t)\,\gamma_{nr}(t_0 - t)\,k_m k_n}\right]\tilde{f}_p'.$$

At least at this point we see that for the general case we cannot expect a result which is easy to comprehend. We, therefore, restrict ourselves to rotational motions in the following and rely for our representation of the deductions on a paper by RÜDIGER [5]. Another possibility of further treating equation (8.76) is the restriction to a first order theory in the mean motion (KRAUSE [2]).

8.5. On the influence of a rotational motion on the correlation tensor

For the rotational motion (8.5) equation (8.76) simply reads

$$\frac{\partial \tilde{v}_i}{\partial t} + \frac{2}{k^2}(k^2\,\delta_{ip} - k_ik_p)\,\bar{u}_{pq}\tilde{v}_q + \nu k^2\tilde{v}_i = (k^2\,\delta_{ip} - k_ik_p)\,\frac{\tilde{f}'_p}{k^2}. \tag{8.77}$$

Introduce the Fourier transforms $\hat{v}_i$, $\hat{f}_i$ with respect to space and time coordinates and write

$$\hat{v}_i^{(0)} = \frac{(k^2\,\delta_{ip} - k_ik_p)\,\hat{f}'_p}{(\nu k^2 - \mathrm{i}\omega)\,k^2}. \tag{8.78}$$

$\boldsymbol{v}^{(0)}$ is clearly that turbulence field which would be excited by the forces $\boldsymbol{f}'$ when there is no mean motion. The forces $\boldsymbol{f}'$ should not depend on the mean motion. In the following we assume that $\boldsymbol{v}^{(0)}$ has a simple structure, e.g. homogeneous isotropic, or inhomogeneous and anisotropic with respect to one preferred direction, and investigate the influence of the rotational motion on such turbulence.

As an equation for $\hat{v}_i$ (8.77) can be cast in the form

$$D_{ij}^{-1}\hat{v}_j = \hat{v}_i^{(0)}, \quad k_i\hat{v}_i = k_i\hat{v}_i^{(0)} = 0, \tag{8.79}$$

with

$$D_{ij}^{-1} = \delta_{ij} - \frac{1}{\nu k^2 - \mathrm{i}\omega}\,\frac{2\boldsymbol{\Omega}\cdot\boldsymbol{k}}{k^2}\,\varepsilon_{ijk}k_k. \tag{8.80}$$

The system (8.79) can be reversed by

$$\hat{v}_i = D_{ij}\hat{v}_j^{(0)},\ k_i\hat{v}_i = k_i\hat{v}_i^{(0)} = 0, \tag{8.81}$$

with

$$D_{ij} = \delta_{ij} + \frac{1}{\nu k^2 - \mathrm{i}\omega}\,\frac{2\boldsymbol{\Omega}\cdot\boldsymbol{k}}{k^2}\,\varepsilon_{ijk}k_k \tag{8.82}$$

in first order with respect to $\boldsymbol{\Omega}$. D_{ij} is an even function of $\boldsymbol{k}$. For later use we note the relations

$$D_{ik}D_{jk}^* = \delta_{ij} + \frac{4\mathrm{i}\omega\boldsymbol{\Omega}\cdot\boldsymbol{k}}{k^2(\nu^2k^4 + \omega^2)}\,\varepsilon_{ijk}k_k, \tag{8.83}$$

$$D_{ik}k_k = D_{ik}^*k_k = k_i. \tag{8.84}$$

From (8.53) follows

$$Q_{ij}(\boldsymbol{x}, \xi, t, \tau \mid \overline{\boldsymbol{u}}) = \iint \hat{Q}_{ij}(\boldsymbol{x}, \boldsymbol{k}, t, \tau \mid \overline{\boldsymbol{u}})\, \mathrm{e}^{\mathrm{i}(\boldsymbol{k}\cdot\xi - \omega\tau)}\, \mathrm{d}\boldsymbol{k}\, \mathrm{d}\omega\,, \tag{8.85}$$

with

$$\hat{Q}_{ij}(\boldsymbol{x}, \boldsymbol{k}, t, \omega \mid \overline{\boldsymbol{u}}) = \iint \overline{\hat{v}_i(\boldsymbol{k}', \omega')\, \hat{v}_j(\boldsymbol{k}, \omega)}\, \mathrm{e}^{\mathrm{i}((\boldsymbol{k}+\boldsymbol{k}')\boldsymbol{x} - (\omega+\omega')t)}\, \mathrm{d}\boldsymbol{k}'\, \mathrm{d}\omega'. \tag{8.86}$$

Let $\hat{Q}_{ij}^{(0)}(\boldsymbol{x}, \boldsymbol{k}, t, \omega)$ be accordingly defined as $\hat{Q}_{ij}$ in (8.86), then follows from (8.81)

$$\hat{Q}_{ij}(\boldsymbol{x}, \boldsymbol{k}, t, \omega \mid \overline{\boldsymbol{u}}) = \frac{1}{(2\pi)^4} \iint D_{ik}(\boldsymbol{k}', \omega')\, D_{jl}(\boldsymbol{k}, \omega)$$

$$\hat{Q}_{kl}^{(0)}(\boldsymbol{x} + \boldsymbol{x}', \boldsymbol{k}, t + t', \omega)\, \mathrm{e}^{-\mathrm{i}((\boldsymbol{k}+\boldsymbol{k}')\boldsymbol{x}' - (\omega+\omega')t')}\, \mathrm{d}\boldsymbol{k}'\, \mathrm{d}\omega'\, \mathrm{d}\boldsymbol{x}'\, \mathrm{d}t'. \tag{8.87}$$

In the case of homogeneous steady turbulence $\boldsymbol{v}^{(0)}$ we obtain

$$\hat{Q}_{ij}(\boldsymbol{k}, \omega \mid \overline{\boldsymbol{u}}) = D_{ik}^{*} D_{jl} \hat{Q}_{kl}^{(0)}(\boldsymbol{k}, \omega). \tag{8.88}$$

Let $\hat{Q}_{kl}^{(0)}$ depend linearly on $\boldsymbol{x}$, thus we can write

$$\hat{Q}_{kl}^{(0)}(\boldsymbol{x} + \boldsymbol{x}', \boldsymbol{k}, \omega) = \hat{Q}_{kl}^{(0)}(\boldsymbol{x}, \boldsymbol{k}, \omega) + x_m' \frac{\partial \hat{Q}_{kl}^{(0)}(\boldsymbol{x}, \boldsymbol{k}, \omega)}{\partial x_m}. \tag{8.89}$$

Then from (8.87) follows

$$\hat{Q}_{ij}(\boldsymbol{x}, \boldsymbol{k}, \omega \mid \overline{\boldsymbol{u}}) = D_{ik}^{*}(\boldsymbol{k}, \omega)\, D_{jl}(\boldsymbol{k}, \omega)\, \hat{Q}_{kl}^{(0)}(\boldsymbol{x}, \boldsymbol{k}, \omega) + \mathrm{i}\, \frac{\partial D_{ik}^{*}(\boldsymbol{k}, \omega)}{\partial k_m}\, D_{jl}(\boldsymbol{k}, \omega)\, \frac{\partial \hat{Q}_{kl}^{(0)}(\boldsymbol{x}, \boldsymbol{k}, \omega)}{\partial x_m}. \tag{8.90}$$

Both the relations, (8.88) and (8.90), solve our problem, i.e. the determination of the influence of the mean motion on the correlation tensor, for the special turbulences considered here. It should be noted in particular that we now find more realistic turbulences with helicity. For, let $\hat{Q}_{kl}^{(0)} = \hat{Q}(k^2\, \delta_{kl} - k_k k_l)$, so that

$$\hat{Q}_{ij}(\boldsymbol{x}, \boldsymbol{k}, \omega \mid \overline{\boldsymbol{u}}) = \hat{Q} \left[(k^2\, \delta_{ij} - k_i k_j) + \frac{4\mathrm{i}\omega \boldsymbol{\Omega} \cdot \boldsymbol{k}}{\nu^2 k^4 + \omega^2}\, \varepsilon_{ijk} k_k \right], \tag{8.91}$$

because of (8.88), (8.83), (8.84). The last term indicates, indeed, helicity, i.e. the imbalance of the two kinds of screw motions. However, in this case the integral helicity,

$$h = \mathrm{i}\, \varepsilon_{ijl} \iint k_l \hat{Q}_{ij}(\boldsymbol{x}, \boldsymbol{k}, \omega \mid \overline{\boldsymbol{u}})\, \mathrm{d}\boldsymbol{k}\, \mathrm{d}\omega\,, \tag{8.92}$$

vanishes since $\hat{Q}$ is an even function of ω because of (6.20). If the turbulence $\boldsymbol{v}^{(0)}$ is still incompressible but anisotropic then $\hat{Q}$ has to be exchanged for $\hat{Q}(k, \omega) + \hat{Q}'(k, \omega)\, \boldsymbol{g} \cdot \boldsymbol{k}$, where $\hat{Q}'$ is an odd function of ω because of (6.20). Now we have a non-vanishing integral helicity (8.92).

CHAPTER 9

THE TURBULENT ELECTROMOTIVE FORCE IN THE CASE OF ROTATIONAL MEAN MOTION

9.1. Illustrating examples

We closed the foregoing chapter by showing that turbulences which have originally no helicity, do have helicity if they undergo the influence of a rotational motion. We could see this in both the cases of an originally isotropic and anisotropic turbulence. We assumed incompressibility, but that was for technical reasons only. A consideration of compressible turbulence would require more mathematical effort.

We shall present here physical arguments in order to show that it is quite natural for turbulences on a rotating body to have helicity. Figure 9.1 illustrates the situation in a compressible medium; it reflects the conditions in the convection zone of the Sun in the northern hemisphere. Rising matter will expand and rotate because of the action of Coriolis forces, thus providing for a left-handed helical motion. Sinking matter is compressed and is forced by the Coriolis forces to rotate in the opposite direction, again a left-handed helical motion. We realize that the convection in the northern hemisphere of the Sun shows a higher probability of left-handed helical motions than of right-handed ones. Obviously, in the southern hemisphere the right-handed helical motions will prevail.

Figure 9.2 illustrates the conditions for an incompressible fluid. A balance of left-handed and right-handed helical motions will exist in the medium layer

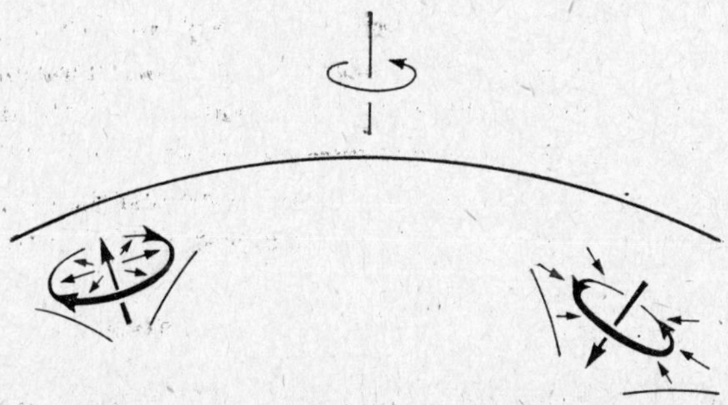

Fig. 9.1. Schematic drawing which explains that in the northern hemisphere of the Sun left-handed helical motions are more probable than right-handed ones

(the intersection of a and b) only if there is no gradient in the turbulence intensity, since one kind of helical motions comes from below (layer a) and the other from above (layer b). Consequently, a gradient in the turbulence intensity in a turbulent medium on a rotating body provides for helicity under the conditions of a flat geometry as considered in figure 9.2. For a spherical geometry this imbalance will occur for a special dependence of the turbulence intensity on the distance from the centre, generally a nonvanishing helicity has to be expected.

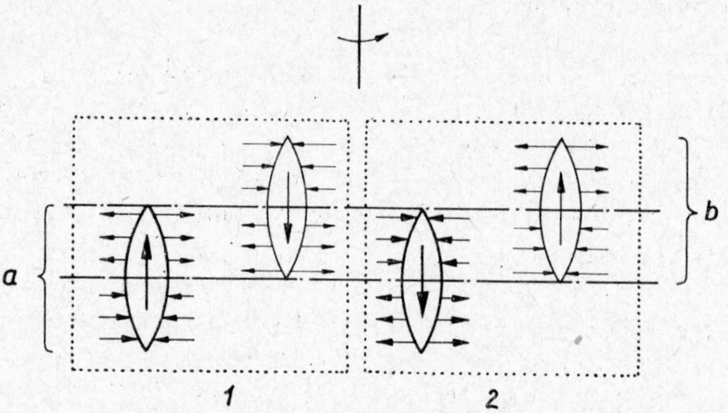

Fig. 9.2. In an incompressible medium the imbalance of right-handed and left-handed helical motions can be due to a gradient of the turbulence intensity $\overline{u'^2}$

9.2. The correlation tensor of an inhomogeneous turbulence

The investigations presented here are already restricted within the limits of two-scale turbulence. Therefore, considering inhomogeneous turbulence we assume a weak dependence of the mean values on the space coordinates, i.e. in our special concern we assume the correlation tensor $\hat{Q}_{ij}^{(0)}$ linearly dependent on $\boldsymbol{x}$ and not dependent on t. Moreover, we shall aim at a consideration of the convective layers of cosmical objects. There the radial direction is the direction of anisotropy and the only one in which the mean quantities vary. Therefore, we assume the correlation tensor $\hat{Q}_{ij}^{(0)}$ to be anisotropic with respect to a certain direction $\boldsymbol{g}$, to depend on $\boldsymbol{g} \cdot \boldsymbol{x}$, and to be linear in $\boldsymbol{g}$ and $\boldsymbol{x}$. We write

$$\hat{Q}_{ij}^{(0)}(\boldsymbol{x}, \boldsymbol{k}, \omega) = \hat{Q}_{ij}^{(00)}(\boldsymbol{k}, \omega) + \frac{i}{2} g_m \frac{\partial \hat{Q}_{ij}^{(01)}(\boldsymbol{k}, \omega)}{\partial k_m} + (\boldsymbol{g} \cdot \boldsymbol{x}) \, \hat{Q}_{ij}^{(01)}(\boldsymbol{k}, \omega), \tag{9.1}$$

where we shall immediately see the reasons for which we introduced the second summand on the right.

Because of the definition (4.3) of the correlation tensor we have the identity

$$Q_{ij}^{(0)}(\boldsymbol{x} - \xi, \xi, \tau) = Q_{ji}^{(0)}(\boldsymbol{x}, -\xi, -\tau), \tag{9.2}$$

which furnishes the condition

$$\hat{Q}_{ij}^{(00)}(\boldsymbol{k}, \omega) + \frac{\mathrm{i}}{2} g_m \frac{\partial \hat{Q}_{ij}^{(01)}(\boldsymbol{k}, \omega)}{\partial k_m} + g_m x_m \hat{Q}_{ij}^{(01)}(\boldsymbol{k}, \omega) - \mathrm{i} g_m \frac{\partial \hat{Q}_{ij}^{(01)}(\boldsymbol{k}, \omega)}{\partial k_m}$$

$$= \hat{Q}_{ji}^{(00)}(-\boldsymbol{k}, -\omega) + \frac{\mathrm{i}}{2} g_m \left[\frac{\partial \hat{Q}_{ji}^{(01)}(\boldsymbol{k}, \omega)}{\partial k_m}\right]_{-\boldsymbol{k}, -\omega} + g_m x_m \hat{Q}_{ji}^{(01)}(-\boldsymbol{k}, -\omega), \tag{9.3}$$

if applied to the tensor given in (9.1).

First we obtain

$$\hat{Q}_{ij}^{(01)}(\boldsymbol{k}, \omega) = \hat{Q}_{ji}^{(01)}(-\boldsymbol{k}, -\omega), \tag{9.4}$$

and, as a consequence,

$$\left[\frac{\partial \hat{Q}_{ji}^{(01)}(\boldsymbol{k}, \omega)}{\partial k_m}\right]_{-\boldsymbol{k}, -\omega} = -\frac{\partial \hat{Q}_{ij}^{(01)}(\boldsymbol{k}, \omega)}{\partial k_m}. \tag{9.10}$$

From (9.3) thus follows

$$\hat{Q}_{ij}^{(00)}(\boldsymbol{k}, \omega) = \hat{Q}_{ji}^{(00)}(-\boldsymbol{k}, -\omega). \tag{9.11}$$

It should be noted that $\hat{Q}_{ij}^{(0)}$ has to fulfil a condition corresponding to Bochner's theorem, which can be deduced from (6.1). From this follows in particular that $\hat{Q}_{ij}^{(00)}$ has to be positive semi-definite, and that $\boldsymbol{x}$ is restricted to a bounded region.

Suppose the fluid is incompressible, so that

$$\frac{\partial \hat{Q}_{ij}^{(0)}(\boldsymbol{x}, \boldsymbol{k}, \omega)}{\partial x_i} = \mathrm{i} k_i \hat{Q}_{ij}^{(0)}(\boldsymbol{x}, \boldsymbol{k}, \omega), \quad \hat{k}_j Q_{ij}^{(0)}(\boldsymbol{x}, \boldsymbol{k}, \omega) = 0. \tag{9.12}$$

Inserting (9.1) we find the conditions

$$k_i \hat{Q}_{ij}^{(01)} = 0, \quad k_j \hat{Q}_{ij}^{(01)} = 0, \tag{9.13}$$

$$k_i \hat{Q}_{ij}^{(00)} = -\frac{\mathrm{i}}{2} g_i \hat{Q}_{ij}^{(01)}; \quad k_j Q_{ij}^{(00)} = \frac{\mathrm{i}}{2} g_j \hat{Q}_{ij}^{(01)}. \tag{9.14}$$

$Q_{ij}^{(01)}$ is isotropic because of the required linearity of $Q_{ij}^{(0)}$ with respect to $\boldsymbol{g}$, therefore

$$\hat{Q}_{ij}^{(01)}(\boldsymbol{k}, \omega) = \hat{Q}_1(k, \omega)\,(k^2\,\delta_{ij} - k_i k_j), \tag{9.15}$$

with

$$\hat{Q}_1(k, -\omega) = \hat{Q}_1(k, \omega). \tag{9.16}$$

Take for $Q_{ij}^{(00)}$ the general ansatz, i.e.

$$\hat{Q}_{ij}^{(00)}(\boldsymbol{k}, \omega) = \{A + (\boldsymbol{k} \cdot \boldsymbol{g})\, A'\}\, k^2\, \delta_{ij} + \{B + (\boldsymbol{k} \cdot \boldsymbol{g})\, B'\}\, k_i k_j + \mathrm{i} C k_i g_j + \mathrm{i} C' k_j g_i,$$

and find from (9.14)

$$B = -A = \hat{Q}(k, \omega), \quad A' = -B' = \hat{Q}'(k, \omega),$$
$$C = -C' = -\frac{1}{2}\, \hat{Q}_1(k, \omega), \tag{9.17}$$

where

$$\hat{Q}(k, \omega) = \hat{Q}(k, -\omega), \quad \hat{Q}'(k, \omega) = -\hat{Q}'(k, -\omega), \tag{9.18}$$

because of (9.11).

We finally obtain

$$\begin{aligned} \hat{Q}_{ij}^{(0)}(\boldsymbol{x}, \boldsymbol{k}, \omega) = {} & \{\hat{Q}(k, \omega) + (\boldsymbol{k} \cdot \boldsymbol{g})\, \hat{Q}'(k, \omega)\}\, (k^2\, \delta_{ij} - k_i k_j) \\ & + \frac{i}{2}\, (\boldsymbol{k} \cdot \boldsymbol{g}) \left\{ \frac{1}{k}\, \frac{\partial k^2 \hat{Q}_1(k, \omega)}{\partial k}\, \delta_{ij} - \frac{1}{k}\, \frac{\partial \hat{Q}_1(k, \omega)}{\partial k}\, k_i k_j \right\} \\ & - \mathrm{i} \hat{Q}_1(k, \omega)\, k_i g_j + (\boldsymbol{g} \cdot \boldsymbol{x})\, \hat{Q}_1(k, \omega)\, (k^2\, \delta_{ij} - k_i k_j), \end{aligned} \tag{9.19}$$

where the three functions $\hat{Q}(k, \omega)$, $\hat{Q}'(k, \omega)$ and $\hat{Q}_1(k, \omega)$ satisfy (9.16) and (9.18), and

$$\hat{Q}(k, \omega) + (\boldsymbol{k} \cdot \boldsymbol{g})\, \hat{Q}'(k, \omega) \geqq 0, \tag{9.20}$$

because of Bochner's theorem. There is no such relation concerning $\hat{Q}_1$, however, $\boldsymbol{x}$ must not be too large.

(9.17) is the most general representation of the correlation tensor of an incompressible turbulent flow which is a linear function of the vector $\boldsymbol{g}$ characterizing the direction of anisotropy as well as the direction of inhomogeneity.

The result (9.19) coincides with that derived by Rädler [2, 11] for a turbulence being inhomogeneous because of the gradient of the turbulence intensity in case $\hat{Q}_1$ is taken proportional to $\hat{Q}$.

Finally, we mention that the summands $\hat{Q}(k^2\, \delta_{ij} - k_i k_j)$ and $(\boldsymbol{g} \cdot \boldsymbol{x})\, \hat{Q}_1(k^2\, \delta_{ij} - k_i k_j)$ in (9.19) are even with respect to $\boldsymbol{k}$, the others are odd.

9.3. Determination of the tensor b_{ipq} for an inhomogeneous turbulence influenced by Coriolis forces

For our purposes (8.60) shall be recorded in convolution form

$$b_{ipq}(\boldsymbol{x}) = -\varepsilon_{ijk}\varepsilon_{klm}\varepsilon_{mnp} \iint \frac{k_l}{\eta k^2 - \mathrm{i}\omega} \frac{\partial \hat{Q}_{jn}(\boldsymbol{x}, \boldsymbol{k}, \omega \mid \overline{\boldsymbol{u}})}{\partial k_q} \,\mathrm{d}\boldsymbol{k}\,\mathrm{d}\omega$$

$$= \varepsilon_{ijk}\varepsilon_{klm}\varepsilon_{mnp} \iint \frac{\partial}{\partial k_q}\left(\frac{k_l}{\eta k^2 - \mathrm{i}\omega}\right) \hat{Q}_{jn}(\boldsymbol{x}, \boldsymbol{k}, \omega \mid \overline{\boldsymbol{u}}) \,\mathrm{d}\boldsymbol{k}\,\mathrm{d}\omega. \tag{9.21}$$

Since the even part of Q_{jn} only contributes to the integral in (9.21) we obtain from (8.83), (8.84), (8.90) and (9.19)

$$b_{ipq}(\boldsymbol{x}) = \varepsilon_{ijk}\varepsilon_{klm}\varepsilon_{mnp} \iint \frac{\partial}{\partial k_q}\left(\frac{k_l}{\eta k^2 - \mathrm{i}\omega}\right) D_{jr}^*(\boldsymbol{k}, \omega)\, D_{ns}(\boldsymbol{k}, \omega)$$

$$\{\hat{Q}(k, \omega) + (\boldsymbol{g} \cdot \boldsymbol{x})\, \hat{Q}_1(k, \omega)\}\, (k^2\, \delta_{rs} - k_r k_s) \,\mathrm{d}\boldsymbol{k}\,\mathrm{d}\omega$$

$$= \varepsilon_{ijk}\varepsilon_{klm}\varepsilon_{mnp} \iint \frac{\partial}{\partial k_q}\left(\frac{k_l}{\eta k^2 - \mathrm{i}\omega}\right)$$

$$\{\hat{Q}(k, \omega) + (\boldsymbol{g} \cdot \boldsymbol{x})\, \hat{Q}_1'(k, \omega)\}\, (k^2\, \delta_{jn} - k_j k_n) \,\mathrm{d}\boldsymbol{k}\,\mathrm{d}\omega$$

$$- 4\mathrm{i}\varepsilon_{ijk}\varepsilon_{klm}\varepsilon_{mnp}\varepsilon_{jnr}\Omega_s \iint \frac{\partial}{\partial k_q}\left(\frac{k_l}{\eta k^2 - \mathrm{i}\omega}\right)$$

$$\omega\, \{\hat{Q}(k, \omega) + \hat{Q}_1(k, \omega)\, (\boldsymbol{g} \cdot \boldsymbol{x})\} \frac{k_r k_s}{\nu^2 k^4 + \omega^2} \,\mathrm{d}\boldsymbol{k}\,\mathrm{d}\omega. \tag{9.22}$$

The first summand is clearly of the form $\beta\varepsilon_{ipq}$ and by multiplying it by ε_{ipq} we obtain

$$\beta = \frac{2}{3} \iint \frac{\hat{Q}(k, \omega) + (\boldsymbol{g} \cdot \boldsymbol{x})\, \hat{Q}_1(k, \omega)}{\eta k^2 - \mathrm{i}\omega}\, k^2 \,\mathrm{d}\boldsymbol{k}\,\mathrm{d}\omega. \tag{9.23}$$

This relation can be rewritten as

$$\beta = \frac{1}{3} \iint \frac{\hat{Q}_{jj}(\boldsymbol{x}, \boldsymbol{k}, \omega \mid \overline{\boldsymbol{u}})}{\eta k^2 - \mathrm{i}\omega} \,\mathrm{d}\boldsymbol{k}\,\mathrm{d}\omega, \tag{9.24}$$

since $\hat{Q}_{jj} - 2k^2(\hat{Q} + (\boldsymbol{g} \cdot \boldsymbol{x})\, \hat{Q}_1)$ is odd with respect to $\boldsymbol{k}$, because of (8.90) and (9.19).

With (9.24) we generalize the representation (7.47) of the turbulent diffusivity to an inhomogeneous incompressible turbulence influenced by a rotational motion.

For the second summand in (9.22), i.e. β_{ipq} with the notation (7.55), we find

$$\beta_{ipq} = -4\mathrm{i}\,(\delta_{il}\delta_{pr} + \delta_{ir}\delta_{pl})\,\Omega_s \iint \frac{(\eta k^2 - \mathrm{i}\omega)\,\delta_{lq} - 2\eta k_l k_q}{(\eta k^2 - \mathrm{i}\omega)^2\,(\nu^2 k^4 + \omega^2)}$$

$$\times \omega k_r k_s \{\hat{Q}(k, \omega) + (\boldsymbol{g}\cdot\boldsymbol{x})\,\hat{Q}_1(k, \omega)\}\,\mathrm{d}\boldsymbol{k}\,\mathrm{d}\omega$$

$$= -\frac{2}{15}\Omega_s \iint \frac{(\delta_{iq}\delta_{ps} + \delta_{is}\delta_{pq})\,(3\eta^2k^4 - 5\omega^2) + 8\eta^2k^4\,\delta_{ip}\delta_{qs}}{(\eta^2k^4 + \omega^2)^2\,(\nu^2k^4 + \omega^2)}$$

$$\omega^2\hat{Q}_{jj}(\boldsymbol{x}, \boldsymbol{k}, \omega \mid \overline{\boldsymbol{u}})\,\mathrm{d}\boldsymbol{k}\,\mathrm{d}\omega\,. \tag{9.25}$$

For the evaluation we used (5.27). This result was already predicted in section 7.5. For the scalars β_1, β_2, β_3 introduced there we find (Roberts and Soward [3]).

$$\beta_1 = \beta_2 = \frac{2}{15}\iint \frac{(3\eta^2k^4 - 5\omega^2)\,\omega^2}{(\eta^2k^4 + \omega^2)^2\,(\nu^2k^4 + \omega^2)}\hat{Q}_{jj}(\boldsymbol{x}, \boldsymbol{k}, \omega \mid \overline{\boldsymbol{u}})\,\mathrm{d}\boldsymbol{k}\,\mathrm{d}\omega\,, \tag{9.26}$$

$$\beta_3 = \frac{16}{15}\iint \frac{\eta^2k^4\omega^2\hat{Q}_{jj}(\boldsymbol{x}, \boldsymbol{k}, \omega \mid \overline{\boldsymbol{u}})}{(\eta^2k^4 + \omega^2)^2\,(\nu^2k^4 + \omega^2)}\mathrm{d}\boldsymbol{k}\,\mathrm{d}\omega\,. \tag{9.27}$$

First expressions for β_1, β_2, β_3 were derived by Rädler [6].

It is essential to note that the quantities β_1, β_2, and β_3 were non-zero already in the case where the undisturbed turbulence is homogeneous and sotropic, i.e. $\hat{Q}' = \hat{Q}_1 = 0$. We find that the inhomogeneity did not produce basically new features.

9.4. Determination of the tensor a_{ip} for an inhomogeneous turbulence influenced by Coriolis forces

We start from relation (8.59) again recorded in convolution form

$$a_{ip} = \mathrm{i}\varepsilon_{ijk}\varepsilon_{klm}\varepsilon_{mnp}\iint \frac{k_l}{\eta k^2 - \mathrm{i}\omega}\hat{Q}_{jn}(\boldsymbol{x}, \boldsymbol{k}, \omega \mid \overline{\boldsymbol{u}})\,\mathrm{d}\boldsymbol{k}\,\mathrm{d}\omega\,. \tag{9.28}$$

For the incompressible case considered here it reduces to

$$a_{ip} = \mathrm{i}\varepsilon_{ijk}\iint \frac{k_p\hat{Q}_{jk}(\boldsymbol{x}, \boldsymbol{k}, \omega \mid \overline{\boldsymbol{u}})}{\eta k^2 - \mathrm{i}\omega}\mathrm{d}\boldsymbol{k}\,\mathrm{d}\omega\,. \tag{9.29}$$

First, suppose the turbulence to be homogeneous so that $\hat{Q}_1 = 0$. (8.83), (8.84), (8.88) and (9.19) give rise to

$$a_{ip} = \mathrm{i}\varepsilon_{ijk}\iint \frac{k_p}{\eta k^2 - \mathrm{i}\omega}\,(D_{jl}^{*}D_{km}\hat{Q}_{lm}^{(0)})\,\mathrm{d}\boldsymbol{k}\,\mathrm{d}\omega$$

$$= -4\varepsilon_{ijk}\varepsilon_{jkl}g_m\Omega_n\iint \frac{\omega k_l k_m k_n k_p\hat{Q}'(k, \omega)}{(\eta k^2 - \mathrm{i}\omega)\,(\nu^2k^4 + \omega^2)}\mathrm{d}\boldsymbol{k}\,\mathrm{d}\omega \tag{9.30}$$

$$= -\frac{8\eta}{15}\iint \frac{\omega k^6\hat{Q}'(k, \omega)}{(\eta^2k^4 + \omega^2)\,(\nu^2k^4 + \omega^2)}\mathrm{d}\boldsymbol{k}\mathrm{d}\omega\,\{\delta_{ip}(\boldsymbol{g}\cdot\boldsymbol{\Omega}) + (\Omega_i g_p + \Omega_p g_i)\}\,.$$

Thus we find (RÜDIGER [5]) for the quantity α introduced in (7.2)

$$\alpha = -\frac{8\eta}{9}\iint \frac{\omega k^6 \hat{Q}'(k, \omega)}{(\eta^2 k^4 + \omega^2)(\nu^2 k^4 + \omega^2)} \,\mathrm{d}\boldsymbol{k}\,\mathrm{d}\omega\,(\boldsymbol{g}\cdot\boldsymbol{\Omega}). \tag{9.31}$$

In order to appreciate this result it shall be noted that this is the first point where we find the quantity α to be non-zero for a real turbulence, in our case a homogeneous anisotropic turbulence influenced by a rotational motion. This has to be underlined because of the importance of α in connection with dynamo excitation as shown in section 7.3.

α proves to be proportional to the scalar product $\boldsymbol{g}\cdot\boldsymbol{\Omega}$, that is, if we consider a convective shell on a rotating body, α is antisymmetric with respect to the equatorial plane.

Finally we note that the tensor a_{ip} in (9.30) is symmetric, in accordance with our results in section 7.1.

The restriction to homogeneity will now be lifted, i.e. $\hat{Q}_1 \neq 0$ is assumed.

In order to investigate at first the antisymmetric part we introduce the vector $\gamma^*\boldsymbol{g}^*$ by

$$\gamma^* g_k^* = \frac{1}{2}\varepsilon_{ipk}a_{ip} = \frac{\mathrm{i}}{2}\iint \frac{k_j \hat{Q}_{jk}(\boldsymbol{x}, \boldsymbol{k}, \omega \mid \overline{\boldsymbol{u}})}{\eta k^2 - \mathrm{i}\omega}\,\mathrm{d}\boldsymbol{k}\,\mathrm{d}\omega. \tag{9.32}$$

According to (8.84), (8.90) and (9.12) we have

$$k_j \hat{Q}_{jk} = -\,\mathrm{i}D_{nl}^* D_{km}\frac{\partial \hat{Q}_{lm}^{(0)}}{\partial x_n}, \tag{9.33}$$

and from (9.19) we thus find

$$\begin{aligned}
\gamma^* g_k^* &= \frac{1}{2} g_n \iint \frac{k^2\,\delta_{kn} - k_k k_n}{\eta k^2 - i\omega}\hat{Q}_1(k, \omega)\,\mathrm{d}\boldsymbol{k}\,\mathrm{d}\omega \\
&\quad + 2 g_n \Omega_m \varepsilon_{knp}\iint \frac{\mathrm{i}\omega k_m k_p}{(\eta k^2 - \mathrm{i}\omega)(\nu^2 k^4 + \omega^2)}\hat{Q}_1(k, \omega)\,\mathrm{d}\boldsymbol{k}\,\mathrm{d}\omega \\
&= \frac{1}{3}\iint \frac{\eta k^4}{\eta^2 k^4 + \eta\,\omega^2}\hat{Q}_1(k, \omega)\,\mathrm{d}\boldsymbol{k}\,\mathrm{d}\omega\, g_k \\
&\quad - \frac{2}{3}\iint \frac{\omega^2 k \hat{Q}_1(k, \omega)}{(\eta^2 k^4 + \omega^2)(\nu^2 k^4 + \omega^2)}\,\mathrm{d}\boldsymbol{k}\,\mathrm{d}\omega\;\varepsilon_{klm} g_l \Omega_m.
\end{aligned} \tag{9.34}$$

This result clearly shows that for an inhomogeneous turbulence the tensor a_{ip} is non-symmetric. Its antisymmetric part shall be recorded in the form

$$a_{ip}^{(a)} = \varepsilon_{ipk}(\gamma g_k + \gamma' \varepsilon_{klm} g_l \Omega_m), \tag{9.35}$$

with

$$\gamma = \frac{1}{3} \iint \frac{\eta k^4}{\eta^2 k^4 + \omega^2} \hat{Q}_1(k, \omega) \, \mathrm{d}\boldsymbol{k} \, \mathrm{d}\omega,$$

$$\gamma' = -\frac{2}{3} \iint \frac{\omega^2 k^2 \hat{Q}_1(k, \omega)}{(\eta^2 k^4 + \omega^2)(\nu^2 k^4 + \omega^2)} \, \mathrm{d}\boldsymbol{k} \, \mathrm{d}\omega. \tag{9.36}$$

In contrast to the cases considered in section 7.1., in the case of an inhomogeneous incompressible turbulence a pumping effect occurs, even if there is no mean motion. The mean motion only modifies the direction in which the mean magnetic field is transported.

In order to determine the symmetric part α_{ip} of the tensor α_{ip} we recall that the general form of this tensor is already known, given in (5.43). We thus can write

$$\alpha_{ip} = -\alpha_1 (\boldsymbol{g} \cdot \boldsymbol{\Omega}) \, \delta_{ip} - \alpha_2 (\Omega_i g_p + \Omega_p g_i), \tag{9.37}$$

or

$$\alpha_{ip} = I_{ipqr} g_q \Omega_r, \tag{9.38}$$

with the isotropic tensor of rank four

$$I_{ipqr} = -\alpha_1 \, \delta_{ip} \delta_{qr} - \alpha_2 (\delta_{iq} \delta_{pr} + \delta_{ir} \delta_{pq}). \tag{9.39}$$

Let

$$D_{ij} = \delta_{ij} + D \Omega_l \varepsilon_{ijk} k_k k_l, \tag{9.40}$$

and

$$\hat{Q}^{(0)}_{ij} = \hat{q}_{ij} + g_k \left[\hat{q}_{ijk} + x_k \hat{Q}_1 (k^2 \, \delta_{ij} - k_i k_j) \right], \tag{9.41}$$

where

$$D = \frac{2}{k^2} \frac{1}{\nu k^2 - \mathrm{i}\omega}, \tag{9.42}$$

and

$$\hat{q}_{ijk} = \left(\hat{Q}' + \frac{\mathrm{i}}{2k} \frac{\partial \hat{Q}_1}{\partial k} \right) (k^2 \, \delta_{ij} - k_j k_i) \, k_k + \mathrm{i} \hat{Q}_1 (k_k \, \delta_{ij} - k_i \, \delta_{jk}), \tag{9.43}$$

because of (8.82) and (9.19).

From (8.90), (9.29), (9.40) and (9.41) we thus find

$$I_{ipqr} = \frac{\mathrm{i}}{2} \iint \frac{k_p \varepsilon_{ijk} + k_i \varepsilon_{pjk}}{\eta k^2 - \mathrm{i}\omega} \Big\{ (D^* \varepsilon_{jls} k_s \, \delta_{km} + D \varepsilon_{kms} k_s \, \delta_{jl}) \, k_r \hat{q}_{lmq} \tag{9.44}$$

$$+ \mathrm{i} \left[2 \frac{\partial D^*}{\partial (k^2)} k_q \varepsilon_{jls} k_s k_r + D^* \varepsilon_{jlq} k_r + D^* \varepsilon_{jls} k_s \delta_{qr} \right] \hat{Q}_1 (k^2 \, \delta_{kl} - k_k k_l) \Big\} \, \mathrm{d}\boldsymbol{k} \, \mathrm{d}\omega.$$

Our problem is the determination of the scalars α_1 and α_2, or equivalent quantities. Those are, e.g., the scalars I_{iiqq} and I_{ipip} which can be derived from (9.44) by contraction. Since

$$k_q \hat{q}_{lmq} = \left[k^2 \hat{Q}' + \frac{\mathrm{i}}{2}\left(k\frac{\partial \hat{Q}_1}{\partial k} + 2\hat{Q}_1\right)\right](k^2\,\delta_{lm} - k_l k_m), \tag{9.45}$$

$$\varepsilon_{ijm}\hat{q}_{lmi} = \left(k^2\hat{Q}' + \mathrm{i}\hat{Q}_1 + \frac{\mathrm{i}}{2}k\frac{\partial \hat{Q}_1}{\partial k}\right)\varepsilon_{ijl}k_i, \tag{9.46}$$

$$\varepsilon_{ilk}\hat{q}_{lmi} = \left(k^2\hat{Q}' + 2\mathrm{i}\hat{Q}_1 + \frac{\mathrm{i}}{2}k\frac{\partial \hat{Q}_1}{\partial k}\right)\varepsilon_{imk}k_i, \tag{9.47}$$

because of (9.43), we obtain from (9.44)

$$\begin{aligned} I_{iiqq} = &-\iint \frac{k^4}{\eta k^2 - \mathrm{i}\omega}\left\{(D^* - D)\left(-2\mathrm{i}k^2\hat{Q}' + 2\hat{Q}_1 + k\frac{\partial \hat{Q}_1}{\partial k}\right)\right. \\ &\left. + \left[4k^2\frac{\partial D^*}{\partial (k^2)} + 8D^*\right]\hat{Q}_1\right\}\mathrm{d}\boldsymbol{k}\,\mathrm{d}\omega \\ = &-\iint \frac{k^4}{\eta k^2 - \mathrm{i}\omega}\left\{(D^* - D)\left(-2\mathrm{i}k^2\hat{Q}' - 5\hat{Q}_1 + \frac{2\eta k^2}{\eta k^2 - \mathrm{i}\omega}\hat{Q}_1\right)\right. \\ &\left. + \left[2k^2\frac{\partial}{\partial (k^2)}(D^* + D) + 8D^*\right]\hat{Q}_1\right\}\mathrm{d}\boldsymbol{k}\,\mathrm{d}\omega, \end{aligned} \tag{9.48}$$

$$\begin{aligned} I_{ipip} = &-\iint \frac{k^4}{\eta k^2 - \mathrm{i}\omega}\left\{(D^* - D)\left(-2\mathrm{i}k^2\hat{Q}' + 2\hat{Q}_1 + k\frac{\partial \hat{Q}_1}{\partial k}\right)\right. \\ &\left. + \left[4k^2\frac{\partial D^*}{\partial (k^2)} - D + 5D^*\right]\hat{Q}_1\right\}\mathrm{d}\boldsymbol{k}\,\mathrm{d}\omega \\ = &-\iint \frac{k^4}{\eta k^2 - \mathrm{i}\omega}\left\{(D^* - D)\left(-2\mathrm{i}k^2\hat{Q}' - 5\hat{Q}_1 + \frac{2\eta k^2}{\eta k^2 - \mathrm{i}\omega}\hat{Q}_1\right)\right. \\ &\left. + \left[2k^2\frac{\partial}{\partial (k^2)}(D^* + D) - D + 5D^*\right]\hat{Q}_1\right\}\mathrm{d}\boldsymbol{k}\,\mathrm{d}\omega. \end{aligned} \tag{9.49}$$

Our main interest was always directed to the determination of the quantity α, one third of the trace of the tensor a_{ip}, for which we find from (9.37), (9.38) and (9.39)

$$\alpha = \frac{1}{3}I_{iiqr}g_q\Omega_r = -\alpha_0(\boldsymbol{g}\cdot\boldsymbol{\Omega}), \tag{9.50}$$

with

$$\alpha_0 = \alpha_1 + \frac{2}{3}\alpha_2 = -\frac{1}{9}I_{iiqq}. \tag{9.51}$$

Thus it becomes clear that we already determined α_0 in (9.48). Inserting the explicit expressions for D from (9.42) we obtain after some reductions (RÜDIGER [5])

$$\alpha_0 = -\frac{8\eta}{9}\iint \frac{\omega k^6 \hat{Q}'(k, \omega)\, \mathrm{d}\boldsymbol{k}\, \mathrm{d}\omega}{(\eta^2 k^4 + \omega^2)(\nu^2 k^4 + \omega^2)}$$

$$+ \frac{4}{9}\iint \frac{k^2 \hat{Q}_1(k, \omega)}{(\eta^2 k^4 + \omega^2)^2 (\nu^2 k^4 + \omega^2)^2} \{\eta\nu(4\eta^2 + 3\eta\nu)\, \omega^2 k^8 \tag{9.52}$$

$$+ (3\eta^2 + 4\eta\nu - \nu^2)\, \omega^4 k^4 - \omega^6\}\, \mathrm{d}\boldsymbol{k}\, \mathrm{d}\omega\,.$$

The first summand is already known to us from (9.31). In order to derive a second equation for the scalars α_1 and α_2 we recall that in the case of homogeneous turbulence both are equal, therefore the difference can be expected to be a simple expression. From (9.39) follows that

$$\alpha_1 - \alpha_2 = \frac{1}{6}(I_{ipip} - I_{iiqq})\,. \tag{9.53}$$

On inserting (9.48) and (9.49), we find (RÜDIGER [5])

$$\alpha_1 - \alpha_2 = \frac{1}{6}\iint \frac{k^4}{\eta k^2 - \mathrm{i}\omega}(D + 3D^*)\, \hat{Q}_1(k, \omega)\, \mathrm{d}\boldsymbol{k}\, \mathrm{d}\omega \tag{9.54}$$

$$= \frac{2}{3}\iint \frac{k^2(2\eta\nu k^4 + \omega^2)}{(\eta^2 k^4 + \omega^2)(\nu^2 k^4 + \omega^2)}\, \hat{Q}_1(k, \omega)\, \mathrm{d}\boldsymbol{k}\, \mathrm{d}\omega\,.$$

With this result we have a complete determination of the tensor a_{ip} for an incompressible inhomogeneous turbulent flow undergoing the influence of a rotational motion. We especially see that the inhomogeneity, in addition to the anisotropy, provides for a non-vanishing α-effect.

9.5. Discussion of the tensor a_{ip}

We have considered an inhomogeneous incompressible turbulence, where, by assumption, the inhomogeneity consists of a linear dependence of the mean quantities of the turbulent velocity field on $(\boldsymbol{g} \cdot \boldsymbol{x})$. Under those conditions we found the turbulent electromotive force to be

$$\mathfrak{E} = -\gamma \boldsymbol{g} \times \overline{\boldsymbol{B}}, \tag{9.55}$$

with γ given by (9.36).

For an interpretation of this result we conclude from (8.85) and (9.19) that

$$\overline{u'^2} = \iint 2\,[\hat{Q}(k, \omega) + (\boldsymbol{g} \cdot \boldsymbol{x})\, \hat{Q}_1(k, \omega)]\, k^2\, \mathrm{d}\boldsymbol{k}\, \mathrm{d}\omega\,, \tag{9.56}$$

and obtain

$$\nabla \overline{u'^2} = 2 \iint \hat{Q}_1(k, \omega)\, k^2\, \mathrm{d}\boldsymbol{k}\, \mathrm{d}\omega\, \boldsymbol{g}. \tag{9.57}$$

If $\iint \hat{Q}_1 k^2 \,\mathrm{d}\boldsymbol{k}\, \mathrm{d}\omega \neq 0$, we can rewrite (9.55) in the form

$$\mathfrak{E} = -\frac{\eta}{6} \frac{\displaystyle\iint \frac{k^4 \hat{Q}_1(k, \omega)}{\eta^2 k^4 + \omega^2}\, \mathrm{d}\boldsymbol{k}\, \mathrm{d}\omega}{\displaystyle\iint k^2 \hat{Q}_1(k,\omega)\, \mathrm{d}\boldsymbol{k}\, \mathrm{d}\omega} (\nabla \overline{u'^2} \times \overline{\boldsymbol{B}}). \tag{9.58}$$

We now see that the mean magnetic field is transported out of the region in which there is turbulence, provided both the integrals in (9.58) have the same sign. This behaviour of the mean magnetic field was called "turbulent diamagnetism" by RÄDLER [2, 4]. RÄDLER discussed the case where the inhomogeneity is due to the gradient of $\overline{u'^2}$ only and found $\hat{Q}_1$ to be proportional to $\hat{Q}$ and, consequently, $\hat{Q}_1$ to have always the same sign. Then the ratio of the integrals in (9.58) is positive. However, in our slightly more general case we have no condition concerning the sign of $\hat{Q}_1$. It is even possible that $\overline{u'^2} = \text{const}$ and still $\gamma \neq 0$.

Relation (9.35) shows that an additional transport of the mean magnetic field occurs in the direction $\boldsymbol{g} \times \boldsymbol{\Omega}$, which is due to the influence of the mean rotational motion. Since $\boldsymbol{g} \times \boldsymbol{\Omega}$ is parallel to $\overline{\boldsymbol{u}}$ this term can also be interpreted as that given by the scalar γ'' in (5.45). We thus see that the scalars α_2', α_3', γ'' in (5.45) do not describe independent effects.

In order to appreciate the derivation of the symmetric part of the tensor a_{ip} we would like to emphasize here that the essential result is to be seen in the fact that there is, indeed, a non-zero α-effect shown to exist for a real turbulence. It is left as a point of discussion whether these values α_0, α_1, α_2 can be used for a special investigation or whether an alternative derivation can be carried out.

For the purpose of an estimation of the order of magnitude we refer to relation (3.30) which was derived for the high-conductivity limit. We obtain

$$\alpha \approx \tau_{\mathrm{cor}} \left| \overline{\boldsymbol{u}' \cdot \operatorname{curl} \boldsymbol{u}'} \right|. \tag{9.59}$$

α proves to be proportional to the turbulence intensity, $\overline{u'^2}$, divided by a certain length scale, say λ_0. This length scale λ_0 can be derived from the gradient of $\overline{u'^2}$ or another mean quantity of the turbulent field; or from the gradient of the mass density or the pressure, e.g. the scale-height. We know, furthermore, that a non-vanishing helicity can be expected only if there is the influence of a mean rotational motion. Therefore an estimation of α must involve the product $\Omega\tau_{\text{cor}}$. All these arguments lead us to the relation

$$|\alpha| \approx \frac{\overline{u'^2}\tau_{\text{cor}}^2}{\lambda_0}\,\Omega\,. \tag{9.60}$$

In the case the Strouhal number S is assumed to be approximately unity, i.e. $\tau_{\text{cor}}u' \approx \lambda_{\text{cor}}$, we have the estimation

$$|\alpha| \approx \frac{\lambda_{\text{cor}}}{\lambda_0}\,u'\Omega\tau_{\text{cor}}\,. \tag{9.61}$$

9.6. Further results concerning the tensor a_{ip}

In our considerations up to now we have realized how important the tensor a_{ip} and, in particular the quantity α, is for the solution of the dynamo problem. This fact shall increasingly be brought out in the following chapters on the generation of the magnetic fields of cosmical bodies. On account of this close connection one should expect to find first results concerning a_{ip} or α in early works about dynamo theory. First of all one should mention the pioneering investigation by PARKER [2] (1955); see also PARKER [7, 8, 10]. Considering an axisymmetric geometry with respect to the mean fields Parker found that a mean poloidal magnetic field is produced from a toroidal one by means of the so-called "cyclonic turbulence", i.e. a motion in convective cells undergoing the influence of Coriolis forces. In terms of our notation this effect can be looked upon as being a special realisation of the α-effect.

Similar results, although based on a quite different approach, were arrived at by BRAGINSKIJ [1, 2, 3]. He considered mean fields, defined to be the averages over the azimuth, and derived the governing equations in the limit of high magnetic Reynolds numbers. A direct comparison between the approach presented here in this book and that of BRAGINSKIJ cannot readily be made. Both of them, however, lead to similar equations for the mean fields. Above all, an effect corresponding to the α-effect appears in BRAGINSKIJ's theory too. (See also TOUGH [1], TOUGH and GIBSON [2], SOWARD [1, 2]).

First determinations of the tensor a_{ip} and of α were actually carried out by STEENBECK, KRAUSE and RÄDLER [2] in the low-conductivity limit and for turbulent velocity fields $\boldsymbol{u}^{(0)}$ of the type

$$\boldsymbol{u}^{(0)} = \mathrm{e}^{a\boldsymbol{g}\cdot\boldsymbol{x}}\,\mathrm{curl}\,\mathrm{curl}\,(\boldsymbol{A}(\boldsymbol{x}, t)\,\mathrm{e}^{-(a+b)\boldsymbol{g}\cdot\boldsymbol{x}}),$$

where $\boldsymbol{A}$ is homogeneous and steady. $a\boldsymbol{g}$ is clearly the density gradient and $b\boldsymbol{g}$ the gradient of $\boldsymbol{u}^{(0)}$. There are some other restrictions within the deductions which are too complex to be described here. The result shows a non-zero α-effect due to the interaction of one of the gradients with the rotational motion. For the ratio $\alpha_2:\alpha_1$ the value $-\frac{1}{4}$ was found.

Without restrictions concerning the conductivity and with inclusion of a mean shear analoguous calculations were carried out by KRAUSE [1]. In addition to the confirmation of the above-mentioned results an electromotive force due to the rate of strain tensor $\frac{\partial u_i}{\partial x_j} + \frac{\partial u_j}{\partial x_i}$ was determined.

On the basis of a technique of double Fourier transformations ROBERTS and SOWARD [3] deduced a number of results concerning the pseudo-tensors a_{ij}, b_{ijk}. Some of their results are identical with those presented here, as mentioned above. They, too, found the value $-\frac{1}{4}$ for the ratio $\alpha_2:\alpha_1$.

A model largely identical with that presented here was treated in KRAUSE and RÄDLER [2], however, the calculations were carried out in the (x, t)-space thus requiring more mathematical effort.

MOFFATT [3] and subsequently SOWARD [5] determined the tensor a_{ip} for a random superposition of inertial waves in a rotating conducting fluid. MOFFATT's investigations are not restricted to small rotational rates and the same is true for RÜDIGER's [4]. It is interesting to note that for fast rotation the tensor a_{ip} becomes two-dimensional with respect to the axis of rotation, i.e. it takes the form

$$a_{ip} \approx a\left(\delta_{ip} - \frac{\Omega_i\Omega_p}{\Omega^2}\right), \quad \text{if} \quad \Omega \to \infty, \tag{9.62}$$

where $a = O(1)$.

HELMIS [1, 2, 3] carried out investigations on the basis of Ohm's law with the inclusion of a term due to the Hall effect. For a homogeneous isotropic non-mirrorsymmetric turbulence the parameters α and β are found to be decreasing functions of the Hall parameter, but the influence of the Hall effect is reduced by increasing electrical conductivity.

We would also like to refer to investigations in which these problems are not treated in a general way but rather on the basis of a concept of more specialized models. LEIGHTON [1, 2] described the turbulent diffusion of magnetic regions in the solar convection zone by a random walk process under the influence of the shear mechanism due to the differential rotation. In the investigations of YOSHIMURA [2, 3, 4] we find a rederivation of the basic equations for the mean fields.

CHAPTER 10

ON THE BACK-REACTION OF THE MAGNETIC FIELD ON THE MOTIONS

10.1. Introductory remarks

The considerations up to this point were based on the assumption that the motion exists uninfluenced by the magnetic field. The velocity was treated as a given quantity and we studied its influence on the magnetic field. This situation is characterized by the inequality

$$\frac{\overline{\boldsymbol{B}^2}}{2\mu} \ll \frac{\varrho}{2}\,\overline{\boldsymbol{u}^2}, \tag{10.1}$$

i.e. the mean magnetic energy is small compared with the kinetic energy.

It is the objective of this chapter to give some insight into situations, where the back-reaction of the magnetic field on the motion becomes noticeable. The relations then clearly take a nonlinear form, in particular, the turbulent electromotive force $\mathfrak{E}$ is no longer a linear functional of the mean magnetic field. The pseudo-tensors a_{ik}, b_{ikl}, for example, become functions of $\overline{\boldsymbol{B}}$.

It is generally expected that the turbulence is suppressed by the magnetic field, and that those characteristic parameters like α and β are decreasing functions of $\overline{B}$. But this is not correct in all cases. As we shall see in the following the turbulence intensity $\overline{\boldsymbol{u}'^2}$ can grow for small $\overline{B}$, although it tends to zero if $\overline{B} \to \infty$. Moreover, experiments with liquid sodium have shown that the turbulence is not simply suppressed by an applied magnetic field but is transformed to take a two-dimensional structure as described by Kit and Tsinober [1], where the motions take place within the planes orthogonal to the applied magnetic field and do not vary along the magnetic field. This result is rather obvious: Let

$$\boldsymbol{B} = (0, 0, B),\ B = \text{const},$$

and

$$\boldsymbol{u} = \left(\frac{\partial A}{\partial y}, -\frac{\partial A}{\partial x}, 0\right),\ A = A(x, y),$$

then

$$\boldsymbol{u}\times\boldsymbol{B} = \left(-B\frac{\partial A}{\partial x}, -B\frac{\partial A}{\partial y}, 0\right) = -\text{grad }(AB), \tag{10.2}$$

a relation showing that the electromotive force $\boldsymbol{u}\times\boldsymbol{B}$ is a gradient and thus will be compensated by space charges. Consequently, no interaction exists between a homogeneous magnetic field and a two-dimensional incompressible flow.

Investigations of turbulent conducting fluids undergoing the influence of a magnetic field were initiated by LEHNERT [1] (1955), further works were carried out by DEISSLER [1], NESTLERODE and LUMLEY [1], MOFFATT [1], MOREAU [1, 2], VAINSHTEIN [3], RÄDLER [12], RÜDIGER [2], ROBERTS and SOWARD [3]. The course we shall follow here is mainly that elaborated by RÜDIGER [2]. Our considerations are based on the second order correlation approximation thus implying that the fluctuating magnetic field here is small compared with the mean magnetic field, i.e.

$$\overline{B'^2} \ll \overline{B}^2. \tag{10.3}$$

From this and (7.150) we see that, in addition, to (3.12) we have to demand

$$\min\,(\mathrm{Rm}, \mathrm{Rm}\;\mathrm{S}) \ll 1. \tag{10.4}$$

The restriction to the second order correlation approximation also exclude considerations of those situations which are mainly an interplay of the fluctuating magnetic field and the fluctuating velocity field. Therefore, we do not provide for any statement concerning the hypothesis by BATCHELOR [1],

$$\frac{\overline{\boldsymbol{B}'^2}}{2\mu} \approx \frac{\varrho}{2}\overline{\boldsymbol{u}'^2}, \tag{10.5}$$

which is often assumed to be valid. Such investigations need higher correlation approximations and a certain closure procedure, or have to be carried out by using other methods. We refer here to the papers of FRISCH, LÉORAT and POUQUET [1], and that of FYFE and MONTGOMERY [1].

10.2. The influence of a uniform magnetic field on the correlation tensor

We consider an incompressible turbulent flow undergoing the influence of a magnetic field, $\overline{\boldsymbol{B}}$ shall be constant and $\overline{\boldsymbol{u}} = 0$. The velocity field is governed by the Navier-Stokes equation (8.61) with $\boldsymbol{F}$ specified to be the sum of the Lorentz-force divided by the mass density and a force $\boldsymbol{F}_0$ producing the turbulent motion, i.e.

$$\boldsymbol{F} = \frac{1}{\mu\varrho}\,\text{curl}\,\boldsymbol{B}\times\boldsymbol{B} + \boldsymbol{F}_0. \tag{10.6}$$

Restriction to the second order correlation approximation gives (8.66). Assuming, in addition, (10.3) we obtain the equations

$$\frac{\partial u_i'}{\partial t} - \nu \Delta u_i' = -\frac{1}{\varrho}\frac{\partial p'}{\partial x_i} - \frac{1}{\mu\varrho}\frac{\partial \overline{B}_j B_j'}{\partial x_i} + \frac{1}{\mu\varrho}\overline{B}_j \frac{\partial B_i'}{\partial x_j}, \tag{10.7}$$

$$\operatorname{div} \boldsymbol{u}' = \frac{\partial u_i'}{\partial x_i} = 0. \tag{10.8}$$

Using Fourier-transformed quantities we have

$$(-\mathrm{i}\omega + \nu k^2)\, \hat{u}_i = -\mathrm{i}k_i \left(\frac{\hat{p}}{\varrho} - \frac{1}{\mu\varrho}\overline{B}_j \hat{B}_j\right) + \mathrm{i}(\boldsymbol{k}\cdot\overline{\boldsymbol{B}})\, \hat{B}_i + \hat{F}_{0i}, \tag{10.9}$$

$$k_i \hat{u}_i = 0, \tag{10.10}$$

and find

$$(-\mathrm{i}\omega + \nu k^2)\, \hat{u}_i = \frac{\mathrm{i}}{\mu\varrho}(\boldsymbol{k}\cdot\overline{\boldsymbol{B}})\, \hat{B}_i + \left(\delta_{ij} - \frac{k_i k_j}{k^2}\right)\hat{F}_{0j}, \tag{10.11}$$

if the pressure is eliminated.

For a vanishing mean magnetic field (10.11) reads

$$(-\mathrm{i}\omega + \nu k^2)\, \hat{u}_i^{(0)} = \left(\delta_{ij} - \frac{k_i k_j}{k^2}\right)\hat{F}_{0j}. \tag{10.12}$$

Assume now the force driving the turbulent motion is independent of the applied magnetic field. Then we can write (10.11) in the form

$$(-\mathrm{i}\omega + \nu k^2)\, (\hat{u}_i - \hat{u}_i^{(0)}) = \frac{\mathrm{i}}{\mu\varrho}(\boldsymbol{k}\cdot\overline{\boldsymbol{B}})\, \hat{B}_i, \tag{10.13}$$

and thus have an equation which describes the distortion of the turbulence $\boldsymbol{u}^{(0)}$ caused by the mean magnetic field.

The set of equations governing our system is completed by the induction equation which, in our case, reads

$$(-\mathrm{i}\omega + \eta k^2)\, \hat{B}_i = \mathrm{i}(\boldsymbol{k}\cdot\overline{\boldsymbol{B}})\, \hat{u}_i. \tag{10.14}$$

Combining (10.13) and (10.14) we obtain

$$\left\{1 + \frac{(\boldsymbol{k}\cdot\overline{\boldsymbol{B}})^2}{\mu\varrho}\frac{1}{(-\mathrm{i}\omega + \eta k^2)(-\mathrm{i}\omega + \nu k^2)}\right\} \hat{u}_i(\boldsymbol{k}, \omega) = \hat{u}_i^{(0)}(\boldsymbol{k}, \omega). \tag{10.15}$$

Writing down this equation once more with the index j and the arguments $\boldsymbol{k}'$, ω' and multiplying both these equations we obtain from (6.18) after integrating over $\boldsymbol{k}$ and ω

$$\left\{1 + \frac{(\boldsymbol{k} \cdot \overline{\boldsymbol{B}})^2}{\mu\varrho} \frac{1}{(-\mathrm{i}\omega + \eta k^2)(-\mathrm{i}\omega + \nu k^2)}\right\}$$
$$\times \left\{1 + \frac{(\boldsymbol{k} \cdot \overline{\boldsymbol{B}})^2}{\mu\varrho} \frac{1}{(\mathrm{i}\omega + \eta k^2)(\mathrm{i}\omega + \nu k^2)}\right\} \hat{Q}_{ij}(\boldsymbol{k},\omega) = \hat{Q}_{ij}^{(0)}(\boldsymbol{k}, \omega), \tag{10.16}$$

or, more explicitely (RÜDIGER [3]),

$$\hat{Q}_{ij}(\boldsymbol{k}, \omega) = \frac{\hat{Q}_{ij}^{(0)}(\boldsymbol{k}, \omega)}{1 + \dfrac{(\boldsymbol{k} \cdot \overline{\boldsymbol{B}})^2}{\mu\varrho} \dfrac{2\eta\nu k^4 - 2\omega^2 + (\boldsymbol{k} \cdot \overline{\boldsymbol{B}})^2/\mu\varrho}{(\eta^2 k^4 + \omega^2)(\nu^2 k^4 + \omega^2)}}. \tag{10.17}$$

Both tensors, $\hat{Q}_{ij}$, $\hat{Q}_{ij}^{(0)}$ are positive semi-definite when one of them is positive semi-definite, since the factor in (10.16) and (10.17) is clearly positive.

A very important although also quite obvious result is the fact that according to (10.17) the correlation tensor of a two-dimensional turbulence as given in (7.76) is not altered by the magnetic field in the case when $\boldsymbol{e}$ is parallel to $\overline{\boldsymbol{B}}$.

10.3. Discussion of the result

We first consider the case of a weak magnetic field where (10.17) yields

$$\hat{Q}_{ij}(\boldsymbol{k}, \omega) = \hat{Q}_{ij}^{(0)}(\boldsymbol{k}, \omega) \left\{1 - \frac{(\boldsymbol{k} \cdot \overline{\boldsymbol{B}})^2}{\mu\varrho} 2 \frac{\eta\nu k^4 - \omega^2}{(\eta^2 k^4 + \omega^2)(\nu^2 k^4 + \omega^2)}\right\}. \tag{10.18}$$

For the turbulence intensity we thus find

$$\overline{u'^2} = \overline{u_0'^2} - 2\frac{\overline{B}_i \overline{B}_j}{\mu\varrho} \iint \frac{k_i k_j (\eta\nu k^4 - \omega^2)}{(\eta^2 k^4 + \omega^2)(\nu^2 k^4 + \omega^2)} \hat{Q}_{kk}^{(0)}(\boldsymbol{k}, \omega) \, \mathrm{d}\boldsymbol{k} \, \mathrm{d}\omega. \tag{10.19}$$

It is quite surprising to see that the turbulence is not necessarily damped, since the first factor in the integrand has no definite sign. (10.19) can be rewritten in the form

$$\overline{u'^2} = \overline{u_0'^2} + \frac{\overline{B}_p \overline{B}_q}{\mu\varrho} \iint k_p k_q \frac{\partial(\hat{Q}_{ii}^{(0)}(\boldsymbol{k}, \omega) + \hat{Q}_{ii}^{(0)}(\boldsymbol{k}, -\omega))}{\partial\omega} \frac{\arctan \dfrac{(\eta - \nu) k^2}{\eta\nu k^4 + \omega^2}}{(\eta - \nu) k^2} \, \mathrm{d}\boldsymbol{k} \, \mathrm{d}\omega, \tag{10.20}$$

which shows that the turbulence intensity is always diminished for turbulences where $\hat{Q}_{ii}^{(0)}(\boldsymbol{k}, \omega) + \hat{Q}_{ii}^{(0)}(\boldsymbol{k}, -\omega)$ is a decreasing function for positive ω. For an increase of $\overline{u'^2}$ a growth of this function is necessary in a sufficiently large ω-range.

The quantity α is found to be

$$\alpha = \alpha_0 - \frac{2\mathrm{i}}{3} \frac{B_l B_m}{\mu\varrho} \int\int \frac{k_l k_m(\eta\nu k^4 - \omega^2)}{(\eta^2 k^4 + \omega^2)(\nu^2 k^4 + \omega^2)} \frac{\varepsilon_{ijk} k_i \hat{Q}_{jk}^{(0)}(\boldsymbol{k}, \omega)}{\eta k^2 - \mathrm{i}\omega} \mathrm{d}\boldsymbol{k}\, \mathrm{d}\omega, \tag{10.21}$$

and we see, again, that α also is not necessarily reduced by the action of the magnetic field. In the low-conductivity limit, however, both quantities, $\overline{u'^2}$ as well as α, will be reduced, as is to be seen in (10.19) and (10.21) for $\eta \to \infty$. This result is, as far as $\overline{u'^2}$ is concerned, in agreement with experimental findings by ALEMANY, MOREAU, SULEM and FRISCH [1].

$\overline{B}$ shall now be large and the case of two-dimensional turbulence shall be excluded. We consider the mean square of the velocity fluctuations and from (10.17) we find

$$\overline{u'^2} = 2\pi \int_{-\infty}^{\infty}\int_0^{\infty}\int_0^{\pi} \frac{\hat{Q}_{ii}(k, \omega)\, k^2\, \mathrm{d}k \sin\vartheta\, \mathrm{d}\vartheta\, \mathrm{d}\omega}{(1 + A\overline{B}^2 \cos^2\vartheta)(1 + A^*\overline{B}^2 \cos^2\vartheta)}, \tag{10.22}$$

with

$$A = \frac{k^2}{\mu\varrho} \frac{1}{(\eta k^2 - \mathrm{i}\omega)(\nu k^2 - \mathrm{i}\omega)}. \tag{10.23}$$

Substituting $s = -\overline{B}\cos\vartheta$ we obtain

$$\overline{u'^2} = \frac{2\pi}{\overline{B}} \int_{-\infty}^{\infty}\int_0^{\infty}\int_{-\overline{B}}^{\overline{B}} \frac{\hat{Q}_{ii}(k, \omega)\, k^2\, \mathrm{d}k\, \mathrm{d}s\, \mathrm{d}\omega}{(1 + As^2)(1 + A^* s^2)}. \tag{10.24}$$

Considering now the limit $\overline{B} \to \infty$ we realize that the integral takes a finite value thus showing that

$$\overline{u'^2} = O\left(\frac{1}{\overline{B}}\right), \text{ if } \overline{B} \to \infty. \tag{10.25}$$

To a certain degree this is a surprising result since inspecting (10.22) one would expect at first glance a decrease of $\overline{u'^2}$ with $\overline{B}^{\,4}$. Obviously, the much slower decay is due to the modes with $\boldsymbol{k} \cdot \overline{\boldsymbol{B}} \approx 0$.

On account of this slow decay we have to conclude that the turbulent electromotive force $\mathfrak{E}$ must not tend to zero if $\overline{B} \to \infty$, since it is connected with the product $\hat{Q}_{ij}\overline{B}_k$. However, considering dynamo excitation one expects the motion to stop being capable of further generating the magnetic field by which it is influenced.

Let us consider homogeneous isotropic turbulence with its correlation tensor given by (6.43). We saw in section 3.9. that the self-excitation is due to the α-effect, i.e. to the ability of the motion to create an electromotive force parallel to the magnetic field being excited. Let us, therefore, ask how the component of $\mathfrak{E}$ parallel to $\overline{\boldsymbol{B}}$ behaves if $\overline{B} \to \infty$. From (5.35) and (7.1) we find

$$\mathfrak{E}_{\parallel} = \frac{\overline{B}_i \mathfrak{E}_i}{\overline{B}} = \mathrm{i}\varepsilon_{ilq} \frac{\overline{B}_i \overline{B}_j}{\overline{B}} \iint \frac{k_j \hat{Q}_{lq}(\boldsymbol{k}, \omega)}{\eta k^2 - \mathrm{i}\omega} \, \mathrm{d}\boldsymbol{k} \, \mathrm{d}\omega \,. \tag{10.26}$$

We take $\hat{Q}_{ij}^{(0)}$ from (6.43) and with (10.17) we obtain

$$\mathfrak{E}_{\parallel} = -4\pi\overline{B} \int_{-\infty}^{\infty}\int_{0}^{\infty}\int_{0}^{\pi} \frac{k^3 \cos^2\vartheta \sin\vartheta}{(\eta k^2 - \mathrm{i}\omega)(1 + A\overline{B}^2\cos^2\vartheta)(1 + A^*\overline{B}^2\cos^2\vartheta)} \frac{\partial \hat{C}}{\partial k} \, \mathrm{d}k \, \mathrm{d}\vartheta \, \mathrm{d}\omega \tag{10.27}$$

$$= -\frac{4\pi}{\overline{B}^2} \int_{-\infty}^{\infty}\int_{0}^{\infty}\int_{-\overline{B}}^{+\overline{B}} \frac{k^3 u^2 \, \mathrm{d}u}{(\eta k^2 - \mathrm{i}\omega)(1 + Au^2)(1 + A^*u^2)} \frac{\partial \hat{C}}{\partial k} \, \mathrm{d}k \, \mathrm{d}\omega \,.$$

Again, the u-integral converges to a finite value if $\overline{B} \to \infty$. Consequently, it becomes clear that

$$\mathfrak{E}_{\parallel} = O\left(\frac{1}{\overline{B}^2}\right), \text{ if } \overline{B} \to \infty \,, \tag{10.28}$$

i.e. the component of the turbulent electromotive force $\mathfrak{E}$, which is parallel to the mean magnetic field, tends to zero if the mean magnetic field tends to infinity. This result, indeed, indicates that the motion has no longer the ability of generating the magnetic field if the generated field becomes larger and larger.

10.4. Two-dimensional turbulence

We mentioned in section 10.1. that experiments with liquid sodium show turbulence taking a two-dimensional structure if subjected to the influence of a strong magnetic field. First results of this kind were published by KIT and TSINOBER [1], further work was done by KOLESNIKOV and TSINOBER [1, 2], VOTSISH and KOLESNIKOV [1, 2]. The transition of homogeneous turbulence from an initially isotropic three-dimensional state to a two-dimensional state was also demonstrated by a numerical simulation (SCHUMANN [1]).

In section 7.6. we derived the two-dimensional magnetic diffusivity. The result (7.94) shows that there is no difference in the order of magnitude if compared with the three-dimensional case. However, the derivation was carried out within the frame of mean-field electrodynamics, i.e. without taking into account the back-reaction of the magnetic field on the motion, and we do not know, therefore, whether this result can be applied to situations with strong magnetic fields.

In order to provide more insight in this subject we recall relation (10.2) which shows that no currents are induced by the interaction of a constant magnetic field with an incompressible two-dimensional flow. Let us now proceed to a more general model (KRAUSE and RÜDIGER [1]): The magnetic field shall be parallel to a certain direction which we identify with the z-direction of a cartesian coordinate system x, y, z; apart from this it can have an arbitrary dependence on the space coordinates and the time, i.e.

$$\boldsymbol{B} = (0, 0, B(x, y, t)). \tag{10.29}$$

There is no dependence of B on z in virtue of div $\boldsymbol{B} = 0$. The two-dimensional flow, $\boldsymbol{u}$, shall always lie completely in the planes orthogonal to $\boldsymbol{B}$ and not vary along the direction of $\boldsymbol{B}$. Hence it can be represented by

$$\boldsymbol{u} = \operatorname{curl} \boldsymbol{A}, \text{ with } \boldsymbol{A} = (0, 0, A(x, y, t)). \tag{10.30}$$

First we shall note that the interaction of $\boldsymbol{u}$ and $\boldsymbol{B}$ provides for currents since

$$\operatorname{curl} (\boldsymbol{u} \times \boldsymbol{B}) = \operatorname{grad} A \times \operatorname{grad} B. \tag{10.31}$$

These currents lie completely in the planes orthogonal to the z-direction and do not vary with z. Consequently, the induced magnetic field is also of the type (10.29), i.e. it does not vary with z and is parallel to the z-direction.

Let us now consider the dependence on time. An electromotive force is induced according to Maxwell's first law,

$$\operatorname{curl} \boldsymbol{E} = -\dot{\boldsymbol{B}}. \tag{10.32}$$

The currents driven by this electromotive force lie, again, in the planes orthogonal to $\boldsymbol{B}$ and do not vary along $\boldsymbol{B}$. Thus, our analysis leads to the conclusion: A magnetic field of type (10.29) continues to be of this type if it interacts with a two-dimensional incompressible flow of the type (10.30).

The Lorentz-force

$$\boldsymbol{j}\times\boldsymbol{B}=\frac{1}{\mu}\operatorname{curl}\boldsymbol{B}\times\boldsymbol{B}=-\frac{1}{2\mu}\operatorname{grad}B^2+\frac{1}{\mu}(\boldsymbol{B}\cdot\operatorname{grad})\,\boldsymbol{B}, \tag{10.33}$$

of the magnetic field (10.29) is not zero, but is a gradient since $(\boldsymbol{B}\cdot\operatorname{grad})\,\boldsymbol{B}=0$. Consequently, the Lorentz force of the magnetic field considered is compensated by pressure and does not influence the motion.

Thus we find the remarkable result that a two-dimensional flow of the type (10.30) can influence and deform a magnetic field of the type (10.29) without being influenced by the magnetic field, even if this is very strong.

Let us apply these findings to the case where $\boldsymbol{u}=\boldsymbol{u}'$ is a two-dimensional isotropic random flow as considered in section 7.6. As before we split B into its mean and its fluctuating part,

$$B(x, y, t)=\overline{B}(x, y, t)+B'(x, y, t), \tag{10.34}$$

and assume the scales of the mean part to be large compared with those of the fluctuating part. Then according to the results deduced in section 7.6. the action of the turbulent motion on the mean magnetic field is a turbulent decay, with the two-dimensional turbulent magnetic diffusivity, $\beta^{(2)}$, given by (7.83) and (7.94). This result holds, in contrast to the decay caused by a three-dimensional isotropic tubulence, for a mean magnetic field of the type (10.29) of arbitrary field strength.

Thus we see that even a strong magnetic field cannot prevent its decay by suppressing the turbulence. It can only transform the turbulence to take a two-dimensional structure. However, these motions destroy the magnetic field without being influenced by it.

In view of applications to sunspots we note that the above results also hold for stratified media with $\varrho=\varrho(z)$, since the motions of the type (10.30) fulfil the equation of continuity $\operatorname{div}\varrho\boldsymbol{u}=0$ also in this case.

10.5. Applications to the decay of sunspots

According to the data given in section 3.7. for the solar convection zone we have

$$\eta=\frac{1}{\mu\sigma}\approx 10^3\,\frac{\mathrm{m}^2}{\mathrm{s}}, \tag{10.35}$$

and for a sunspot with a diameter $\overline{\lambda}=10^4$ km we find the decay time

$$\overline{\tau}=\frac{\overline{\lambda}^2}{\eta}\approx 3\cdot 10^3\ \text{years}, \tag{10.36}$$

a result which clearly contradicts the observations. We should, however, interpret this phenomenon as a turbulent decay. The data in section 3.7. further give

$$\eta_T \approx \beta \approx 10^7 \frac{\text{m}^2}{\text{s}} \approx 10^4 \, \eta. \tag{10.37}$$

Deriving the decay time of a sunspot from

$$\bar{\tau} = \frac{\bar{\lambda}^2}{\eta} \approx 0.3 \text{ years} \approx 4 \text{ month}, \tag{10.38}$$

we find, indeed, the correct order of magnitude observed for long-lived, recurrent spots.

This analysis, however, suffers from an important gap, since the action of the strong magnetic field in the sunspot is not taken into account. Therefore, the deductions of the foregoing section are of importance here. Although the magnetic field influences the turbulence, forcing it to be two-dimensional, it cannot prevent its own decay, for the two-dimensional turbulence enhances the decay to the same order of magnitude provided the scales of the two-dimensional turbulence are of the same order of magnitude as those of the three-dimensional turbulence uninfluenced by the magnetic field. Thus the estimate (10.38) remains correct if η_T is replaced by $\eta_T^{(2)}$.

It is worth noting that the assumption of a suppressed convection provides an explanation of the darkness of sunspots but not for the rapid decay within some weeks or months. The assumption of two-dimensional turbulence in a sunspot explains both the rapid decay and the darkness, since a two-dimensional turbulence does not allow a convective transfer of energy along the magnetic field either.

A consideration of a simple model shows that the conception of a turbulent decay of the magnetic fields in sunspots can also be used for the explanation of further details found by observations (Meyer and Schmidt [1], Krause and Rüdiger [1], Meyer, Schmidt, Weiss and Wilson [1]).

For $B(x, y, t)$ in (10.29) we take

$$B(x, y, t) = \Phi_0 \frac{1}{4\eta_T^{(2)} t} \exp\left(-r^2/4\eta_T^{(2)} t\right), \quad r^2 = x^2 + y^2, \tag{10.39}$$

thus describing a decaying magnetic field in z-direction, where the flux, Φ_0, penetrating the (x, y)-plane is independent of time and concentrated at $r = 0$ for $t = 0$.

The umbra of the spot shall be represented by the circular region, where the magnetic field strength is larger than a critical value B_c. Let the area of this region be A_u and its radius r_u. A_u and r_u are functions of time. $\bar{\tau}$ denotes the lifetime of a spot. At $t = \bar{\tau}$ the maximum value of B is B_c, and A_u will vanish. Introducing B_c and $\bar{\tau}$ into (10.39) we obtain

$$B(x, y, t) = B_c \frac{\bar{\tau}}{t} \exp\left(-\frac{r^2}{4\eta_T^{(2)} t}\right), \tag{10.40}$$

and find the relation

$$1 = \frac{\bar{\tau}}{t} \exp\left(-\frac{r_u^2}{4\eta_T^{(2)} t}\right), \tag{10.41}$$

and, consequently, for the model umbra

$$A_u(t) = 4\pi\eta_T^{(2)} t\,(\log \bar{\tau} - \log t). \tag{10.42}$$

We see that $A_u(t)$ has a maximum at $t_{\max} = \dfrac{\bar{\tau}}{\mathrm{e}}$ which is

$$A_{u,\max} = \frac{4\pi}{\mathrm{e}} \eta_T^{(2)} \bar{\tau}, \tag{10.43}$$

and, furthermore, that at $t = \bar{\tau}$ the area $A_u(t)$ has the decay rate

$$\left(\frac{\mathrm{d}A_u}{\mathrm{d}t}\right)_{t=\bar{\tau}} = -4\pi\eta_T^{(2)}. \tag{10.44}$$

From the analysis of a large number of observations WALDMEIER derived the rule

$$\bar{\tau} = 0.1 A_{u,\max}, \tag{10.45}$$

where the lifetime is measured in days and the area in millionths of the visible hemisphere of the sun. Interpreting (10.43) as WALDMEIER's rule we obtain the estimate

$$\frac{4\pi}{\mathrm{e}} \eta_T^{(2)} = 10 \cdot 10^{-6} \frac{(\text{hemisphere})_{\text{sun}}}{\text{day}}, \tag{10.46}$$

i.e., the turbulent diffusivity is

$$\eta_T^{(2)} \approx 7 \cdot 10^7 \frac{\mathrm{m}^2}{\mathrm{s}}. \tag{10.47}$$

The agreement with (10.37) is fairly satisfactory. The fact that the value $\eta_T^{(2)}$ derived from WALDMEIER's rule is larger than that derived from convection will become more understandable by taking into account the more detailed investigations of BUMBA [1]. BUMBA presents the results of a statistical analysis of the time dependence of sunspot areas. He finds two categories of sunspot groups: (A) groups with rapid area decrease, (B) groups with a slow part in the curve of decrease. All groups show rapid growth and decrease in the first days, thus belonging to category (A). Some of them, the long-lived ones, especially the recurrent groups, become slow at the end of their lifetime, thus belonging to category (B).

Bumba's analysis shows that the long-lived spots, category (B), decay at a constant rate towards the end of their lifetime independent of their size. This result is in agreement with relation (10.44). The decay rate found from the observations is

$$\left(\frac{\mathrm{d}A_u}{\mathrm{d}t}\right)_{t=\tau} = -4.2 \cdot 10^{-6}\,\frac{(\text{hemisphere})_{\text{Sun}}}{\text{day}}, \tag{10.48}$$

and from (10.44) it follows that

$$\eta_T^{(2)} \approx 10^7\,\frac{\text{m}^2}{\text{s}}, \tag{10.49}$$

a result which is in striking agreement with the value (10.37) derived from the data of the observed convection.

Let us now compare the curves of our model (figure 10.1) with the curves derived from observations of the growth and the early decrease of sunspot group areas as described in category (A). Figure 10.1 gives two examples, where

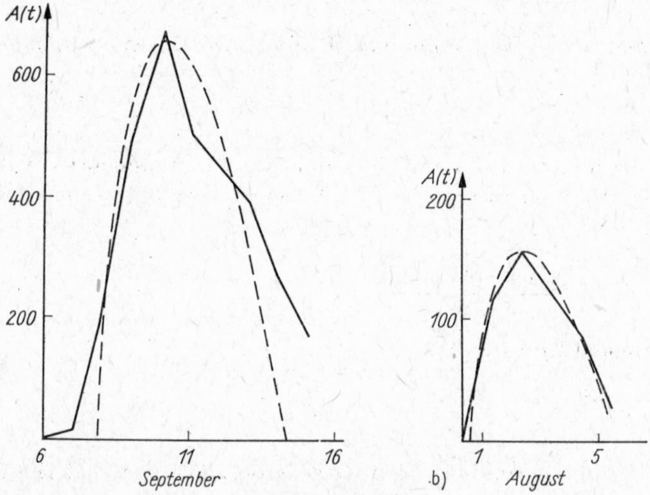

Fig. 10.1. Model curves (dashed lines) fitted to the time development curves of the spot group areas for the non-recurrent groups No. 18195, (a), and No. 18143, (b), according to Bumba. Growth and decay correspond to values of the turbulent diffusivity, $\eta_T^{(2)}$, which are larger than that of the convection in non-active regions by a factor 70 in case (a), and 20 in case (b)

models are fitted to two spot groups with quite different values for the maximum of the areas and for $\eta_T^{(2)}$ too. From the estimation of the area decay rate by BUMBA for a large number of sunspot groups of category (A) follows that

$$(\eta_T^{(2)})_{(A)} = 25(\eta_T^{(2)})_{(B)}. \tag{10.50}$$

Now we can see why the value of $\eta_T^{(2)}$ derived from WALDMEIER's rule, (10.45), (10.47), lies between the value of category (A) and that of (B): WALDMEIER's rule is deduced by averaging over all sunspot groups.

Combining (7.94) and (10.50) we are led to the relation

$$(\overline{u'^2}\tau_{\mathrm{cor}})_{(A)} \approx 25(\overline{u'^2}\tau_{\mathrm{cor}})_{(B)}, \tag{10.51}$$

which suggests an explanation for the different behaviours of the spots of categories (A) and (B), which is based on different scales of the turbulence. Indeed, the emerging of a sunspot is a violent process, which provides for additional causes of turbulence. A turbulent velocity enhanced by a factor 5 would explain (10.51). Furthermore, the emerging of a large spot group is a more violent process than that of a small group. This might provide an explanation of the considerable difference between the values of $\eta_T^{(2)}$ for the two examples in figure 10.1.

CHAPTER 11

THE DYNAMO PROBLEM OF MAGNETOHYDRODYNAMICS

11.1. The question of the origin of cosmical magnetic fields

The dynamo theory to which the following chapters will be devoted has been motivated mainly by the desire to find an explanation of the origin and the behaviour of the magnetic fields of the Earth, the Sun and other cosmical objects.

Let us recall the basic knowledge of these fields in very general terms (for more detailed information see sections 17.1., 17.3., and 17.5.). The Earth shows a magnetic field which, at the surface and in outer space, has a dipole-like structure. This field is to some extent symmetric with respect to the axis of rotation. Over long time intervals it may be considered as steady. At the Sun a great number of magnetic phenomena has been observed. They are related to a general magnetic field, which mainly consists of two oppositely orientated belts under the surface, one in each hemisphere, and a dipole-like part, all with some symmetry with respect to the axis of rotation. This general field oscillates with the period of the solar acitivity cycle, i.e. 22 years. At several stars too magnetic fields have been observed. In many cases one has to assume that these fields show considerable deviations from symmetry with respect to the axis of rotation. For a co-rotating observer they may be steady.

In the following we want to analyse possible causes for such magnetic fields. As is well known, magnetic fields are either due to magnetization of matter originating from magnetic moments localized in atomic or molecular regions or due to electric currents caused by the motion of free charged particles.

A magnetization which is supposed to be the origin of a magnetic field must be of ferromagnetic nature. The existence of ferromagnetism, however, is subject to special conditions, e.g. with respect to the temperature, which are scarcely fulfilled in the cosmical bodies mentioned. In all probability the causes of the Earth's magnetic field are located in its core, the temperature of which is so high that any ferromagnetism has to be excluded. As for the Sun, an additional difficulty arises from the fact that an alternating field has to be explained. By these and other reasons, ferromagnetism can be ruled out as an origin for the magnetic fields of the objects considered.

The existence of electric currents requires both electrically conducting matter and an electric field or electromotive forces. The Earth's and the planetary interiors consist of matter with metallic properties, and in the Sun and in normal stars the matter is in the plasma state, i.e. in all these objects high electric conductivity may be assumed.

We first discuss the case in which an electric field has been created by some kind of cause and then exists in absence of electromotive forces. In any case this electric field, and thus also the electric currents and the magnetic field connected with it, are bound to decay. The energy stored in these fields is converted into heat. For the Earth the decay time has been estimated to be of the order of only 10^5 years (cf. section 11.3.). Therefore, the presence of the geomagnetic field, which proved to be constant over much longer intervals, cannot be understood without the assumption of suitable electromotive forces. For the Sun and stars of comparable dimensions the decay times may reach or exceed the order of 10^{10} years and, hence, the presumed ages of these objects (cf. section 11.3.). Yet, this does not provide for an explanation of the solar magnetic field, since it is periodic in time. Stellar magnetic fields not considerably varying in time could be thought of as relics from earlier stages of the stellar evolution, and may now exist without the assistance of any electromotive force. Of course, the early processes which led to the generation of these fields still remain to be discussed.

As far as the electromotive forces are concerned we shall distinguish between those of non-electromagnetic and those of electromagnetic nature depending on whether the force on the charged particles exists already without or only due to the presence of electromagnetic fields.

There is a series of electromotive forces of non-electromagnetic nature which may occur in cosmical bodies. The different masses of positively and negatively charged particles lead to different inertial behaviours when being accelerated, especially in the case of rotation. The result is an electromotive force analogous to that occurring with the Tolman effect observed in rigid conductors. Inhomogeneities in composition and in temperature of the matter may also cause electromotive forces which correspond to galvanic or thermoelectric effects. A detailed analysis shows that all electromotive forces of that kind are hardly suitable for an explanation of the magnetic fields of the cosmical bodies considered. In the case of the Earth the fields to be generated in this way differ from the observed one by many orders of magnitude. Likewise, the Sun's magnetic field can scarcely be understood on this basis simply because of its time dependence. For some stars the electromotive force due to different inertial behaviour of differently charged particles may lead to fields with flux densities of some 10^2 G but with geometrical structures not concurring with the observations. In any case we may state that also the electromotive forces of non-electromagnetic nature cannot provide a general explanation of magnetic fields of cosmical objects.

Let us now turn to electromotive forces of electromagnetic nature. The simplest of them is that which is crucial for all magnetohydrodynamics: the electromotive force which acts in matter moving in the presence of a magnetic field and which is described by the Lorentz term in Ohm's law, i.e. the $\boldsymbol{u} \times \boldsymbol{B}$ term in the last equation (2.2). As is well known, this electromotive force is also responsible for the action of dynamo machines used for the production of electric current. In the externally excited dynamo machine the magnetic field necessary for the appearance of the electromotive force in the moving anchor exists independently of this force and the currents produced by it. In a self-excited dynamo machine, as first proposed by von Siemens (1866), the magnetic field is generated with the help of those currents. Only at the first instant, when the machine shall commence to work, a magnetic field produced by some other source is necessary, but this seed field may be arbitrarily small. It only initiates a process in which the electromotive force, the electric currents, and the magnetic fields permanently grow due to their mutual influence. Then, of course, also the Lorentz forces grow, which act against the motion of the anchor. If the mechanical forces available are no longer able to maintain the motion, or if any other limiting effect occurs, the growth will be restricted. Depending on the special construction of the machines both direct and alternating currents and corresponding magnetic fields can be generated.

This leads us to the question whether the magnetic fields of cosmical bodies could be produced by a process which corresponds to that in a self-excited dynamo machine. This possibility was, with respect to the Sun, first proposed by Larmor [1] (1919). As we have seen before, seed fields are always available. Besides, in all cosmical bodies under consideration internal motions may be assumed. There is, however, a drastic difference between the situations in a dynamo machine and that of a cosmical body. A dynamo machine consists not only of conducting but also of insulating parts, especially conducting wires enveloped by insulating material; thus the conducting matter covers a region with a multiple connection. Special arrangements of the wires offer the possibilities to determine the current paths and therefore also shape and magnitude of the magnetic field in order to meet the requirements of the dynamo principle. In a cosmical body the conducting matter generally fills a certain region completely which is to be considered as simply-connected. If an initial magnetic field and the motion are given, the distribution of the electric current and the magnetic field generated by this motion is already fixed. It can by no means be taken for granted that initial fields and motions are to be found which enable processes like those in a self-excited dynamo machine. Only after considerable research it did become clear that magnetic fields can indeed be generated in this way. The dynamo principle has proved to be the most attractive and most convincing possibility of an all-embracing explanation of the origin of the magnetic fields observed at cosmical bodies.

11.2. General view of the dynamo problem

The considerations on the origin of cosmical magnetic fields given in the foregoing chapter lead to the dynamo problem of magnetohydrodynamics. In this context we speak of a "dynamo" if a magnetic field is generated or maintained due to motions inside an electrically conducting fluid medium which fills a simply-connected region. Sometimes it is also called "homogeneous dynamo" in order to emphasize that the medium does not contain any insulating part. Some general aspects of the dynamo problem will now be discussed briefly.

Let us consider a body of a fluid electrically conducting medium occupying a simply-connected region and surrounded by vacuum. We exclude any magnetization of this medium and any electromotive force except those described by the Lorentz term in Ohm's law. At a given instant a weak magnetic field shall be present. In absence of motions it is, of course, bound to decay. Its energy is, due to Ohmic dissipation, converted into heat. At first we have to ask the crucial question: Is it at all possible to find internal motions such that the magnetic field does not decay or that it even grows? Of course, this presupposes a permanent input of mechanical energy which is converted into magnetic energy and, finally, into heat. The question posed here can easily be reduced to: Is it possible to find a magnetic field and a motion such that this magnetic field, which may be stationary or oscillatory, can be maintained by this motion?

Let us now assume that the question is answered positively and that corresponding magnetic fields and motions are known. Then, of course, a further question arises: Under what conditions and by what kind of forces can such motions be generated?

A magnetic field which is growing at a given instant, inevitably continues to grow as long as the motions remain unchanged. In reality, however, the motions are influenced by the Lorentz forces which appear as a consequence of the magnetic field. As long as the magnetic field is sufficiently weak this influence may be neglected. If the field becomes stronger this influence will be of crucial importance. It prevents the incessant increase of the field and determines its magnitude in the final state. A complete treatment of the dynamo problem requires that this complicated interaction between magnetic field and motions is taken into account.

We shall speak of the kinematic dynamo problem, and of kinematic dynamo models, if the motion is fixed without regard to dynamic aspects. This means that any back-reaction of the magnetic field on the motions is ignored. A solution of the kinematic dynamo problem provides the excitation condition of a magnetic field but no information about its magnitude. The study of the kinematic problem has to be understood as a first step towards the solution of the full magnetohydrodynamic problem, with the dynamics of the motion being taken into account, i.e. where the motion is considered as a consequence of other factors, especially forces including those due to the magnetic field. Only by solving this full magnetohydrodynamic problem the magnitude of the magnetic field can be determined.

11.3. Mathematical formulation of the dynamo problem and simple consequences

The dynamo problem as discussed in the foregoing section is equivalent to a mathematical problem which shall be formulated now.

As before, a body of a fluid electrically conducting medium is considered which occupies a simply-connected bounded region. All space outside this body is assumed to show the electromagnetic properties of vacuum. We denote that region by V, its surface by S, and the space outside by V', furthermore the unit outward normal to S by $\boldsymbol{n}$, and a jump of any quantity, F, across S by $[F]$.

Again, the electrically conducting medium is assumed to show no magnetization and no other electromotive forces than those described by the Lorentz term in Ohm's law. Starting from the basic electromagnetic equations in the magnetohydrodynamic approximation, for the magnetic flux density, $\boldsymbol{B}$, we obtain the equations

$$\frac{1}{\mu}\operatorname{curl}\left(\frac{1}{\sigma}\operatorname{curl}\boldsymbol{B}\right)-\operatorname{curl}(\boldsymbol{u}\times\boldsymbol{B})+\frac{\partial\boldsymbol{B}}{\partial t}=0, \tag{11.1a}$$

$$\operatorname{div}\boldsymbol{B}=0, \qquad \text{in } V. \tag{11.1b}$$

For the velocity field, $\boldsymbol{u}$, inside the medium no restrictions are made at first. Contrary to the permeability, μ, which is identical with that of vacuum, the electrical conductivity, σ, may depend on position. As for the space outside the body, in accordance with the magnetohydrodynamic approximation only quasi-steady electromagnetic processes are taken into account, so that

$$\operatorname{curl}\boldsymbol{B}=0, \quad \operatorname{div}\boldsymbol{B}=0, \text{ in } V'. \tag{11.2a, b}$$

Since there is no reason to include surface currents, we arrive at the condition

$$[\boldsymbol{B}]=0, \quad \text{on } S. \tag{11.3a, b}$$

It implies that $\boldsymbol{n} \cdot \operatorname{curl} \boldsymbol{B} = 0$, on S. In order to ensure that the shape of the body is conserved we require

$$\boldsymbol{n} \cdot \boldsymbol{u} = 0, \text{ on } S. \tag{11.3c}$$

Finally, only magnetic fields due to self-excitation by processes inside the body shall be considered. In order to exclude external excitation by causes located at infinity, it is claimed that the magnetic field vanishes at infinity at least like a dipole field, i.e.

$$\boldsymbol{B} = \mathrm{O}(r^{-3}), \text{ as } r \to \infty, \tag{11.4}$$

where r means a distance from a point inside the body.

The problem of constructing a kinematic dynamo model consists in a suitable specification of the $\boldsymbol{u}$-field. It has to be chosen in such a way that the equations and conditions formulated here allow a $\boldsymbol{B}$-feild which does not vanish as $t \to \infty$. If we restrict attention to a steady state it is sufficient to find a non-trivial $\boldsymbol{B}$-field.

It is useful to consider the energy balance too. Departing again from the basic equations for the electromagnetic fields in the magnetohydrodynamic approximation we obtain

$$\frac{\mathrm{d}}{\mathrm{d}t} \int\limits_{V+V'} \frac{\boldsymbol{B}^2}{2\mu} \,\mathrm{d}\boldsymbol{x} = - \int\limits_{V} \frac{\boldsymbol{j}^2}{\sigma} \,\mathrm{d}\boldsymbol{x} - \int\limits_{V} \boldsymbol{u} \cdot (\boldsymbol{j} \times \boldsymbol{B}) \,\mathrm{d}\boldsymbol{x}. \tag{11.5}$$

Here $\boldsymbol{B}^2/2\mu$ represents the magnetic energy density, $\boldsymbol{j}^2/\sigma$ the Ohmic heat production in a unit volume, and $\boldsymbol{u} \cdot (\boldsymbol{j} \times \boldsymbol{B})$ the work done by the Lorentz force in a unit volume during a time unit. It can easily be shown that

$$- \int\limits_{V} \boldsymbol{u} \cdot (\boldsymbol{j} \times \boldsymbol{B}) \,\mathrm{d}\boldsymbol{x}$$

$$= \frac{1}{\mu} \int\limits_{V} \left(e_{ij} - \frac{1}{2} \operatorname{div} \boldsymbol{u} \cdot \delta_{ij} \right) B_i B_j \,\mathrm{d}\boldsymbol{x} - \frac{1}{\mu} \int\limits_{S} (\boldsymbol{u} \cdot \boldsymbol{B}) (\boldsymbol{B} \cdot \mathrm{d}\boldsymbol{f}), \tag{11.6}$$

where e_{ij} is the rate of strain tensor defined by

$$e_{ij} = \frac{1}{2} \left(\frac{\partial u_i}{\partial x_j} + \frac{\partial u_j}{\partial x_i} \right), \tag{11.7}$$

and $\mathrm{d}\boldsymbol{f}$ the surface element parallel to the outward normal. The last term on the right-hand side of (11.6) is zero not only if $\boldsymbol{u}$ vanishes on S but also if it corresponds to a rigid rotation of S. As is to be seen from (11.5) and (11.6) a conversion of mechanical into magnetic energy, or vice versa, requires a deformation of the conducting medium, which is, of course, always connected with a deformation of the magnetic field lines. The magnetic energy grows in the case of stretching and decreases in the case of contraction of the field lines.

Let us at first consider the case of a medium at rest. Then the behaviour of the magnetic field is governed by Ohmic dissipation only. As can be seen from (11.5), any magnetic field is bound to decay. On the basis of the usual scaling laws the decay time, T, can be estimated from (11.1) or (11.5). Supposing that σ is constant we obtain

$$T = \mu\sigma L^2. \tag{11.8}$$

Here L represents a characteristic scale of the magnetic field the order of which, at least in simple cases, will coincide with a characteristic dimension of the conducting body. As we shall see in section 14.4., for a spherical body with the radius R we should put $L = R/\pi$. Then T coincides with the time in which the slowest decaying mode is weakened by a factor 1/e.

At this point we are able to justify some data used in section 11.1. For the Earth's core we may accept $\mu = \mu_0 = 4\pi \cdot 10^{-7}\ \mathrm{VsA^{-1}\ m^{-1}}$, $\sigma = 3 \cdot 10^5\ \Omega^{-1}\ \mathrm{m}^{-1}$ and $L = R/\pi$ with $R = 3 \cdot 10^6$ m. Hence, $T \approx 1 \cdot 10^5$ a. With regard to the Sun we choose $\mu = \mu_0$, $\sigma = 10^7\ \Omega^{-1}\ \mathrm{m}^{-1}$, and $L = R/\pi$ with $R = 7 \cdot 10^8$ m, thus obtaining $T \approx 2 \cdot 10^{10}$ a.

A necessary condition for the existence of a dynamo is a sufficiently intense motion with a proper geometrical structure so that the induction effect of this motion balances or exceeds the dissipative effect considered before. As explained above, the motion must result in a deformation of the medium by which the magnetic field lines are essentially stretched.

We compare the magnitude of the induction term and the dissipative term in (11.1) or in (11.5). If σ is again assumed to be constant, its ratio can be estimated to be smaller than or of the order of the magnetic Reynolds number, Rm, defined by

$$\mathrm{Rm} = \mu\sigma UL\,, \tag{11.9}$$

where U is a characteristic value of the velocity and L has the same meaning as above. Only in the case of a proper geometrical structure of the velocity field, the ratio mentioned can indeed reach the order of Rm. In an estimation referring to (11.6), a modified magnetic Reynolds number, Rm^+, appears instead of the usual one, namely

$$\mathrm{Rm}^+ = \mu\sigma U^+ L^2\,, \tag{11.10}$$

with U^+ being a characteristic value of that part of the velocity gradient which corresponds to a deformation of the medium, and L again as above. It has to be expected that in the case of a dynamo both numbers, Rm and Rm^+, have to exceed certain bounds in the order of unity. As shall be seen in section 11.4., for a spherical body with the radius R it is reasonable to put again $L = R/\pi$ in both (11.9) and (11.10).

For cosmical bodies the values of Rm and Rm^+ can generally be assumed to be much higher than unity. This mainly appears as a consequence of the large dimensions of these objects. In order to illustrate this we again consider the Earth's core. Taking the same numerical values for σ and L as above we only need $U \approx 3 \cdot 10^{-6}\,\mathrm{ms^{-1}}$ and $U^+R \approx 9 \cdot 10^{-6}\,\mathrm{ms^{-1}}$ in order to get $\mathrm{Rm} \approx 1$ and $\mathrm{Rm}^+ \approx 1$, respectively. There are, however, certainly internal motions with much higher values of U and U^+R so that also much higher values of Rm and Rm^+ appear. Likewise, for the Sun and similar objects values of Rm and Rm^+ have to be expected which exceed unity considerably.

It would be desirable that we could reproduce in the laboratory a dynamo process as assumed to take place at a cosmical body. A suitable medium for such experiments is liquid sodium for which $\mu = \mu_0$ and $\sigma = 6 \cdot 10^6\,\Omega^{-1}\,\mathrm{m}^{-1}$. We consider a spherical volume of sodium with a diameter of 1 m, putting $L = R/\pi$ with $R = 0.5$ m. Then we need already $U \approx 0.8\,\mathrm{ms^{-1}}$ and $U^+R \approx 2.6\,\mathrm{ms^{-1}}$ in order to get $\mathrm{Rm} \approx 1$ and $\mathrm{Rm}^+ \approx 1$. We may conclude that the experimental realization of magnetic field generation due to the motion of a conducting fluid seems to be possible. It requires, however, enormous technical efforts. Up to now no experiment of this kind has been carried out.

In this context it should be mentioned that in special types of thermonuclear reactors, in particular fast breeders, the cooling is due to the circulation of liquid metals, e.g. liquid sodium. The possibility of operating trouble due to the excitation of magnetic fields must be taken into account as far as very large cooling systems are concerned involving very high velocities. We refer to investigations by Bevir [1] and Pearson [1].

So far we have only dealt with the kinematic aspects of the dynamo problem. When the full magnetohydrodynamic problem is treated, the equations given above have to be supplemented by equations governing the behaviour of the velocity field $\boldsymbol{u}$. We only mention the Navier-Stokes equation completed by the Lorentz force, and the continuity equation,

$$\varrho\left(\frac{\partial \boldsymbol{u}}{\partial t} + (\boldsymbol{u} \cdot \nabla)\,\boldsymbol{u}\right) = -\nabla p + \frac{1}{\mu}\,\mathrm{curl}\,\boldsymbol{B} \times \boldsymbol{B} + \varrho \boldsymbol{F}, \tag{11.11a}$$

$$\frac{\partial \varrho}{\partial t} + \mathrm{div}\,(\varrho \boldsymbol{u}) = 0, \qquad \text{in } V. \tag{11.11b}$$

Here ϱ is the mass density, p the hydrodynamic pressure, and $\boldsymbol{F}$ stands for several forces like those due to gravity, internal friction etc. In dependence on the special model considered, these equations have to be completed by other ones, e.g. the equations of state or of internal energy, and by suitable boundary conditions.

11.4. Some necessary conditions for dynamos

The difficulties in the construction of dynamo models result from the geometrical structure of both the magnetic and the velocity field rather than from the magnitude of the velocity field. Some theorems shall be introduced which contain necessary conditions especially for the geometrical structures of the fields.

A remarkable theorem of this kind has been proved by COWLING [1] (1934) and in several respects generalized by some other authors. According to this theorem a steady magnetic field being symmetric with respect to a given axis cannot be maintained by a steady velocity field which is symmetric with respect to the same axis. It can easily be shown that such a magnetic field cannot be maintained by any other velocity field either. In this way COWLING's theorem leads to the conclusion that a steady axisymmetric magnetic field can never be maintained by dynamo action.

For spherical models with constant electrical conductivity BRAGINSKIJ [1] (1964) has generalized this theorem by including non-steady magnetic fields. He has shown that any field which is symmetric with respect to a given axis is bound to decay if the velocity field is solenoidal and symmetric with respect to the same axis. We can add that this statement remains valid for all solenoidal velocity fields regardless of their symmetry properties.

While dynamos with steady axisymmetric magnetic fields turn out to be impossible, those with a steady axisymmetric velocity field cannot be excluded. GAILITIS [2] presented an example for the maintenance of a non-axisymmetric magnetic field by steady axisymmetric motions.

Besides COWLING's theorem, the essence of which can be understood as a condition for the magnetic field, there are other theorems which contain conditions for the motions. A theorem of this type traces back to ELSASSER [1] (1946) and BULLARD and GELLMAN [1] (1954) and has been reconsidered by MOFFATT [8]. It presupposes constant electrical conductivity. It states that a magnetic field cannot be maintained by solenoidal motions without radial components, i.e. by a solenoidal flow with streamlines which lie in concentric spherical surfaces. This implies that a magnetic field may exist due to motions of an incompressible medium only if these motions show radial components too. Furthermore, maintenance of a magnetic field turns out to be impossible even in the case of a compressible medium if, e.g., only rotational motions are present regardless of the dependence of the angular velocity on radius or latitude. A lower bound for the magnitude of the radial motion necessary for dynamo action has been given by BUSSE [4] (1975).

A theorem by BACKUS [2] (1958) concerns both the geometrical structure and the magnitude of the velocity field. It poses a condition for the rate of strain tensor e_{ij} defined by (11.7). Again a solenoidal motion, i.e. div $\boldsymbol{u} = 0$, is supposed so that the highest eigenvalue of this tensor is positive. Of course, like the tensor itself, also this highest eigenvalue varies in space. BACKUS' theorem requires that its maximum value occurring inside the body, say $U^+_{\max}$, cannot be smaller than the lowest decay rate of the magnetic field in absence of motion. This can be formulated as a condition for a modified magnetic Reynolds number like (11.10). We restrict ourselves to a spherical model with constant electric conductivity. Then this condition reads

$$\mu\sigma U^+_{\max} R^2/\pi^2 \geqq 1 , \tag{11.12}$$

where, as before, R denotes the radius of the conducting body.

Finally, we mention a theorem by CHILDRESS [1] (1969) concerning the magnitude of the velocity field. It makes more precise the observation that in the case of a dynamo a magnetic Reynolds number like (11.9) has to exceed a certain bound. For several models such bounds were estimated. In the case of a spherical model with constant electrical conductivity and solenoidal motions one obtains

$$\mu\sigma U_{\max} R/\pi \geqq 1 , \tag{11.13}$$

where $U_{\max}$ is the maximum relative velocity inside the body and R again its radius.

11.5. Successful attempts to construct kinematic dynamo models

There have been numerous attempts to construct kinematic dynamo models, i.e. to find solutions of the corresponding mathematical problem. Many of them failed because of the great difficulties arising for several reasons, especially as a consequence of the necessary conditions discussed above.

HERZENBERG [1] (1958) was the first to construct a kinematic dynamo model. His solution of the governing equations also played the role of an existence proof for dynamos. In this model the conducting medium occupies a sphere. Apart from two smaller separate spherical regions inside this sphere the medium is at rest. In each of the two regions it rotates like a rigid body. For special rates of rotation and a special arrangement of the axes of rotation self-excitation occurs. As GAILITIS [3] has shown not only steady but also oscillatory magnetic fields are possible. Similar models with three spherical rotors, the action of which is in some respect easier to comprehend, have been investigated by GIBSON [1, 2] and by KROPACHEV, GORSHKOV and SEREBRYANAYA [1, 2].

A modification of HERZENERG's dynamo model where the two spherical rotors are replaced by two cylindrical ones has been constructed in the laboratory and successfully put into action by LOWES and WILKINSON [1, 2]. Both the nonmoving and the moving parts were made from solid electrically conducting material, and the contact between these parts was provided by a mercury film.

Other examples of kinematic dynamos have been given by Lortz [2, 3] (1968, 1972). His first model supposes a conducting medium of infinite spatial extent. He showed that special non-axisymmetric steady motions within a cylindrical region, the stream lines of which are helices, are able to maintain a steady magnetic field, which also shows some helical structure. In a second model the conducting medium only occupies a torus which is surrounded by vacuum. Also in this case special helical motions may maintain a magnetic field which passes as an axisymmetric field into the vacuum.

Ideas developed by Tverskoj [1] (1966) led to a further interesting kinematic dynamo model elaborated by Gailitis [2] (1970). A modification of it has been investigated by Gailitis and Freiberg [1]. Again, the conducting medium is assumed to be of infinite spatial extent. The motion is restricted to a pair of thin vortex rings. It is symmetric with respect to a given axis, to a plane perpendicular to this axis and to all planes containing this axis. By this simple kind of motion magnetic fields can indeed be maintained. In agreement with Cowling's theorem these fields lack symmetry with respect to the axis mentioned.

Remarkable investigations on kinematic dynamo models were carried out by Bullard and Gellman [1] (1954). In their models the conducting medium occupies a sphere which is embedded in vacuum. The authors established a mathematical method which allows to reduce the basic equations without any restriction on magnetic field or motion to an infinite system of differential equations for scalar functions; in the steady case ordinary differential equations for these functions appear. For a special model, i.e. a special pattern of motions, Bullard and Gellman tried to solve this system of equations numerically. However, their conclusion that this model may act as a dynamo was rejected by Gibson and Roberts [2] who pointed out that the numerical method applied does not converge. A modification of this model by Lilley [1] was refuted by Pekeris, Accad and Shkoller [1] who, in turn, proposed a model which indeed seems to work as a dynamo. In this model, which is adapted to the situation inside the Earth's core, the motions form spherical convective cells.

The hitherto mentioned dynamo models, apart from the last one, scarcely apply to situations as they occur at the Earth, at the Sun or at stars. There are, however, two approaches to the understanding of dynamo processes and to the construction of dynamo models which are of interest for these cosmical bodies. The basic features of both approaches are very similar. In both cases the magnetic field and the velocity field are each split up into a mean and a residual field. The mean fields are defined by a proper averaging procedure which ensures relatively simple geometrical structures and time dependencies, whereas the residual fields necessarily show a more complicated behaviour. Instead of the complete magnetic field only its mean part is considered, the behaviour of which under the influence of a given motion is determined by the mean velocity field and some averages taken over quantities which depend on the remaining velocity field.

The first of these two approaches mentioned consists in the theory of the "nearly symmetric dynamo" established by BRAGINSKI [1, 2, 3] (1964). In this case the mean fields are those parts of the original fields which are symmetric with respect to a given axis. In this frame spherical dynamo models have been considered in which the magnetic field occurs as a consequence of the simultaneous presence of a mean, i.e. axisymmetric motion consisting of differential rotation and meridional circulation and some special non-axisymmetric motions. Models of that kind provide for one of the possibilities to explain the existence and some fundamental properties of the magnetic field of the Earth.

The second approach is based on mean-field magnetohydrodynamics as introduced in the foregoing chapters, and it is often referred to as the theory of the "turbulent dynamo". In this case the mean fields are usually defined by statistical or space or time averages and they are not necessarily axisymmetric. As we have seen in the foregoing chapters a homogeneous isotropic non-mirror-symmetric turbulence, which leads to an α-effect, is able to maintain a mean magnetic field. Furthermore, it is also shown that turbulent motions in rotating bodies, i.e. under the influence of Coriolis forces, also give rise to an α-effect. Spherical dynamo models have been proved to be possible where the maintenance of the magnetic field is only due to the α-effect. Also such spherical dynamo models have been constructed where the magnetic field is a consequence of both mean motions, e.g. differential rotation, and the α-effect, or other effects caused by turbulent motions. A series of models of that kind have been investigated up to now, which also provide explanations of the origin and the behaviour of magnetic fields of some cosmical bodies, especially of the Earth and the Sun.

In the following chapters we shall deal with this special branch of dynamo theory, the theory of the turbulent dynamo, and with some of its applications with respect to cosmical bodies.

CHAPTER 12

FUNDAMENTALS OF THE THEORY OF THE TURBULENT DYNAMO

12.1. Basic concept

We shall now consider the dynamo problem as discussed in the foregoing chapter within the framework of mean-field magnetohydrodynamics. As we have seen, in the case of a dynamo generally both the magnetic field and the motion will have rather complicated geometrical structures. We now do not ask for all details of the magnetic field but only for a mean magnetic field with a more simple structure, defined by a proper averaging procedure. Of course, the behaviour of this mean magnetic field depends not only on the mean velocity field but also on the electromotive force caused by the residual part of the velocity field. In this way we arrive at a special approach to the dynamo problem. Since this approach also leads to dynamos with turbulent motions and hence turbulent magnetic fields, it has become known as the theory of the turbulent dynamo.

We point out the fact that the general concept of mean-field magnetohydrodynamics applies to all cases in which an averaging procedure is devised for electromagnetic and hydrodynamic fields, no matter whether or not these fields correspond to any definition of turbulence. Only the validity of the Reynolds relations (cf. section 2.4.) must be required. Likewise, the general considerations of this chapter do not only apply to cases with the presence of some kind of turbulence. We may also think of more or less regular flow patterns. Nevertheless, we shall use the terminology adapted to situations with turbulence.

Until further notice we assume that the mean fields are defined by a statistical average, for which the Reynolds relations are automatically fulfilled. In addition, we adopt all assumptions and notations introduced in the mathematical formulation of the dynamo problem given in section 12.2.

By virtue of the fact that only statistical averages will be used the basic electromagnetic equations for mean fields, (2.5) and (2.6), may be accepted for all space; of course, in vacuum the electric current vanishes. Starting from these equations, or taking the average of the relations (11.1)...(11.4), we may easily derive the basic equations for the turbulent dynamo. The equations governing the mean magnetic flux density, $\overline{\boldsymbol{B}}$, inside the conducting medium read

$$\frac{1}{\mu}\operatorname{curl}\left(\frac{1}{\sigma}\operatorname{curl}\overline{\boldsymbol{B}}\right)-\operatorname{curl}\left(\overline{\boldsymbol{u}}\times\overline{\boldsymbol{B}}+\mathfrak{E}\right)+\frac{\partial\overline{\boldsymbol{B}}}{\partial t}=0, \tag{12.1a}$$

$$\operatorname{div}\overline{\boldsymbol{B}}=0, \qquad \text{in } V, \tag{12.1b}$$

where $\mathfrak{E}$ is again the electromitive force caused by the fluctuating motions, i.e. is identical with $\overline{\boldsymbol{u}'\times\boldsymbol{B}'}$. For several purposes it is convenient to put (12.1) in the form

$$\frac{1}{\mu}\operatorname{curl}\left(\frac{1}{\sigma_T}\operatorname{curl}\overline{\boldsymbol{B}}\right)-\operatorname{curl}\left(\overline{\boldsymbol{u}}\times\overline{\boldsymbol{B}}+\mathfrak{E}^+\right)+\frac{\partial\overline{\boldsymbol{B}}}{\partial t}=0, \tag{12.2a}$$

$$\operatorname{div}\overline{\boldsymbol{B}}=0, \qquad \text{in } V, \tag{12.2b}$$

with the turbulent conductivity, σ_T, and a reduced electromotive force, $\mathfrak{E}^+$, defined by

$$\sigma_T=\sigma/(1+\mu\sigma\beta), \quad \mathfrak{E}^+=\mathfrak{E}+\beta\operatorname{curl}\overline{\boldsymbol{B}}, \tag{12.2c, d}$$

where β remains to be fixed suitably. In addition, we have the equations

$$\operatorname{curl}\overline{\boldsymbol{B}}=0, \quad \operatorname{div}\overline{\boldsymbol{B}}=0, \text{ in } V', \tag{12.3a, b}$$

and the conditions

$$[\overline{\boldsymbol{B}}]=0, \quad \boldsymbol{n}\cdot\overline{\boldsymbol{u}}=0, \text{ on } S, \tag{12.4a, b}$$

and

$$\overline{\boldsymbol{B}}=O(r^{-3}), \text{ as } r\to\infty. \tag{12.5}$$

As explained in section 2.3., the electromotive force may be comprehended as a functional of $\overline{\boldsymbol{u}}$, $\boldsymbol{u}'$, and $\overline{\boldsymbol{B}}$ which proves to be linear in $\overline{\boldsymbol{B}}$. Correspondingly the above equations have to be supplemented by a relation

$$\mathfrak{E}=\mathfrak{E}(\overline{\boldsymbol{B}}), \tag{12.6}$$

the form of which is determined by $\overline{\boldsymbol{u}}$ and $\boldsymbol{u}'$. For all cases considered in the following we shall assume that $\mathfrak{E}$ is zero if $\overline{\boldsymbol{B}}$ vanishes everywhere and that $\mathfrak{E}$ at a given point may be expressed by $\overline{\boldsymbol{B}}$ and its first spatial derivatives in this point. Then (12.6) takes the special form

$$\mathfrak{E}_i=a_{ij}\overline{B}_j+b_{ijk}\frac{\partial\overline{B}_j}{\partial x_k}, \tag{12.7}$$

with a_{ij} and b_{ijk} depending only on $\overline{\boldsymbol{u}}$ and $\boldsymbol{u}'$.

In order to construct a kinematic dynamo model in addition to the $\overline{\boldsymbol{u}}$-field those properties of the $\boldsymbol{u}'$-field relevant for the electromotive force $\mathfrak{E}$ have to be specified in such way that the equations formulated above allow $\overline{\boldsymbol{B}}$-fields which do not vanish as $t \to \infty$.

For many purposes it will be of interest to have an energy equation analogous to (11.5). The mean energy density $\overline{\boldsymbol{B}^2}/2\mu$ of the magnetic field can always be split in the two parts $\overline{\boldsymbol{B}}^2/2\mu$ and $\overline{\boldsymbol{B}'^2}/2\mu$ related to the mean and the fluctuating part of the magnetic field. We are only interested in the first part. Starting from the basic equations (2.5) and (2.6) for the mean fields we obtain

$$\frac{\mathrm{d}}{\mathrm{d}t} \int\limits_{V+V'} \frac{\overline{\boldsymbol{B}}^2}{2\mu}\,\mathrm{d}\boldsymbol{x} = -\int\limits_{V} \frac{\overline{\boldsymbol{j}}^2}{\sigma}\,\mathrm{d}\boldsymbol{x} - \int\limits_{V} \overline{\boldsymbol{u}} \cdot (\overline{\boldsymbol{j}} \times \overline{\boldsymbol{B}})\,\mathrm{d}\boldsymbol{x} + \int\limits_{V} \overline{\boldsymbol{j}} \cdot \mathfrak{E}\,\mathrm{d}\boldsymbol{x}. \tag{12.8}$$

This relation retains its validity if σ and $\mathfrak{E}$ are replaced by σ_T and $\mathfrak{E}^+$.

If instead of a kinematic dynamo model, a true magnetohydrodynamic model is treated we have to take into account that both the mean and the fluctuating part of the motion are influenced by the magnetic field. In this way we are confronted with a rather complex problem. We at first may supplement the equations (12.1) for $\overline{\boldsymbol{B}}$ by corresponding equations for $\overline{\boldsymbol{u}}$. Departing from (11.11) and restricting ourselves to incompressible media, i.e. $\varrho = \text{const}$, we obtain

$$\varrho \left(\frac{\partial \overline{\boldsymbol{u}}}{\partial t} + (\overline{\boldsymbol{u}} \cdot \nabla)\, \overline{\boldsymbol{u}} \right) = -\nabla \overline{p} + \frac{1}{\mu} \operatorname{curl} \overline{\boldsymbol{B}} \times \overline{\boldsymbol{B}} + \varrho(\overline{\boldsymbol{F}} + \mathfrak{F}), \tag{12.9a}$$

$$\operatorname{div} \overline{\boldsymbol{u}} = 0, \qquad \text{in } V, \tag{12.9b}$$

with

$$\mathfrak{F} = -\overline{(\boldsymbol{u}' \cdot \nabla)\, \boldsymbol{u}'} + \frac{1}{\mu\varrho} \overline{\operatorname{curl} \boldsymbol{B}' \times \boldsymbol{B}'}. \tag{12.9c}$$

Analogous to $\mathfrak{E}$ now a further quantity, $\mathfrak{F}$, occurs which is determined by the fluctuations $\boldsymbol{u}'$ and $\boldsymbol{B}'$. The equations for these fluctuations can be derived from (11.1) and (11.11), and also contain $\overline{\boldsymbol{u}}$ and $\overline{\boldsymbol{B}}$. We therefore may try to find relations connecting $\mathfrak{E}$ and $\mathfrak{F}$ with $\overline{\boldsymbol{u}}$, $\overline{\boldsymbol{B}}$, and, e.g., the forces which cause the turbulence. This presupposes that suitable solutions of the equations mentioned are available. If we are indeed successful in establishing relations of that kind for $\mathfrak{E}$ and $\mathfrak{F}$ we may insert them into equations (12.1) and (12.9). In this way we then arrive at a system of partial differential equations for $\overline{\boldsymbol{u}}$ and $\overline{\boldsymbol{B}}$, which remains to be integrated. Up to now the problem sketched here has been far from being completely solved.

12.2. Remarks concerning averaging procedures and the scales of mean and fluctuating quantities

When the concept explained above is applied to special problems, some points require special attention.

Up to now statistical averages have been used. In view of a comparison between theoretical and observational results it is desirable to admit also space or time averages. In doing so we have to put up with some restrictions concerning the scales of the mean and fluctuating fields. For the following considerations we denote the characteristic length and time scales of the mean fields by $\bar{\lambda}$ and $\bar{\tau}$, and those of the fluctuations by λ_{cor} and τ_{cor}.

Let us first deal with space averages defined by an integration over a certain three-dimensional range. The scale of the averaging range is denoted by λ_{av}. In order to ensure the validity of the Reynolds relations we have to require that the condition

$$\bar{\lambda} \gg \lambda_{\text{av}} \gg \lambda_{\text{cor}} \tag{12.10}$$

is fulfilled. Generally, $\bar{\lambda}$ cannot exceed the characteristic dimensions of the conducting body. Therefore λ_{cor} must be much smaller than these dimensions. In a model of the Earth's core, e.g., an irregular motion with a length scale which is not much smaller than the radius of the core can therefore never be treated as a fluctuation. At the Sun the same difficulty arises for the motions occurring with the giant cells of the convection zone.

We want to mention a further difficulty. In the case of space averages Ohm's law, i.e. the last equation (2.2), and therefore the induction equations (12.1 a) and (12.2 a) for mean fields, can only be justified if the spatial variations of the electrical conductivity, σ, are negligible within the averaging range. In general, there will be a jump in the electrical conductivity at the boundary. For points whose averaging range contains this jump the equations (12.1 a), (12.2 a), and also (12.3 a) are not exactly justified. Possible errors would then have to be estimated for each special situation.

There is, however, a special version of a space average which is defined by an integration over a single coordinate. For this type of average, at least under special suppositions, the restrictions discussed before do not hold. We refer to a cylindrical or a spherical coordinate system with the azimuthal coordinate φ. The average mentioned of a scalar function is defined as a suitably normalized integral of this function over all φ, and the average of a vector as the result of averaging its components related to one of these coordinate systems. Then, of course, all averaged fields are automatically symmetric with respect to the polar axis of the coordinate system. For this special average the Reynolds relations hold exactly. Let us also suppose the shape of the conducting body and the distribution of the electrical conductivity to be symmetric with respect to the polar axis. Then the basic equations for the mean fields, (2.1) and (2.2), and the relations (12.1)...(12.5) hold without any restriction.

The average explained is used throughout in the theory of the nearly symmetric dynamo, and may be used here too. In applying it, however, we must bear in mind that even in the case of a field with large-scale structures the corresponding mean field may vanish. This always happens if this field contains no part which is symmetric with respect to the axis mentioned. In so far the use of the average explained implies a restriction to special situations here. It should be mentioned that if this average is used some of the relations for averaged quantities given in the foregoing chapters have to be modified.

We now consider time averages which are defined by means of an integration over certain time intervals. The averaging time is denoted by τ_{av}. With respect to the Reynolds relations we have to require that

$$\bar{\tau} \gg \tau_{\mathrm{av}} \gg \tau_{\mathrm{cor}}\,. \tag{12.11}$$

In comparison with the space averages defined by integrations over a three-dimensional range, generally the time average seems to be more suited for models which are to be applied to cosmical objects. When dealing with a model for the Earth we may utilize the fact that the spectrum of the most important time scales of the relevant quantities shows a clear gap separating regions with large and small scales. The large scales are greater than or in the order of that belonging to the westward drift of the magnetic field, i.e. at least 10^3 years. All other important scales are smaller than 1 year. We then may choose an averaging interval of a few decades. In this way (12.11) is fairly satisfied. All quantities with the large scales mentioned appear as mean fields, and all with small scales as fluctuations, regardless of their spatial extent. In the case of the Sun the essential time scales are greater than or equal to the period of the solar cycle, i.e. 22 years, or they are comparable with or smaller than a rotational period, i.e. one month. An averaging interval of one or two years seems to be appropriate. Then the above considerations apply analogously.

In the case of a time average, provided the electrical conductivity does not vary in time, the basic equations for the mean fields (2.1) and (2.2) and the relations (12.1)...(12.5) can readily be justified.

We now turn to the assumptions under which the electromotive force $\mathfrak{E}$ takes the special form (12.7), which is supposed for all models considered in the following.

At first, the existence of any part of $\mathfrak{E}$ is ignored which is independent of $\overline{\boldsymbol{B}}$, i.e. which is non-zero if $\overline{\boldsymbol{B}}$ vanishes. As explained in section 2.4. a part of $\mathfrak{E}$

being independent of $\overline{\boldsymbol{B}}$ is to be expected only if fluctuating motions in certain small regions are already able to constitute a dynamo. We do not want to deal with this possibility but refer instead to a discussion given by RÄDLER [15].

Furthermore, $\mathfrak{E}$ at a given point generally depends on the $\overline{\boldsymbol{B}}$-field in a certain range around this point. The relation (12.7) presupposes a sufficiently weak variability of the $\overline{\boldsymbol{B}}$-field so that its behaviour within this range can be expressed by $\overline{\boldsymbol{B}}$ and its first spatial derivatives in the given point. Thus regardless of the conditions resulting from the averaging procedure, further conditions for the scales of mean and fluctuating quantities appear. Let us denote the length and time scales of $\overline{\boldsymbol{B}}$ again by $\bar{\lambda}$ and $\bar{\tau}$, those of the range mentioned by λ^+ and τ^+, and those of $\boldsymbol{u}'$ by λ_{cor} and τ_{cor}. Clearly, we have to require that

$$\bar{\lambda} \gg \lambda^+, \quad \bar{\tau} \gg \tau^+. \tag{12.12a, b}$$

According to our general considerations about $\mathfrak{E}$, the quantities λ^+ and τ^+ cannot exceed the orders of λ_{cor} and τ_{cor}, respectively. They may even be much smaller than λ_{cor} and τ_{cor}. From the considerations of sections 4.6. and 4.7. we may conclude that in the second order correlation approximation

$$\lambda^+ = \min\left(\lambda_{cor}, \sqrt{\tau_{cor}/\mu\sigma}\right), \tag{12.13a}$$

$$\tau^+ = \min\left(\tau_{cor}, \mu\sigma\lambda_{cor}^2\right). \tag{12.13b}$$

For higher approximations similar relations have to be expected.

Let us again consider the situation in the Earth. We use a time average for which (12.11) holds. Then (12.12b) is also fulfilled. Only (12.12a) remains to be discussed. It is reasonable to assume that $\bar{\lambda}$ is in the order of the radius of the core, i.e. $3 \cdot 10^6$ m. With respect to λ^+ we accept (12.13a) and conclude $\lambda^+ \leqq \sqrt{\tau_{cor}/\mu\sigma}$. If $\tau_{cor} < 1$ year and $\sigma = 3 \cdot 10^5\ \Omega^{-1}\ \mathrm{m}^{-1}$ (cf. section 11.3.) we find $\lambda^+ < 1 \cdot 10^4$ m. Thus (12.12a) too may be considered as fulfilled. An analogous situation appears at the Sun. By reasons which will be explained later on (cf. section 17.4.) we may restrict our considerations to the solar convection zone. We assume $\bar{\lambda}$ in the order of the thickness of the convection zone, i.e. in the order of 10^8 m. We again use $\lambda^+ \leqq \sqrt{\tau_{cor}/\mu\sigma}$. With $\tau_{cor} < 1$ month and $\sigma = 3 \cdot 10^3\ \Omega^{-1}\ \mathrm{m}^{-1}$ (cf. section 17.4.) we obtain $\lambda^+ < 3 \cdot 10^4$ m. Again, (12.12a) may be considered as fulfilled.

CHAPTER 13

TOROIDAL AND POLOIDAL VECTOR FIELDS

13.1. Preliminary remarks

For the treatment of spherical dynamo models it has proved to be advantageous to represent vector fields, like the magnetic flux density, as a sum of a toroidal and a poloidal field and to profit from the special properties of these fields. On this basis, e.g., BULLARD and GELLMAN [1] reduced the relevant vector equations to a system of equations for scalar functions. In a similar way we may proceed in the theory of the turbulent dynamo too.

In this chapter we shall explain the definition and some properties of toroidal and poloidal vector fields. In doing so we shall follow the ideas of a paper by RÄDLER [10]. We also want to refer to other introductions to this topic, e.g. ROBERTS [1].

The subject of our following considerations is a vector field, $\boldsymbol{U}$, which is defined inside a certain simply-connected region of the Euclidean space, whose shape will be specified later on. As to the regularity of $\boldsymbol{U}$ we assume the existence of the spatial derivatives up to the second order.

We introduce a spherical polar coordinate system with the coordinates r, ϑ, and φ, the corresponding unit vectors $\boldsymbol{e}_r$, $\boldsymbol{e}_\vartheta$, and $\boldsymbol{e}_\varphi$, and the radius vector $\boldsymbol{r}$. For any scalar F we then have

$$\nabla F = \frac{\partial F}{\partial r}\,\boldsymbol{e}_r + \frac{1}{r}\,\frac{\partial F}{\partial \vartheta}\,\boldsymbol{e}_\vartheta + \frac{1}{r \sin\vartheta}\,\frac{\partial F}{\partial \varphi}\,\boldsymbol{e}_\varphi, \tag{13.1a}$$

$$\Delta F = \frac{1}{r^2}\left(\frac{\partial}{\partial r}\left(r^2\,\frac{\partial F}{\partial r}\right) + \Omega F\right), \tag{13.1b}$$

$$\Omega F = \frac{1}{\sin\vartheta}\,\frac{\partial}{\partial \vartheta}\left(\sin\vartheta\,\frac{\partial F}{\partial \vartheta}\right) + \frac{1}{\sin^2\vartheta}\,\frac{\partial^2 F}{\partial \varphi^2}, \tag{13.1c}$$

where $(1/r^2)\,\Omega$ is the Laplacean operator for a surface $r = \text{const}$. Concerning integrals over such surfaces we use the notation

$$\oint\!\!\oint F\,\mathrm{d}\tilde{\omega} = \int_{\vartheta=0}^{\pi}\int_{\varphi=0}^{2\pi} F \sin\vartheta\,\mathrm{d}\vartheta\,\mathrm{d}\varphi. \tag{13.2}$$

Denoting the components of $\boldsymbol{U}$ by U_r, U_ϑ, and U_φ we have

$$\operatorname{div} \boldsymbol{U} = \frac{1}{r^2} \frac{\partial}{\partial r} (r^2 U_r) + \frac{1}{r \sin \vartheta} \frac{\partial}{\partial \vartheta} (\sin \vartheta U_\vartheta) + \frac{1}{r \sin \vartheta} \frac{\partial U_\varphi}{\partial \varphi}, \tag{13.3a}$$

$$\begin{aligned} \operatorname{curl} \boldsymbol{U} = {} & \frac{1}{r \sin \vartheta} \left(\frac{\partial}{\partial \vartheta} (\sin \vartheta U_\varphi) - \frac{\partial U_\vartheta}{\partial \varphi} \right) \boldsymbol{e}_r \\ & + \frac{1}{r} \left(\frac{1}{\sin \vartheta} \frac{\partial U_r}{\partial \varphi} - \frac{\partial}{\partial r} (r U_\varphi) \right) \boldsymbol{e}_\vartheta \\ & + \frac{1}{r} \left(\frac{\partial}{\partial r} (r U_\vartheta) - \frac{\partial U_r}{\partial \vartheta} \right) \boldsymbol{e}_\varphi . \end{aligned} \tag{13.3b}$$

13.2. Toroidal and poloidal vector fields in the axisymmetric case

Let us at first assume that the vector field $\boldsymbol{U}$, and also the region of definition, are axisymmetric and the coordinate system fixed in such way that its polar axis coincides with the axis of symmetry. Then the components U_r, U_ϑ, and U_φ do not depend on φ.

We now consider the field $\boldsymbol{U}$ as a sum of a toroidal field, $\boldsymbol{U}_\mathrm{t}$, and a poloidal field, $\boldsymbol{U}_\mathrm{p}$, so that

$$\boldsymbol{U} = \boldsymbol{U}_\mathrm{t} + \boldsymbol{U}_\mathrm{p} \tag{13.4}$$

and define

$$\boldsymbol{U}_\mathrm{t} = U_\varphi \boldsymbol{e}_\varphi, \quad \boldsymbol{U}_\mathrm{p} = U_r \boldsymbol{e}_r + U_\vartheta \boldsymbol{e}_\vartheta. \tag{13.5a, b}$$

Hence, we can state:

(i) If, at a point, $\boldsymbol{U} = 0$ then also $\boldsymbol{U}_\mathrm{t} = \boldsymbol{U}_\mathrm{p} = 0$.

(ii) If a is any scalar, $a\boldsymbol{U}_\mathrm{t}$ is toroidal and $a\boldsymbol{U}_\mathrm{p}$ is poloidal.

(iii) If $\boldsymbol{U}^+$ is any axisymmetric vector field, at any fixed point $\boldsymbol{U}_\mathrm{t}^+ \times \boldsymbol{U}_\mathrm{t}$ is zero, $\boldsymbol{U}_\mathrm{t}^+ \times \boldsymbol{U}_\mathrm{p}$ and $\boldsymbol{U}_\mathrm{p}^+ \times \boldsymbol{U}_\mathrm{t}$ are poloidal, and $\boldsymbol{U}_\mathrm{p}^+ \times \boldsymbol{U}_\mathrm{p}$ is toroidal; especially $\boldsymbol{r} \times \boldsymbol{U}_\mathrm{t}$ is poloidal and $\boldsymbol{r} \times \boldsymbol{U}_\mathrm{p}$ is toroidal.

(iv) $\operatorname{div} \boldsymbol{U}_\mathrm{t} = 0$.

(v) $\operatorname{curl} \boldsymbol{U}_\mathrm{t}$ is poloidal, $\operatorname{curl} \boldsymbol{U}_\mathrm{p}$ is toroidal.

(vi) If, throughout the whole region, $\operatorname{curl} \boldsymbol{U}_\mathrm{t} = 0$ then $\boldsymbol{U}_\mathrm{t} = 0$.

(vii) At any fixed point $\boldsymbol{U}_\mathrm{t}$ and $\boldsymbol{U}_\mathrm{p}$ are orthogonal, i.e. $\boldsymbol{U}_\mathrm{t} \cdot \boldsymbol{U}_\mathrm{p} = 0$.

Most of the foregoing statements are self-evident; we mention them with a view to generalizing them later. Statements (iv) and (v) become clear by considering the special representations of $\operatorname{div} \boldsymbol{U}_\mathrm{t}$, $\operatorname{curl} \boldsymbol{U}_\mathrm{t}$, and $\operatorname{curl} \boldsymbol{U}_\mathrm{p}$ in spherical coordinates, where, of course, all derivatives with respect to φ vanish. As for (vi), the simple-connectedness of the region permits to conclude from curl $\boldsymbol{U}_\mathrm{t} = 0$ that $\boldsymbol{U}_\mathrm{t}$ is a gradient of a single-valued potential. Of course, this gradient must be axisymmetric. Such gradients, however, are always poloidal. Hence, on account of (i) we have $\boldsymbol{U}_\mathrm{t} = 0$.

In connection with statements (iv) and (v) we should take a brief look at the quantity $\Delta \boldsymbol{U}$ which may be written as $\Delta \boldsymbol{U}_\mathrm{t} + \Delta \boldsymbol{U}_\mathrm{p}$. We have

$$\Delta \boldsymbol{U}_\mathrm{t} = -\mathrm{curl}\,\mathrm{curl}\,\boldsymbol{U}_\mathrm{t}, \tag{13.6a}$$

$$\Delta \boldsymbol{U}_\mathrm{p} = -\mathrm{curl}\,\mathrm{curl}\,\boldsymbol{U}_\mathrm{p} + \nabla\,\mathrm{div}\,\boldsymbol{U}_\mathrm{p}. \tag{13.6b}$$

Clearly, $\Delta \boldsymbol{U}_\mathrm{t}$ proves to be toroidal and $\Delta \boldsymbol{U}_\mathrm{p}$ to be poloidal.

We now assume the axisymmetric field $\boldsymbol{U}$ to be solenoidal, i.e. satisfying

$$\mathrm{div}\,\boldsymbol{U} = 0. \tag{13.7}$$

As $\boldsymbol{U}_\mathrm{t}$ is solenoidal in any case, $\boldsymbol{U}_\mathrm{p}$ must be solenoidal too. Hence, $\boldsymbol{U}_\mathrm{p}$ can be replaced by curl $\hat{\boldsymbol{U}}$ with $\hat{\boldsymbol{U}}$ being a vector potential. Assuming that $\hat{\boldsymbol{U}}$ is also axisymmetric we may represent it by $\hat{\boldsymbol{U}}_\mathrm{t} + \hat{\boldsymbol{U}}_\mathrm{p}$. Yet only curl $\hat{\boldsymbol{U}}_\mathrm{t}$ is poloidal; curl $\hat{\boldsymbol{U}}_\mathrm{p}$ is toroidal and therefore has to vanish. Thus we may write

$$\boldsymbol{U}_\mathrm{p} = \mathrm{curl}\,\hat{\boldsymbol{U}}_\mathrm{t}. \tag{13.8}$$

Like $\boldsymbol{U}_\mathrm{t}$, also $\boldsymbol{U}_\mathrm{p}$ can be expressed by one scalar quantity.

The assumption concerning $\hat{\boldsymbol{U}}$ does not imply a serious restriction. If a $\hat{\boldsymbol{U}}$ exists at all, the components of which are single-valued functions of φ, an averaging with respect to φ readily provides another one which is axisymmetric. As long as $\boldsymbol{U}_\mathrm{p}$ is supposed to be continuous in all infinite space the existence of such $\hat{\boldsymbol{U}}$ can easily be proved. If, however, $\boldsymbol{U}_\mathrm{p}$ shows this property only in a bounded region, there are exceptional cases where no such $\hat{\boldsymbol{U}}$ are to be found. For instance, $\boldsymbol{U}_\mathrm{p} = \boldsymbol{r}/r^3$ is continuous in all regions not containing the origin $r = 0$, but it cannot be represented by a $\hat{\boldsymbol{U}}$ with the properties assumed above.

13.3. A special representation of a vector field

Let us now deal with an arbitrary field $\boldsymbol{U}$ defined inside a sphere or a spherical shell. The centre of the sphere or shell is taken as the origin of the spherical coordinate system. The sphere may also be extended to the infinite space.

We start from the fact that any continuously differentiable vector field $\boldsymbol{U}$ can be represented in the form

$$\boldsymbol{U} = \boldsymbol{r} \times \nabla U + \boldsymbol{r} V + \nabla W \tag{13.9}$$

by three single-valued scalar fields U, V, and W. In components this reads

$$U_r = rV + \frac{\partial W}{\partial r}, \tag{13.10a}$$

$$U_\vartheta = -\frac{1}{\sin\vartheta}\frac{\partial U}{\partial\varphi} + \frac{1}{r}\frac{\partial W}{\partial\vartheta}, \tag{13.10b}$$

$$U_\varphi = \frac{\partial U}{\partial\vartheta} + \frac{1}{r\sin\vartheta}\frac{\partial W}{\partial\varphi}. \tag{13.10c}$$

In order to justify this representation we consider at first the field $U_\vartheta \boldsymbol{e}_\vartheta + U_\varphi \boldsymbol{e}_\varphi$. As is well known, any such field can be represented at a spherical surface $r = \text{const}$ by means of a single-valued stream function, U, and a single-valued potential, W, in such a way that (13.10b) and (13.10c) hold. For a field $U_r\boldsymbol{e}_r$ we are always entitled to write (13.10a), with a further single-valued function, V. In this way, for the sum $U_r\boldsymbol{e}_r + U_\vartheta\boldsymbol{e}_\vartheta + U_\varphi\boldsymbol{e}_\varphi$ of these fields, we obtain (13.10).

We note that the determination of the functions U, V, and W generally requires the integration of a system of partial differential equations, namely of the system (13.10) where U_r, U_ϑ, and U_φ are to be understood as inhomogeneities. The main difficulty is the determination of U and W. From (13.10b) and (13.10c) we get

$$\Omega U = \frac{1}{\sin\vartheta}\left(\frac{\partial}{\partial\vartheta}(\sin\vartheta U_\varphi) - \frac{\partial U_\vartheta}{\partial\varphi}\right), \tag{13.11a}$$

$$\Omega W = \frac{r}{\sin\vartheta}\left(\frac{\partial}{\partial\vartheta}(\sin\vartheta U_\vartheta) + \frac{\partial U_\varphi}{\partial\varphi}\right). \tag{13.11b}$$

If W is known, (13.10a) immediately provides V.

The representation (13.9) for $\boldsymbol{U}$ implies the possibility of gauge transformations of U, V, and W. Obviously, $\boldsymbol{U}$ is invariant under replacement of U, V, and W by $\tilde{U}$, $\tilde{V}$, and $\tilde{W}$ provided that

$$U - \tilde{U} = g, \quad V - \tilde{V} = -\frac{1}{r}\frac{\mathrm{d}h}{\mathrm{d}r}, \quad W - \tilde{W} = h, \tag{13.12a, b, c}$$

where g and h may arbitrarily depend on r but never on ϑ and φ. The transformations given by (13.12) turn out to be the only possible ones. In order to see this we make sure that (13.10) with $U_r = U_\vartheta = U_\varphi = 0$ admits only $U = g$, $V = -\dfrac{1}{r}\dfrac{\mathrm{d}h}{\mathrm{d}r}$, and $W = h$ with g and h as above. From (13.10b, c) we obtain $\Omega U = 0$ and therefore $U = g$. Hence, (13.10b, c) immediately leads to $W = h$, and (13.10a) further to $V = -\dfrac{1}{r}\dfrac{\mathrm{d}h}{\mathrm{d}r}$.

It should be pointed out that not only $\boldsymbol{U}$ but also $\boldsymbol{r}\times\nabla U$ and $\boldsymbol{r}V + \nabla W$ remain unchanged under gauge transformations. Consequently, any field $\boldsymbol{U}$ can be split in a unique manner in two parts of the form $\boldsymbol{r}\times\nabla U$ and $\boldsymbol{r}V + \nabla W$.

For several purposes it is useful to remove the non-uniqueness of U, V, and W by a normalization condition. In that sense U, V, and W may be fixed by the requirement that, for all r,

$$\oiint U\,\mathrm{d}\tilde{\omega} = \oiint W\,\mathrm{d}\tilde{\omega} = 0. \tag{13.13}$$

As is easily seen, this condition can always be fulfilled. If, instead of functions U, V, and W satisfying (13.13), other functions, $\tilde{U}$, $\tilde{V}$, and $\tilde{W}$, are given, we may put

$$\oiint \tilde{U}\,\mathrm{d}\tilde{\omega} = -4\pi g, \quad \oiint \tilde{W}\,\mathrm{d}\tilde{\omega} = -4\pi h. \tag{13.14a, b}$$

Then (13.12) provides U, V, and W in agreement with (13.13). Since (13.12) are the only possible gauge transformations, (13.13) indeed fixes U, V, and W uniquely.

If only U and V are to be fixed, conditions (13.13) may be replaced by

$$\oiint U\,\mathrm{d}\tilde{\omega} = \oiint V\,\mathrm{d}\tilde{\omega} = 0. \tag{13.15}$$

These conditions, too, can always be fulfilled. Then U and V are really uniquely fixed, whereas W is determined only up to an additive constant.

When using the representation (13.9), we should also keep in mind some relations of the vector analysis. For any scalar F we have

$$\operatorname{curl}(\boldsymbol{r}F) = -\boldsymbol{r}\times\nabla F, \tag{13.16a}$$

$$\operatorname{curl}\operatorname{curl}(\boldsymbol{r}F) = -\operatorname{curl}(\boldsymbol{r}\times\nabla F) = -\boldsymbol{r}\,\Delta F + \nabla\frac{\partial}{\partial r}(rF), \tag{13.16b}$$

$$\operatorname{curl}\operatorname{curl}\operatorname{curl}(\boldsymbol{r}F) = \Delta(\boldsymbol{r}\times\nabla F) = -\operatorname{curl}(\boldsymbol{r}\,\Delta F) = \boldsymbol{r}\times\nabla\,\Delta F, \tag{13.16c}$$

$$\boldsymbol{r}\times\operatorname{curl}(\boldsymbol{r}F) = -\boldsymbol{r}\times(\boldsymbol{r}\times\nabla F) = -\frac{\boldsymbol{r}}{r}\frac{\partial}{\partial r}(r^2F) + \nabla(r^2F). \tag{13.16d}$$

In connection with the normalization conditions it may be of interest that, again for any F,

$$\oiint \Omega F\,\mathrm{d}\tilde{\omega} = 0. \tag{13.17}$$

With the help of (13.1b) we conclude that

$$\oiint \Delta F\,\mathrm{d}\tilde{\omega} = 0, \text{ if } \oiint F\,\mathrm{d}\tilde{\omega} = 0. \tag{13.18}$$

13.4. Toroidal and poloidal vector fields in the general case

We shall now consider an arbitrary vector field $\boldsymbol{U}$ defined in a sphere or a spherical shell. As we have done in the axisymmetric case, in this general case too, we understand the field $\boldsymbol{U}$ as a sum of a toroidal field, $\boldsymbol{U}_t$, and a poloidal field, $\boldsymbol{U}_p$, so that

$$\boldsymbol{U} = \boldsymbol{U}_t + \boldsymbol{U}_p. \tag{13.19}$$

We define $\boldsymbol{U}_t$ and $\boldsymbol{U}_p$ by requiring that they can be represented as

$$\boldsymbol{U}_t = \boldsymbol{r}\times\nabla U, \tag{13.20a}$$

$$\boldsymbol{U}_p = \boldsymbol{r}V + \nabla W, \tag{13.20b}$$

or, in component form,

$$\boldsymbol{U}_t = -\frac{1}{\sin\vartheta}\frac{\partial U}{\partial\varphi}\boldsymbol{e}_\vartheta + \frac{\partial U}{\partial\vartheta}\boldsymbol{e}_\varphi, \tag{13.21a}$$

$$\boldsymbol{U}_p = \left(rV + \frac{\partial W}{\partial r}\right)\boldsymbol{e}_r + \frac{1}{r}\frac{\partial W}{\partial\vartheta}\boldsymbol{e}_\vartheta + \frac{1}{r\sin\vartheta}\frac{\partial W}{\partial\varphi}\boldsymbol{e}_\varphi. \tag{13.21b}$$

According to the foregoing considerations any vector field $\boldsymbol{U}$ can indeed be represented in this manner, and $\boldsymbol{U}_t$ and $\boldsymbol{U}_p$ are uniquely fixed. Obviously, this definition of $\boldsymbol{U}_t$ and $\boldsymbol{U}_p$ implies that given in the axisymmetric case. Because in a suitably chosen coordinate system we have in this case $\partial U/\partial\varphi = \partial W/\partial\varphi = 0$ so that (13.5) and (13.21) are in agreement.

In the general case the splitting of a given field $\boldsymbol{U}$ into the parts $\boldsymbol{U}_t$ and $\boldsymbol{U}_p$ is more complicated than in the axisymmetric case. It requires the knowledge of U, V, and W, which, as explained above, generally have to be determined as solutions of a system of partial differential equations. This implies a further difference. In the axisymmetric case the splitting of the field at a given point does not require any information on the behaviour of this field in the neighbourhood of this point. In the general case, however, the splitting at a given point depends on the behaviour of the field at the surface r = const containing this point.

If a field is given in the form of the right-hand side of (13.20a) or (13.20b) it is clearly toroidal or poloidal, respectively. In particular, any gradient of a single-valued function is poloidal.

Now statements similar to those given for the axisymmetric case are possible:

(i) If, at a surface r = const, $\boldsymbol{U} = 0$ then also $\boldsymbol{U}_t = \boldsymbol{U}_p = 0$.

(ii) If a is a scalar depending only on r, but not on ϑ and φ, then $a\boldsymbol{U}_t$ is toroidal and $a\boldsymbol{U}_p$ is poloidal.

(iii) $\boldsymbol{r} \times \boldsymbol{U}_t$ is poloidal and $\boldsymbol{r} \times \boldsymbol{U}_p$ is toroidal.
(iv) div $\boldsymbol{U}_t = 0$.
(v) curl $\boldsymbol{U}_t$ is poloidal, and curl $\boldsymbol{U}_p$ is toroidal.
(vi) If, throughout the whole region, curl $\boldsymbol{U}_t = 0$ then $\boldsymbol{U}_t = 0$.
(vii) $\boldsymbol{U}_t$ and $\boldsymbol{U}_p$ are orthogonal to each other in the sense of $\oiint \boldsymbol{U}_t \cdot \boldsymbol{U}_p \, d\tilde{\omega} = 0$.

Statement (i) can easily be proved with the help of the considerations of the foregoing section according to which $\boldsymbol{U} = 0$ results in $U = g$; this again leads to $\boldsymbol{U}_t = 0$ and, hence, to $\boldsymbol{U}_p = 0$. Statement (ii) is a simple consequence of definition (13.20). As for the statements (iii), (iv), and (v), the proof is given by the vector formulae (13.16) in connection with definition (13.20). In order to justify statement (vi), again referring to the simple-connectedness of the region, we at first conclude from curl $\boldsymbol{U}_t = 0$ that $\boldsymbol{U}_t$ is a gradient of a single-valued potential; because such gradients are always poloidal we then obtain $\boldsymbol{U}_t = 0$. With respect to statement (vii) we note at first that

$$\oiint \boldsymbol{U}_t \cdot \boldsymbol{U}_p \, d\tilde{\omega} = \frac{1}{r} \int_{\vartheta=0}^{\pi} \int_{\varphi=0}^{2\pi} \left(\frac{\partial U}{\partial \vartheta} \frac{\partial W}{\partial \varphi} - \frac{\partial U}{\partial \varphi} \frac{\partial W}{\partial \vartheta} \right) d\vartheta \, d\varphi$$

$$= \frac{1}{r} \left(\int_0^{2\pi} \left[U \frac{\partial W}{\partial \varphi} \right]_{\vartheta=0}^{\vartheta=\pi} d\varphi - \int_0^{\pi} \left[U \frac{\partial W}{\partial \vartheta} \right]_{\varphi=0}^{\varphi=2\pi} d\vartheta \right), \tag{13.22}$$

where the last line is the result of partial integrations. Since $\boldsymbol{U}$ is supposed to be non-singular everywhere, it follows that $\partial W/\partial \varphi = 0$ at $\vartheta = 0$ and $\vartheta = \pi$ and, therefore, the first integral of the last line vanishes. Due to the fact that U and W are single-valued the second integral of this line vanishes too.

The remarks on $\Delta \boldsymbol{U}$ given for the axisymmetric case remain valid, in particular formulae (13.6). Again, $\Delta \boldsymbol{U}_t$ and $\Delta \boldsymbol{U}_p$ turn out to be toroidal and poloidal, respectively.

Dealing with fields $\boldsymbol{U}$ which are solenoidal we may essentially repeat considerations given for the axisymmetric case. As far as $\boldsymbol{U}_p$ can be represented by a vector potential $\hat{\boldsymbol{U}}$ which is single-valued we again arrive at (13.8). Concerning the existence of suitable $\hat{\boldsymbol{U}}$ we refer to the remarks under (13.8).

By contrast with the axisymmetric case, the general case considered here was started from the assumption that the field $\boldsymbol{U}$ is defined in a spherical region. Then for the definition of the toroidal and poloidal parts, $\boldsymbol{U}_t$ and $\boldsymbol{U}_p$, we made reference to a spherical coordinate system, especially of the spherical coordinate surface $r = $ const. The question arises whether it is possible to extend these considerations to other regions or other coordinate systems in a way so that statements similar to (i)...(vii) hold. It is not difficult to represent a field $\boldsymbol{U}$ as a sum of two parts which are defined like $\boldsymbol{U}_t$ and $\boldsymbol{U}_p$ but with respect to other surfaces. However, only in the case of planes or spherical surfaces statements like (iv) and (v) are to be expected; see RÄDLER [10]. Already in the case of ellipsoidal surfaces statements of this kind can no longer be justified.

13.5. Expansions in spherical harmonics

As we have pointed out, the determination of the toroidal and the poloidal parts of a given vector field $\boldsymbol{U}$ generally requires the solution of a system of partial differential equations for the functions U, V, and W. In many cases it is appropriate to expand these functions with respect to spherical harmonics. Thus, the integration of differential equations can be reduced to quadratures. In this section we shall present some relations relevant for such expansions with spherical harmonics.

Let each of the functions U, V, and W be expanded as

$$F = \sum_{n,m} F_n^m(r)\, Y_n^m(\vartheta, \varphi). \tag{13.23}$$

The functions F_n^m are complex quantities, and the spherical harmonics Y_n^m are defined by

$$Y_n^m = P_n^{|m|}(\cos\vartheta)\, \mathrm{e}^{\mathrm{i}m\varphi}, \tag{13.24}$$

with P_n^m being associated Legendre polynomials. The summation is over all n and m with $n \geqq 0$ and $|m| \leqq n$. Since F should be real we have to require that $F_n^{-m} = F_n^{m*}$.

We may, of course, understand the vector field $\boldsymbol{U}$ too as a sum of special fields which are related to the terms in the expansions of U, V, and W. These fields are ascribed to special multipoles. We designate all fields which correspond to terms of U, V, or W with $n = 1$ as dipole fields, if $n = 2$ as quadrupole fields, if $n = 3$ as octupole fields, etc.

As is well known the spherical harmonics Y_n^m satisfy the differential equation

$$\Omega Y_n^m + n(n+1)\, Y_n^m = 0 \tag{13.25}$$

and the orthogonality relation

$$\oiint Y_n^m Y_\nu^{-\mu}\, \mathrm{d}\tilde{\omega} = \begin{cases} A_n^m \text{ if } n = \nu \text{ and } m = \mu \\ 0 \text{ otherwise}. \end{cases} \tag{13.26}$$

Adopting Ferrer's normalization of the P_n^m we have

$$A_n^m = \frac{4\pi(n + |m|)!}{(2n+1)\,(n - |m|)!}. \tag{13.27}$$

With the help of (13.16) from (13.19) and (13.20) we now conclude

$$\Omega U = \boldsymbol{r} \cdot \operatorname{curl} \boldsymbol{U}, \tag{13.28a}$$

$$\Omega V = -\boldsymbol{r} \cdot \operatorname{curl} \operatorname{curl} \boldsymbol{U}, \tag{13.28b}$$

$$\Omega W = \boldsymbol{r} \cdot \operatorname{curl} (\boldsymbol{r} \times \boldsymbol{U}), \tag{13.28c}$$

of which (13.28a) and (13.28c) are only repetitions of (13.11). Expressing U, V, and W by expansions like (13.23), and using (13.25) and (13.26) we then obtain

$$U_n^m = -(n(n+1)\, A_n^m)^{-1} \oiint (\boldsymbol{r} \cdot \operatorname{curl} \boldsymbol{U})\, Y_n^{-m}\, \mathrm{d}\tilde{\omega}, \tag{13.29a}$$

$$V_n^m = (n(n+1)\, A_n^m)^{-1} \oiint (\boldsymbol{r} \cdot \operatorname{curl} \operatorname{curl} \boldsymbol{U})\, Y_n^{-m}\, \mathrm{d}\tilde{\omega}, \tag{13.29b}$$

$$W_n^m = -(n(n+1)\, A_n^m)^{-1} \oiint (\boldsymbol{r} \cdot \operatorname{curl} (\boldsymbol{r} \times \boldsymbol{U}))\, Y_n^{-m}\, \mathrm{d}\tilde{\omega}, \text{ for } n \geqq 1. \tag{13.29c}$$

Analogously it results

$$r^2 V + r \frac{\partial W}{\partial r} = \boldsymbol{r} \cdot \boldsymbol{U} \tag{13.30}$$

and

$$V_n^m + \frac{1}{r} \frac{\partial W_n^m}{\partial r} = (r^2 A_n^m)^{-1} \oiint (\boldsymbol{r} \cdot \boldsymbol{U})\, Y_n^{-m}\, \mathrm{d}\tilde{\omega}, \text{ for } n \geqq 0. \tag{13.31}$$

Owing to the possibility of gauge transformations, U_0^0 is not fixed by these relations, and V_0^0 and W_0^0 are only obliged to satisfy (13.31).

If $\boldsymbol{U}$ is solenoidal, we may use (13.8) instead of (13.20b), and express $\hat{\boldsymbol{U}}_\mathrm{t}$ according to (13.20a) by a function $\hat{U}$. Doing so we find

$$\Omega \hat{U} = \boldsymbol{r} \cdot \boldsymbol{U} \tag{13.32}$$

and

$$\hat{U}_n^m = -(n(n+1)\, A_n^m)^{-1} \oiint (\boldsymbol{r} \cdot \boldsymbol{U})\, Y_n^{-m}\, \mathrm{d}\tilde{\omega}, \text{ for } n \geqq 1. \tag{13.33}$$

In agreement with the possibility of gauge transformations, $\hat{U}_0^0$ may be chosen arbitrarily.

With regard to the relations occuring here we note that, for any $\boldsymbol{U}$ and $\boldsymbol{F}$,

$$\boldsymbol{r} \cdot \operatorname{curl} \boldsymbol{U} = -\operatorname{div} (\boldsymbol{r} \times \boldsymbol{U}) \tag{13.34}$$

and

$$\oiint (\boldsymbol{r} \cdot \operatorname{curl} \boldsymbol{U})\, F\, \mathrm{d}\tilde{\omega} = -\oiint \operatorname{div} (\boldsymbol{r} \times \boldsymbol{U})\, F\, \mathrm{d}\tilde{\omega}$$
$$= -\oiint \boldsymbol{U} \cdot (\boldsymbol{r} \times \nabla F)\, \mathrm{d}\tilde{\omega}. \tag{13.35}$$

CHAPTER 14

A SIMPLE MODEL OF AN α-EFFECT DYNAMO

14.1. Description of the model

In this chapter we shall deal with an extremely simple spherical model of a turbulent dynamo. We consider a sphere of an electrically conducting medium which shows no mean motion but only a homogeneous isotropic non-mirror-symmetric steady turbulence. It is assumed that the influence of this turbulence on the electromagnetic fields is described correctly simply by the turbulent electrical conductivity and the α-effect. This presupposes, in particular, that the scales of the turbulence are sufficiently small compared with those of the mean magnetic field. All space outside the conducting medium is again assumed to be vacuum.

At first the model shall be considered from the kinematic point of view, i.e. no influence of the magnetic field on the motion is taken into account. Only at the end of this chapter we shall modify this model by admitting a back-reaction of the magnetic field on the turbulence and, therefore, on the α-effect.

Up to now the model envisaged has been treated only with regard to axisymmetric magnetic fields. Within this frame KRAUSE and STEENBECK [2] (1967) dealt with the steady case, VOIGTMANN [1] (1968) also studied the non-steady case, and RÜDIGER [1] (1973) made a first step towards including the back-reaction of the magnetic field on the turbulence. In the following we shall generalize these considerations by admitting arbitrary magnetic fields.

Of course, our model is rather unrealistic. It is hard to imagine a mechanism which would create and maintain a homogeneous isotropic non-mirrorsymmetric turbulence. Besides, even if such turbulence exists in some inner regions of the sphere, it must be accompanied by an inhomogeneous anisotropic turbulence near the boundary. In this way other influences on the electromagnetic fields will occur, which are, however, ignored here. Nevertheless it is very instructive to study this model. Since it allows an analytical treatment it is well suited to demonstrate some features of turbulent dynamos. Thus it provides some orientation for the investigation of more realistic models, which in all cases requires extensive numerical computations.

14.2. Basic equations and their reduction to equations for scalar functions

We now specify the basic equations (12.2)...(12.5) for models of turbulent dynamos to the special situations considered here. Again, a spherical coordinate system with the coordinates r, ϑ, and φ and the radius vector $\boldsymbol{r}$ is used. The origin of this system coincides with the centre of the sphere occupied by the conducting medium. The radius of the sphere is denoted by R. As far as the space inside the sphere is concerned we put $\overline{\boldsymbol{u}} = 0$ and $\mathfrak{E}^+ = \alpha \overline{\boldsymbol{B}}$, and we suppose σ_T and α to be constant. In this way we obtain the equations

$$\frac{1}{\mu\sigma_T} \Delta \overline{\boldsymbol{B}} + \alpha \operatorname{curl} \overline{\boldsymbol{B}} - \frac{\partial \overline{\boldsymbol{B}}}{\partial t} = 0, \tag{14.1a}$$

$$\operatorname{div} \overline{\boldsymbol{B}} = 0, \qquad \text{for } r < R. \tag{14.1b}$$

We have to supplement them by the equations

$$\operatorname{curl} \overline{\boldsymbol{B}} = 0, \quad \operatorname{div} \overline{\boldsymbol{B}} = 0, \text{ for } r > R, \tag{14.2a, b}$$

and the conditions

$$[\overline{\boldsymbol{B}}] = 0, \quad \text{on } r = R, \tag{14.3a}$$

$$\overline{\boldsymbol{B}} = O(r^{-3}), \text{ as } r \to \infty. \tag{14.3b}$$

When solving these equations we want to recall the mathematical considerations of the foregoing chapter. We shall utilize the basic ideas of sections 13.3., 13.4., and 13.5., especially the statements (i)...(vii), without special reference. For simplicity, instead of $\overline{\boldsymbol{B}}$ we shall write $\boldsymbol{B}$ in the following.

At first we split $\boldsymbol{B}$ into a toroidal part, $\boldsymbol{B}_t$, and a poloidal part, $\boldsymbol{B}_p$, so that

$$\boldsymbol{B} = \boldsymbol{B}_t + \boldsymbol{B}_p. \tag{14.4}$$

By reason of (14.1b) and (14.2b) we then put

$$\boldsymbol{B}_p = \operatorname{curl} \boldsymbol{A}_t, \tag{14.5}$$

with $\boldsymbol{A}_t$ being a vector potential which is purely toroidal.

We focus attention on equation (14.1a). Inserting (14.4) and separating the toroidal and the poloidal parts of this equation we obtain

$$\frac{1}{\mu\sigma_T} \Delta \boldsymbol{B}_t + \alpha \operatorname{curl} \boldsymbol{B}_p - \frac{\partial \boldsymbol{B}_t}{\partial t} = 0, \tag{14.6a}$$

$$\frac{1}{\mu\sigma_T} \Delta \boldsymbol{B}_p + \alpha \operatorname{curl} \boldsymbol{B}_t - \frac{\partial \boldsymbol{B}_p}{\partial t} = 0, \text{ for } r < R. \tag{14.6b}$$

By virtue of (14.5) the first equation may be rewritten as

$$\frac{1}{\mu\sigma_T}\Delta(\boldsymbol{B}_t - \mu\sigma_T\alpha\boldsymbol{A}_t) - \frac{\partial \boldsymbol{B}_t}{\partial t} = 0, \text{ for } r < R, \tag{14.7a}$$

and the second one allows an integration which results in

$$\frac{1}{\mu\sigma_T}\Delta\boldsymbol{A}_t + \alpha\boldsymbol{B}_t - \frac{\partial \boldsymbol{A}_t}{\partial t} = 0, \text{ for } r < R. \tag{14.7b}$$

Likewise, equation (14.2a) leads to

$$\boldsymbol{B}_t = 0, \quad \Delta\boldsymbol{A}_t = 0, \text{ for } r > R. \tag{14.8a, b}$$

The conditions (14.3) result in

$$\boldsymbol{B}_t = 0, \quad [\text{curl } \boldsymbol{A}_t] = 0, \text{ on } r = R, \tag{14.9a, b}$$

$$\boldsymbol{A}_t = O(r^{-2}), \text{ as } r \to \infty. \tag{14.9c}$$

Let us now represent $\boldsymbol{B}_t$ and $\boldsymbol{A}_t$ by

$$\boldsymbol{B}_t = -\boldsymbol{r}\times\nabla T, \tag{14.10a}$$

$$\boldsymbol{A}_t = -\boldsymbol{r}\times\nabla S, \tag{14.10b}$$

where T and S are single-valued scalar functions normalized by

$$\oiint T \, \mathrm{d}\tilde{\omega} = \oiint S \, \mathrm{d}\tilde{\omega} = 0. \tag{14.11}$$

It should be noted that by combining (14.4), (14.5), and (14.10) one obtains

$$\boldsymbol{B} = -\boldsymbol{r}\times\nabla T - \boldsymbol{r}\,\Delta S + \nabla \frac{\partial}{\partial r}(rS). \tag{14.12}$$

Again we turn to equations (14.7). Using (14.10) we are able to reduce them to

$$\boldsymbol{r}\times\nabla\left\{\frac{1}{\mu\sigma_T}\Delta(T - \mu\sigma_T\alpha S) - \frac{\partial T}{\partial t}\right\} = 0, \tag{14.13a}$$

$$\boldsymbol{r}\times\nabla\left\{\frac{1}{\mu\sigma_T}\Delta S + \alpha T - \frac{\partial S}{\partial t}\right\} = 0, \text{ for } r < R. \tag{14.13b}$$

From this we may at first conclude that the expressions $\{\cdots\}$ are functions of r and t alone. By reason of (13.18) and (14.11) these functions, however, turn out to be zero. Thus we have

$$\frac{1}{\mu\sigma_T}\Delta(T - \mu\sigma_T\alpha S) - \frac{\partial T}{\partial t} = 0, \tag{14.14a}$$

$$\frac{1}{\mu\sigma_T}\Delta S + \alpha T - \frac{\partial S}{\partial t} = 0, \text{ for } r < R. \tag{14.14b}$$

Analogously, from (14.8) we obtain

$$T = 0, \quad \Delta S = 0, \quad \text{for } r > R. \tag{14.15a, b}$$

Finally, (14.9) provides for

$$T = 0, \quad [S] = [\partial S/\partial r] = 0, \text{ on } r = R, \tag{14.16a, b}$$

$$S = O(r^{-2}), \text{ as } r \to \infty. \tag{14.16c}$$

We introduce dimensionless coordinates,

$$x = r/R, \quad \tau = t/\mu\sigma_T R^2, \tag{14.17a, b}$$

and a dimensionless parameter,

$$C = \mu\sigma_T \alpha R, \tag{14.18}$$

which represents a measure for the α-effect. In addition, we define an operator, D_n, by

$$D_n f = \frac{1}{x^2}\left(\frac{\partial}{\partial x}\left(x^2 \frac{\partial f}{\partial x}\right) - n(n+1)\, f\right), \tag{14.19}$$

with f being any function and n an integer.

Furthermore, we represent the functions T and S by series of spherical harmonics,

$$T = \sum_{n,m} T_n^m(x, \tau)\, Y_n^m(\vartheta, \varphi), \tag{14.20a}$$

$$S = R \sum_{n,m} S_n^m(x, \tau)\, Y_n^m(\vartheta, \varphi). \tag{14.20b}$$

The factor R accomplishes the agreement of the dimensions of the T_n^m and S_n^m. As for the Y_n^m we refer to (13.24) ... (13.27). The summation is over all n and m with $n \geqq 1$ and $|m| \leqq n$; because of (14.11) $n = 0$ has to be excluded. Since T and S should be real quantities we require $T_n^{-m} = T_n^{m*}$ and $S_n^{-m} = S_n^{m*}$.

The equations (14.14) for T and S now reduce to

$$D_n(T_n^m - C S_n^m) - \partial T_n^m/\partial\tau = 0, \tag{14.21a}$$

$$D_n S_n^m + C T_n^m - \partial S_n^m/\partial\tau = 0, \qquad \text{for } 0 \leqq x \leqq 1. \tag{14.21b}$$

From (14.15) and (14.16c) follows

$$T_n^m = 0, \quad S_n^m = a_n^m x^{-(n+1)}, \text{ for } x \geqq 1, \tag{14.22a, b}$$

with a_n^m being independent of x. Then (14.16a) is equivalent to

$$T_n^m = 0, \text{ at } x = 1. \tag{14.23}$$

From (14.16b) one gets

$$S_n^m = a_n^m, \quad \partial S/\partial x = -(n+1)\, a_n^m, \text{ at } x = 1, \tag{14.24a, b}$$

and, by eliminating a_n^m,

$$\partial S_n^m/\partial x + (n+1)\, S_n^m = 0, \text{ at } x = 1. \tag{14.25}$$

As a result of our deductions the problem of solving the system of partial differential equations (14.1) and (14.2) with the conditions (14.3) governing the vector field $\boldsymbol{B}$ for all space is now reduced to the problem of solving the differential equations (14.21) with the boundary conditions (14.23) and (14.25) determining the scalar functions T_n^m and S_n^m in $0 \leqq x \leqq 1$.

It is noteworthy that neither the equatons (14.21) nor the conditions (14.23) and (14.25) provide any coupling between pairs of T_n^m and S_n^m which differ in n or m. In addition to this, these equations and conditions do not depend on m.

Finally, we want to point out that the basic equations for $\boldsymbol{B}$ are only justified if the scales of $\boldsymbol{B}$ exceed certain bounds determined by the scales of turbulence. This requires a restriction to sufficiently small n and m. The physical sense of our considerations becomes questionable if n and m are too large.

14.3. The steady case

We ask at first for steady $\boldsymbol{B}$-fields. In this case the equations (14.20) reduce to

$$D_n(T_n^m - C S_n^m) = 0, \tag{14.26a}$$

$$D_n S_n^m + C T_n^m = 0, \qquad \text{for } 0 \leqq x \leqq 1. \tag{14.26b}$$

Again the conditions (14.23) and (14.25) have to be added. In this way an eigenvalue problem results, in which C is the eigenvalue parameter.

Until further notice we consider the T_n^m and S_n^m for an arbitrary but fixed pair of n and m. The case $C = 0$ shall at first be excluded.

From (14.26) we may derive an equation for T_n^m alone, namely

$$(D_n + C^2)\, T_n^m = 0, \text{ for } 0 \leqq x \leqq 1. \tag{14.27}$$

Its only solution without singularities in the interval considered reads

$$T_n^m = (c_n^m/\sqrt{x}) J_{n+1/2}(Cx), \tag{14.28}$$

where c_n^m is a constant and $J_{n+1/2}$ is the Bessel function of the order $n + 1/2$. Condition (14.23) requires $c_n^m = 0$ unless

$$J_{n+1/2}(C) = 0. \tag{14.29}$$

In the case $c_n^m = 0$ we may easily conclude from (14.22)...(14.26) that $T_n^m = S_n^m = 0$ for all x. We pay special attention to the case $c_n^m \neq 0$ which only occurs if equation (14.29) is satisfied. By this equation discrete values of C are defined which represent the eigenvalues of our problem. Until further notice we suppose that C is specified to be one of these eigenvalues. Using now equation (14.26a), and again excluding singularities, we obtain

$$S_n^m = \frac{c_n^m}{C\sqrt{x}} J_{n+1/2}(Cx) + d_n^m x^n, \text{ for } 0 \leqq x \leqq 1, \tag{14.30}$$

where d_n^m is a new constant. In order to satisfy condition (14.25) this constant must be properly fixed. In doing so and using general relations for Bessel functions we get

$$S_n^m = \frac{c_n^m}{C}\left(\frac{1}{\sqrt{x}} J_{n+1/2}(Cx) - \frac{C}{2n+1} J_{n-1/2}(C)\, x^n\right), \quad \text{for} \leqq x \leqq 1. \tag{14.31}$$

Referring to the relations (14.22) and (14.24a), we supplement (14.28), (14.29) and (14.31) with

$$T_n^m = 0, \tag{14.32a}$$

$$S_n^m = -\frac{c_n^m}{2n+1} J_{n-1/2}(C)\, x^{-(n+1)}, \text{ for } x \geqq 1. \tag{14.32b}$$

For the case $C = 0$, which is again included in the following considerations, we easily find that $T_n^m = S_n^m = 0$ for all x.

We return to the series for T and S given by (14.20) where, of course, again all n and m with $n \geqq 1$ and $|m| \geqq n$ are to be considered. As long as C does not coincide with one of the eigenvalues defined by (14.29), T and S vanish everywhere. Only if C equals one of these eigenvalues, T and S may differ from zero. According to the properties of the $J_{n+1/2}$ the eigenvalues for different n are always different from each other. Therefore, in the series for T and S the summation over n can be omitted; there are only contributions from one special n. Clearly, the eigenvalues for a given n but different m coincide, i.e. there is a degeneration with respect to m. Therefore, T and S may remain sums over terms with different m.

We see from these calculations that stationary $\boldsymbol{B}$-fields are only possible if the parameter C, which is a measure for the α-effect, takes special values, namely the eigenvalues mentioned. These $\boldsymbol{B}$-fields correspond to multipoles of a given order which depends on the special eigenvalue. The presence of a $\boldsymbol{B}$-mode of the order n excludes the occurrence of modes with other n. However, the coexistence of modes with the same n but different m is possible.

Discussing now some details, we restrict ourselves at first to the case $\alpha \geqq 0$ so that $C \geqq 0$. We again consider the eigenvalue equation (14.29). According to the properties of the $J_{n+1/2}$, for any given n we have an infinite sequence of eigenvalues which we denote by C_{nl} with $l = 1, 2, \ldots$ We arrange them such that the C_{nl} grow monotonically with l. Again referring to the properties of the $J_{n+1/2}$ we may then state that the C_{nl}, for any fixed l, also grow monotonically with n. Therefore, C_{11} turns out to be the smallest of all values C_{nl}.

Let us discuss the case $n = 1$. The related $\boldsymbol{B}$-modes are dipole fields, where those for $m = 0$ and for $m = \pm 1$ differ only by a rotation about an axis running through the origin of the coordinate system. From (14.29), which for $n = 1$ reduces to

$$(1/C_{1l}) \tan C_{1l} - 1 = 0, \quad C_{1l} > 0, \tag{14.33}$$

we get

$$C_{11} = 4.4934, \quad C_{12} = 7.7253, \tag{14.34a, b}$$

$$C_{13} = 10.9041, \quad C_{14} = 14.0662. \tag{14.34c, d}$$

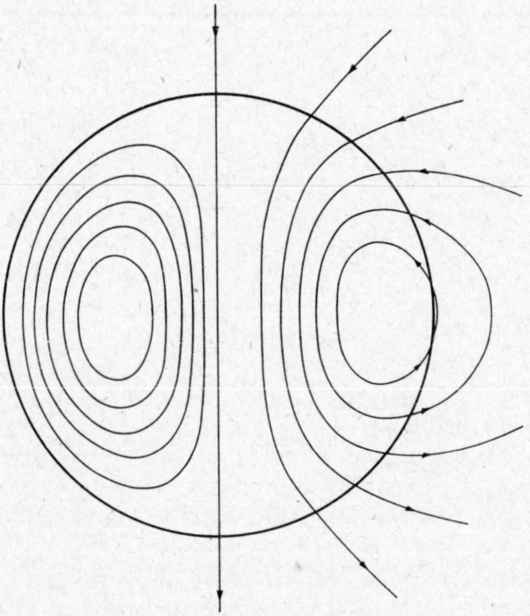

Fig. 14.1. The steady $\boldsymbol{B}$-field in the case $n = l = 1$. The left-hand side of the picture shows the lines of constant magnitude of the toroidal field, the right-hand side field lines of the poloidal field. (After KRAUSE and STEENBECK [2])

By means of (14.12), (14.17), (14.28), (14.31), and (14.32) the corresponding $\boldsymbol{B}$-fields can be calculated. For that belonging to $l = 1$ the field lines are given in figure 14.1. In the case of $l > 1$ the field patterns inside the sphere are more complicated.

We now consider the case of $n > 1$ and, therefore, $\boldsymbol{B}$-modes which correspond to quadrupole fields, octupole fields, etc. Such modes with $m = 0$, $m = \pm 1, \ldots$ possess quite different field structures which can become congruent after rotation only in special cases. In this respect it is remarkable that the eigenvalues are independent of m. On the basis of (14.29) it results

$$C_{21} = 5.7635, \quad C_{22} = 9.0950, \tag{14.35a, b}$$

$$C_{23} = 12.3229, \quad C_{24} = 15.5146. \tag{14.35c, d}$$

Obviously, the equations (14.26) and the corresponding boundary conditions remain valid if, at the same time, the signs of C and of either T or S are changed. Consequently, for each positive eigenvalue C_{nl} there exists a negative one, namely $-C_{nl}$. Starting from a $\boldsymbol{B}$-field belonging to C_{nl} we get a field belonging to $-C_{nl}$ by changing the sign of either $\boldsymbol{B}_\mathrm{t}$ or $\boldsymbol{B}_\mathrm{p}$.

14.4. The non-steady case

Now the initial value problem for the $\boldsymbol{B}$-field shall be considered. We suppose that $\boldsymbol{B}(\boldsymbol{x}, 0)$ is given and ask for $\boldsymbol{B}(\boldsymbol{x}, t)$ for $t > 0$. Due to the foregoing deductions this problem can also be formulated in terms of the T_n^m and S_n^m. We then suppose all $T_n^m(x, 0)$ and $S_n^m(x, 0)$ to be given and ask for $T_n^m(x, \tau)$ and $S_n^m(x, \tau)$ for $\tau > 0$.

In preparation of this, a simpler problem shall be solved. We look for $\boldsymbol{B}$-fields with a purely exponential dependence on time, i.e. for such of the form $\boldsymbol{B}(\boldsymbol{x}, 0)\, \mathrm{e}^{\lambda t}$ with some real constant λ. To this end we put, by way of trial,

$$T_n^m = T_{nl}^m(x) \exp\{\lambda_{nl}\tau\}, \tag{14.36a}$$

$$S_n^m = S_{nl}^m(x) \exp\{\lambda_{nl}\tau\}. \tag{14.36b}$$

This notation implies that, if such T_n^m and S_n^m are possible, the λ_{nl} may not depend on m. The index l will be specified later.

We now refer to the equations (14.21). Inserting (14.36) we obtain

$$D_n(T_{nl}^m - CS_{nl}^m) - \lambda_{nl}T_{nl}^m = 0, \tag{14.37a}$$

$$D_nS_{nl}^m + CT_{nl}^m - \lambda_{nl}S_{nl}^m = 0, \text{ for } 0 \leqq x \leqq 1. \tag{14.37b}$$

The general solution which is non-singular in the interval considered reads

$$T_{nl}^m = \frac{c_{nl}^{(1)m}}{\sqrt{x}} J_{n+1/2}(u_{nl}^{(1)}x) + \frac{c_{nl}^{(2)m}}{\sqrt{x}} J_{n+1/2}(u_{nl}^{(2)}x), \tag{14.38a}$$

$$S_{nl}^m = \frac{c_{nl}^{(1)m}}{u_{nl}^{(1)}\sqrt{x}} J_{n+1/2}(u_{nl}^{(1)}x) + \frac{c_{nl}^{(2)m}}{u_{nl}^{(2)}\sqrt{x}} J_{n+1/2}(u_{nl}^{(2)}x), \text{ for } 0 \leqq x \leqq 1, \tag{14.38b}$$

where $c_{nl}^{(1)m}$ and $c_{nl}^{(2)m}$ are arbitrary constants and

$$u_{nl}^{(1,2)} = \frac{1}{2}\,(C \pm \sqrt{C^2 - 4\lambda_{nl}})\,. \tag{14.39}$$

The conditions (14.23) and (14.25) together with special relations for Bessel functions result in

$$c_{nl}^{(1)m} J_{n+1/2}(u_{nl}^{(1)}) + c_{nl}^{(2)m} J_{n+1/2}(u_{nl}^{(2)}) = 0\,, \tag{14.40a}$$

$$c_{nl}^{(1)m} J_{n-1/2}(u_{nl}^{(1)}) + c_{nl}^{(2)m} J_{n-1/2}(u_{nl}^{(2)}) = 0\,. \tag{14.40b}$$

We only look for non-trivial T_{nl}^m and S_{nl}^m. More precisely, we require T_{nl}^m and S_{nl}^m, i.e. $c_{nl}^{(1)m}$ and $c_{nl}^{(2)m}$, not to vanish simultaneously for any fixed n and m. In this case, (14.40) implies

$$J_{n+1/2}(u_{nl}^{(1)})\, J_{n-1/2}(u_{nl}^{(2)}) - J_{n-1/2}(u_{nl}^{(1)})\, J_{n+1/2}(u_{nl}^{(2)}) = 0\,. \tag{14.41}$$

If supplemented by (14.39), this relation determines an infinite set of values λ_{nl} for each given n and each C. We now number the λ_{nl} and the corresponding T_{nl}^m and S_{nl}^m by $l = 1, 2, \ldots$ in such way that the λ_{nl} for any fixed n grow with l. In order to evaluate the λ_{nl} as functions of C it is useful at first to determine pairs $u_{nl}^{(1)}$ and $u_{nl}^{(2)}$. Then related values of λ_{nl} and C result from the formulae

$$\lambda_{nl} = u_{nl}^{(1)} u_{nl}^{(2)}\,, \tag{14.42a}$$

$$C = u_{nl}^{(1)} + u_{nl}^{(2)}\,, \tag{14.42b}$$

which are a consequence of (14.39).

On this basis we may specify the ***B***-modes with a purely exponential dependence on time.

In the special case $C = 0$ these modes coincide with the modes of free decay. For this case, referring to (14.42b), we put $u_{nl}^{(1)} = -u_{nl}^{(2)} = u_{nl}$, where $u_{nl} > 0$. Regarding the properties of $J_{n-1/2}$ and $J_{n+1/2}$ we then may replace (14.41) by

$$J_{n-1/2}(u_{nl})\, J_{n+1/2}(u_{nl}) = 0\,, \tag{14.43}$$

and (14.42a) by

$$\lambda_{nl} = -(u_{nl})^2 < 0\,. \tag{14.44}$$

For the dipole modes, i.e. $n = 1$, we obtain

$$\sin u_{1l}((1/u_{1l}) \tan u_{1l} - 1) = 0, \tag{14.45}$$

leading to

$$\lambda_{11} = -\pi^2, \qquad \lambda_{12} = -(4.4934)^2, \tag{14.46a, b}$$

$$\lambda_{13} = -(2\pi)^2, \qquad \lambda_{14} = -(7.7253)^2. \tag{14.46c, d}$$

Similarly, for the quadrupole modes, i.e. $n = 2$, we obtain

$$\lambda_{21} = -(4.4934)^2, \qquad \lambda_{22} = -(5.7635)^2, \tag{14.47a, b}$$

$$\lambda_{23} = -(7.7253)^2, \qquad \lambda_{24} = -(9.0950)^2. \tag{14.47c, d}$$

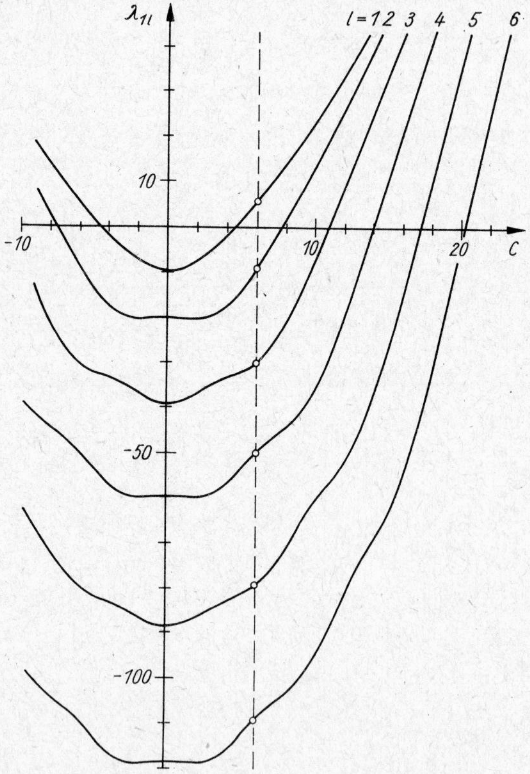

Fig. 14.2. The dependence of the λ_{nl} on C for $n = 1$ and $l = 1, 2, \ldots, 6$. The λ_{nl} are symmetric in C. The dotted line corresponds to a special value of C used in the example considered in the text. (After VOIGTMANN [1])

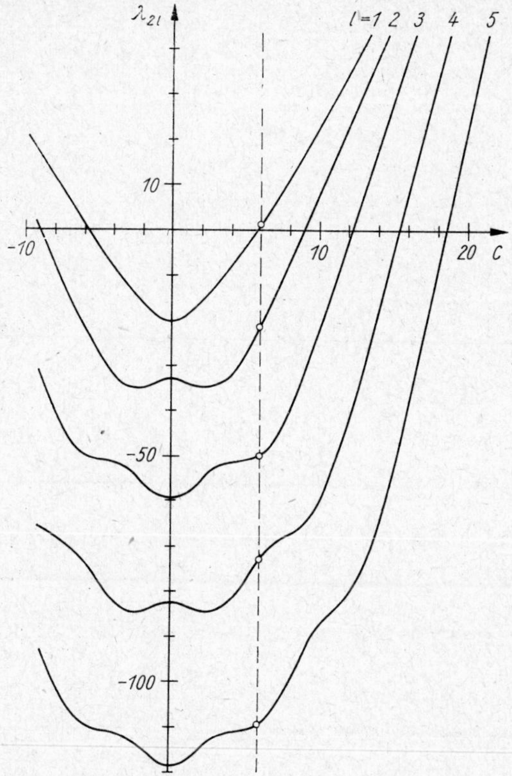

Fig. 14.3. The dependence of the λ_{nl} on C for $n = 2$ and $l = 1, 2, \ldots, 5$. The λ_{nl} are symmetric in C. The dotted line corresponds to a special value of C used in the example considered in the text. (After VOIGTMANN [1])

In the case $C \neq 0$ the $\boldsymbol{B}$-fields are influenced by the α-effect. Then the λ_{nl} are not necessarily negative. For sufficiently large $|C|$ positive λ_{nl} may also occur, and the corresponding modes will grow in time. As for the dipole and quadrupole modes, i.e. $n = 1$ and $n = 2$, the dependence of the λ_{nl} on C is shown in figures 14.2 and 14.3. Obviously, $\lambda_{nl} = 0$ corresponds to the steady modes, for which $C = C_{nl}$. Comparing the λ_{nl} at any fixed C for different modes we find that they are the larger the smaller the $|C_{nl}|$ related to these modes. Therefore a mode belonging to a given $|C_{nl}|$ will grow more slowly, or decay more rapidly, than one with a smaller $|C_{nl}|$.

For any fixed C the λ_{nl} increase if n or l grows. Therefore, the $\boldsymbol{B}$-modes with smaller n and l decay more slowly, or grow more rapidly, than those with higher

n and l. This becomes clear if the connection between n and l and the geometrical structure of the $\boldsymbol{B}$-fields is considered. Obviously, n is a measure for the variations with ϑ and φ. More precisely, for $m = 0$ there are only variations with ϑ, and if n is fixed but $|m|$ grows, those with ϑ become smaller and those with φ larger. Furthermore, l is a measure for the variations with r. If l grows, the dependence on r becomes more complex. The stronger the spatial variations of a $\boldsymbol{B}$-field the more effective is the Ohmic dissipation, i.e. the destructive influence on the field.

We now return to the initial value problem for the $\boldsymbol{B}$-field posed above. It can be shown that it has a unique solution. On grounds of the foregoing considerations all basic relations for $\boldsymbol{B}$, i.e. (14.1)...(14.3), are satisfied if we put

$$T_n^m = \sum_l T_{nl}^m(x) \exp\{\lambda_{nl}\tau\}, \tag{14.48a}$$

$$S_n^m = \sum_l S_{nl}^m(x) \exp\{\lambda_{nl}\tau\}, \text{ for } 0 \leqq x \leqq 1, \tag{14.48b}$$

and supplement them according to (14.20) and (14.22a) with suitable relations for $x > 1$. Obviously, (14.48) already constitutes the solution of the initial value problem provided that $T_n^m(x, 0)$ and $S_n^m(x, 0)$, which must be considered as given but, of course, satisfying (14.23) and (14.25), may be represented by

$$T_n^m(x, 0) = \sum_l T_{nl}^m(x), \tag{14.49a}$$

$$S_n^m(x, 0) = \sum_l S_{nl}^m(x), \text{ for } 0 \leqq x \leqq 1; \tag{14.49b}$$

the relations for $x > 1$ cause no difficulties. In the case $C = 0$ it can easily be seen that the representations (14.49) are always possible. For $C \neq 0$ a proof seems to be more complicated.

Without going into detail we assume that our solution of the initial value problem for the $\boldsymbol{B}$-field indeed applies to all initial fields. We envisage a special value of C so that, e.g., λ_{11} and λ_{21} are positive whereas all other λ_{nl} are negative; this situation corresponds to the dotted lines in figures 14.2 and 14.3. Let us also suppose that in the initial fields particularily those modes are present which belong to λ_{11} and λ_{21}, i.e. special dipole and quadrupole modes. Then only these modes will grow whereas the others will decay. Since $\lambda_{11} > \lambda_{21}$, the modes belonging to λ_{11}, i.e. the dipole modes, will grow more rapidly than those belonging to λ_{21}, i.e. the quadrupole modes. Independent of the inital ratio of the magnitudes, the dipole modes will dominate after a certain time. In other words, those modes will dominate which are related to the smallest value of $|C_{nl}|$.

This finding will be of interest for other models too. As long as it is true that the λ_{nl} for different modes are the larger the smaller the corresponding $|C_{nl}|$, the modes with the smallest $|C_{nl}|$ will generally dominate.

The treatment of the kinematic dynamo model considered so far may be understood as a stability problem. The possibility of growing magnetic fields corresponds to an instability. In terms of the stability theory we have stability if $\lambda_{11} < 0$, marginal stability if $\lambda_{11} = 0$, and instability if $\lambda_{11} > 0$. In this sense C_{11} shall be called the marginal value of C.

14.5. Considerations involving the back-reaction of the magnetic field on the motions

So far our model has been considered only from the kinematic point of view. In particular, the influence of the Lorentz force on the motion of the medium has been ignored. Since the Lorentz force is proportional to the square of the magnetic field it may be neglected as long as the field is sufficiently small, but it becomes very important if the field grows. The mean part of the Lorentz force will give rise to a mean motion so that $\overline{\boldsymbol{u}}$ may no longer be supposed to be zero. The fluctuating part will influence the turbulence and, therefore, the electromotive force $\mathfrak{E}$, which loses its linearity with respect to $\boldsymbol{B}$. Thus we are confronted with a rather complex situation.

In treating our model we do not intend to consider every effect of the magnetic field. In order to point out some special aspects drastic simplifications shall be introduced. As before, no mean motion is taken into account. As for the effects of turbulence we modify earlier assumptions only by admitting that α may depend on $\boldsymbol{B}$, whereas σ_T continues to be treated as a constant. Only the steady case is considered. Then, instead of equation (14.1a), we have

$$\frac{1}{\mu\sigma_{\mathrm{T}}}\Delta\boldsymbol{B} + \operatorname{curl}(\alpha\boldsymbol{B}) = 0, \text{ for } r < R. \tag{14.50}$$

Due to the dependence of α on $\boldsymbol{B}$ this equation is non-linear with respect to $\boldsymbol{B}$. The relations (14.1b)...(14.3b) remain unchanged. We assume that the $\boldsymbol{B}$-field is small, and write

$$\alpha = \alpha_0 - \alpha_B\boldsymbol{B}^2, \tag{14.51}$$

with α_0 and α_B being constants. Presupposing that $|\alpha|$ decreases with growing $|\boldsymbol{B}|$ we require that α_0 and α_B have the same signs; for simplicity we only consider $\alpha_0, \alpha_B > 0$. Furthermore, we only deal with the case where α_0 slightly exceeds α_{11} but no other α_{nl}, where the α_{nl} are defined by $\mu\sigma_{\mathrm{T}}\alpha_{nl}R = C_{nl}$.

Hence,

$$\alpha_0 = \alpha_{11} + \delta\alpha, \tag{14.52}$$

with $\delta\alpha$ being a small positive constant.

Clearly, $\boldsymbol{B}$ must tend to zero as $\delta\alpha$ vanishes. Therefore, it seems to be reasonable to put

$$\boldsymbol{B} = \left(\frac{\delta\alpha}{\alpha_B}\right)^{1/2} \boldsymbol{b} \tag{14.53}$$

and to assume that $\boldsymbol{b}$ can be represented as a power series of the form

$$\boldsymbol{b} = \boldsymbol{b}^{(0)} + \varepsilon \boldsymbol{b}^{(1)} + \varepsilon^2 \boldsymbol{b}^{(2)} + \ldots, \tag{14.54}$$

where ε is a small parameter defined by

$$\varepsilon = \delta\alpha/\alpha_{11}. \tag{14.55}$$

In this way equation (14.50) leads to

$$\frac{1}{\mu\sigma_T} \Delta \boldsymbol{b}^{(0)} + \alpha_{11} \operatorname{curl} \boldsymbol{b}^{(0)} = 0, \tag{14.56a}$$

$$\frac{1}{\mu\sigma_T} \Delta \boldsymbol{b}^{(1)} + \alpha_{11} \operatorname{curl} \boldsymbol{b}^{(1)} = -\alpha_{11} \operatorname{curl}\left((1 - \boldsymbol{b}^{(0)2})\, \boldsymbol{b}^{(0)}\right), \tag{14.56b}$$

$$\frac{1}{\mu\sigma_T} \Delta \boldsymbol{b}^{(2)} + \alpha_{11} \operatorname{curl} \boldsymbol{b}^{(2)}$$
$$= -\alpha_{11} \operatorname{curl}\left((1 - \boldsymbol{b}^{(0)2})\, \boldsymbol{b}^{(1)} - 2(\boldsymbol{b}^{(0)} \cdot \boldsymbol{b}^{(1)})\, \boldsymbol{b}^{(0)}\right). \tag{14.56c}$$

These equations have to be supplemented by corresponding relations resulting from (14.1b)...(14.3c).

At first the equations for $\boldsymbol{b}^{(0)}$ are considered. They possess a non-trivial solution which can be extracted from section 14.3.; it is fixed apart from a constant factor. As for the equations for $\boldsymbol{b}^{(1)}$, which show an inhomogeneity containing $\boldsymbol{b}^{(0)}$, we point out that the corresponding homogeneous equations also possess a non-trivial solution. Therefore, the complete equations allow such solution only if the inhomogeneity satisfies a certain condition which results in a requirement for $\boldsymbol{b}^{(0)}$. In this way $\boldsymbol{b}^{(0)}$ is fixed. Of course, $\boldsymbol{b}^{(1)}$ again contains a free constant. The equations for $\boldsymbol{b}^{(2)}$, however, pose a condition by which $\boldsymbol{b}^{(1)}$ is fixed. In this way, the terms in the expansion (14.54) can be determined successively.

As far as the details of these calculations are concerned we refer to the paper by Rüdiger [1]. Explaining his result we again rely on the representation of the $\boldsymbol{B}$-field given by (14.4), (14.5), (14.10), (14.11), and (14.17). We first consider the zero order approximation, where in the result for $\boldsymbol{B}$ the expansion (14.54) for $\boldsymbol{b}$ is reduced to the term $\boldsymbol{b}^{(0)}$; this corresponds to $\varepsilon = 0$. According to the above assumptions the expansions (14.20) for T and S can only have terms with $n = 1$, and we choose $m = 0$, which is possible here without any loss of generality. Clearly, the $\boldsymbol{B}$-field is a dipole field. Of course, T_1^0 and S_1^0 are again given by (14.28), (14.30), (14.31), and (14.32) with $C = C_{11}$, and the field pattern is identical with that given in figure 14.1. We can now add that

$$c_1^0 = z(\delta\alpha/\alpha_B)^{1/2}, \tag{14.57}$$

with z being a dimensionless constant. The numerical calculations have led to

$$z = 1.1525. \tag{14.58}$$

Let us now take the first order approximation where in the result for $\boldsymbol{B}$ not only the term $\boldsymbol{b}^{(0)}$ but also that with $\boldsymbol{b}^{(1)}$ is included, i.e. small positive ε are admitted. In this case T and S have not only terms with $n = 1$ but also such

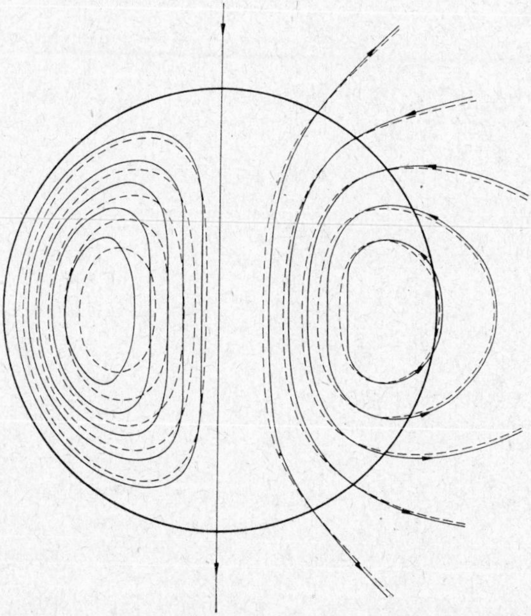

Fig. 14.4. The steady $\boldsymbol{B}$-field in a model with a dependence of α on $\boldsymbol{B}$ in the case $n = l = 1$. The left-hand side of the picture shows the lines of constant magnitude of the toroidal field, the right-hand side field lines of the poloidal field. The dotted lines correspond to $\varepsilon = 0$; they are identical with the full lines of figure 14.1. The full lines correspond to $\varepsilon = 1/C_{11}$. (After RÜDIGER [1])

with $n = 3$, both of which with $m = 0$. Thus the $\boldsymbol{B}$-field is a sum of a dipole and an octupole part. The functions T_1^0, S_1^0, T_3^0, and S_3^0 are already rather complicated and shall not be given here. In figure 14.4 the field pattern for $\varepsilon = 1/C_{11}$, i.e. $\mu\sigma_T\, \delta\alpha R = 1$ is drawn. Compared with the case $\varepsilon = 0$, both the line of maximum toroidal field and the line of zero poloidal field are shifted outward.

The problem considered here, i.e. the case implying a non-linear equation for $\boldsymbol{B}$, has been tackled by other methods too. In a further paper by RÜDIGER [3] the equations for $\boldsymbol{B}$ have been integrated numerically on the basis of a grid-point method.

CHAPTER 15

SPHERICAL MODELS OF TURBULENT DYNAMOS AS SUGGESTED BY COSMICAL BODIES. GENERAL ASPECTS

15.1. General description of the models

We shall now be concerned with spherical dynamo models reflecting essential features of the processes which are assumed to take place in cosmical objects. In all cases again a spherical body consisting of an electrically conducting medium is considered which rotates about a fixed axis running through its centre. Together with the axis of rotation also an equatorial plane is defined. The distribution of the electrical conductivity is supposed to be symmetric with respect to both the axis of rotation and the equatorial plane; later on we shall restrict attention to the case where the conductivity depends on radius only. As far as the mean motion is concerned, in the simplest case only a rigid body rotation is assumed. Yet we also admit a differential rotation with an angular velocity which may depend on radius or latitude, and a meridional circulation. In any case, however, we again require the mean motion to be symmetric with respect to both the axis of rotation and the equatorial plane and, in addition, to be steady. As well as the mean motion, fluctuating or turbulent motions are assumed which result in an electromotive force acting on the mean electromagnetic fields. The distribution of these motions is assumed to show the same symmetries and to be steady. More precisely, we require the invariance of all mean quantities depending on these motions under arbitrary virtual rotations of the fluctuating velocity field about the axis of rotation, under reflection of this field at the equatorial plane, and under time displacements. As for the mean electromotive force we restrict ourselves to such situations in which this quantity is non-zero only in the presence of a mean magnetic field and may be expressed at a given point by the mean magnetic flux density and its first spatial derivatives at this point. All space outside the conducting body is regarded as vacuum.

We shall only deal with kinematic models. The influence of the magnetic field on the motions is ignored. Within this frame we may study the occurrence of dynamos, the excitation conditions and the structures of the magnetic fields in the case of small field strengths. The question of what the behaviour of the dynamos will be at higher field strengths must remain unanswered here.

15.2. Basic equations and some of their symmetry properties

We again start from the general equations (12.1)...(12.5) for models of turbulent dynamos. As in the foregoing chapter a spherical polar coordinate system with the coordinates r, ϑ, and φ and the radius vector $\boldsymbol{r}$ is used. The origin of this system coincides with the centre of the sphere occupied by the conducting medium, and the polar axis with the axis of rotation. In the following, axisymmetry shall always be understood as symmetry with respect to the axis of rotation. The radius of this sphere is denoted by R.

As for the $\overline{\boldsymbol{B}}$-field inside the conducting medium we have the equations

$$\frac{1}{\mu} \operatorname{curl} \left(\frac{1}{\sigma} \operatorname{curl} \overline{\boldsymbol{B}} \right) - \operatorname{curl} (\overline{\boldsymbol{u}} \times \overline{\boldsymbol{B}} + \mathfrak{E}) + \frac{\partial \overline{\boldsymbol{B}}}{\partial t} = 0, \tag{15.1a}$$

$$\operatorname{div} \overline{\boldsymbol{B}} = 0, \text{ for } r < R, \tag{15.1b}$$

where σ, $\overline{\boldsymbol{u}}$ and $\mathfrak{E}$ have to be specified according to the assumptions introduced above. Again, we may replace σ and $\mathfrak{E}$ by σ_T and $\mathfrak{E}^+$. Furthermore, we have the equations

$$\operatorname{curl} \overline{\boldsymbol{B}} = 0, \quad \operatorname{div} \overline{\boldsymbol{B}} = 0, \text{ for } r > R, \tag{15.2a, b}$$

and the conditions

$$[\overline{\boldsymbol{B}}] = 0, \quad \boldsymbol{r} \cdot \overline{\boldsymbol{u}} = 0, \text{ on } r = R, \tag{15.3a, b}$$

$$\overline{\boldsymbol{B}} = O(r^{-3}), \text{ as } r \to \infty. \tag{15.3c}$$

The assumptions on the symmetries of the conductivity distribution and of the motions lead to special properties of the solutions of these equations. Let us at first consider the models dealt with here within the scope of equations (11.1)...(11.4). We pay attention to special rotations and a special reflection of the $\boldsymbol{u}$ and $\boldsymbol{B}$-field occuring there. More precisely, we consider rotations about the polar axis which transform $\boldsymbol{u}$ into $\boldsymbol{u}^{\text{rot}}$, and $\boldsymbol{B}$ into $\boldsymbol{B}^{\text{rot}}$, so that

$$u^{\text{rot}}_{r,\vartheta,\varphi}(r, \vartheta, \varphi) = u_{r,\vartheta,\varphi}(r, \vartheta, \varphi - \varphi_0), \tag{15.4a}$$

$$B^{\text{rot}}_{r,\vartheta,\varphi}(r, \vartheta, \varphi) = B_{r,\vartheta,\varphi}(r, \vartheta, \varphi - \varphi_0), \tag{15.4b}$$

with some angle φ_0, and the reflection at the equatorial plane transforming $\boldsymbol{u}$ into $\boldsymbol{u}^{\text{ref}}$, and $\boldsymbol{B}$ into $\boldsymbol{B}^{\text{ref}}$, so that

$$u^{\text{ref}}_{r,\varphi}(r, \vartheta, \varphi) = u_{r,\varphi}(r, \pi - \vartheta, \varphi), \; u^{\text{ref}}_{\vartheta}(r, \vartheta, \varphi) = -u_{\vartheta}(r, \pi - \vartheta, \varphi), \tag{15.5a, b}$$

$$B^{\text{ref}}_{r,\varphi}(r, \vartheta, \varphi) = -B_{r,\varphi}(r, \pi - \vartheta, \varphi), \; B^{\text{ref}}_{\vartheta}(r, \vartheta, \varphi) = B_{\vartheta}(r, \pi - \vartheta, \varphi). \tag{15.5c, d}$$

Provided that equations (11.1)...(11.4) hold with given $\boldsymbol{u}$ and $\boldsymbol{B}$ they continue to hold if these are replaced by $\boldsymbol{u}^{\mathrm{rot}}$ and $\boldsymbol{B}^{\mathrm{rot}}$ or by $\boldsymbol{u}^{\mathrm{ref}}$ and $\boldsymbol{B}^{\mathrm{ref}}$. Taking the average of (11.1)...(11.4) we get the mean field equations (12.1)...(12.5). Clearly, they must show the same features. According to our assumptions, however, the transition from $\boldsymbol{u}$ to $\boldsymbol{u}^{\mathrm{rot}}$ or $\boldsymbol{u}^{\mathrm{ref}}$ is without any influence on mean fields, in particular on $\overline{\boldsymbol{u}}$, $\mathfrak{E}$, and $\overline{\boldsymbol{B}}$. Consequently, the validity of (12.1)...(12.5) remains unaffected even if only $\boldsymbol{B}$ is replaced by $\boldsymbol{B}^{\mathrm{rot}}$ or $\boldsymbol{B}^{\mathrm{ref}}$. We may suppose that $\boldsymbol{B}$ occurs only in the form of $\overline{\boldsymbol{B}}$; for $\mathfrak{E}$ can be expressed by $\boldsymbol{u}$ and $\overline{\boldsymbol{B}}$. Thus we find that the validity of (12.1)...(12.5) is maintained if $\overline{\boldsymbol{B}}$ is replaced by $\overline{\boldsymbol{B}}^{\mathrm{rot}}$ or $\overline{\boldsymbol{B}}^{\mathrm{ref}}$. Since equations (12.1)...(12.5) essentially coincide with (15.1)...(15.3) we may state that any $\overline{\boldsymbol{B}}$-field which is a solution of equations (15.1)...(15.3) turns again into a solution of these if it is rotated or reflected according to (15.4b) or (15.5c, d).

By reason of the steadiness of the conductivity distribution and the restricted time dependence of the motions an analogous conclusion may be drawn with respect to time displacements.

15.3. Special magnetic field modes

We may decompose any $\overline{\boldsymbol{B}}$-field into special modes. At first we split it in two fields, $\overline{\boldsymbol{B}}_{\mathrm{A}}$ and $\overline{\boldsymbol{B}}_{\mathrm{S}}$, where $\overline{\boldsymbol{B}}_{\mathrm{A}}$ is antisymmetric and $\overline{\boldsymbol{B}}_{\mathrm{S}}$ symmetric with respect to the equatorial plane. Carrying out a harmonic analysis we further decompose each of these fields into modes showing the form of the real or imaginary part of $\boldsymbol{C}^m \mathrm{e}^{\mathrm{i}m\varphi}$ where $\boldsymbol{C}^m$ is a complex axisymmetric vector field, and m is a non-negative integer. Obviously, $m = 0$ corresponds to axisymmetric, $m \neq 0$ to non-axisymmetric fields.

We now assume that the $\overline{\boldsymbol{B}}$-field satisfies the basic equations (15.1)...(15.3). If it is not already in the form of one of these $\overline{\boldsymbol{B}}$-modes it can be decomposed into such modes. Owing to the linearity of these equations and the mentioned properties of their solutions we may conclude that each individual mode represents a solution.

We denote a $\overline{\boldsymbol{B}}$-mode by A or S depending on whether it is antisymmetric or symmetric with respect to the equatorial plane, and we add the number m indicating the dependence on φ. Simple examples of $\overline{\boldsymbol{B}}$-modes of A0, S0, A1, and S1 types may be given by dipole and quadrupole fields. Corresponding field patterns are depicted in figure 15.1. In general, of course, we may not expect those $\overline{\boldsymbol{B}}$-modes which are solutions of (15.1)...(15.3) to be pure dipole or quadrupole fields. These fields will only play the role of leading terms in multipole expansions which also contain higher order terms with the same symmetry properties.

Let us now bear in mind that the $\overline{\boldsymbol{B}}$-field generally varies in time. Extending its decomposition in this respect we arrive at $\overline{\boldsymbol{B}}$-modes showing the form of the

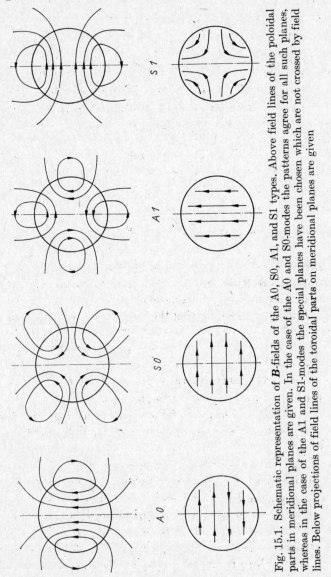

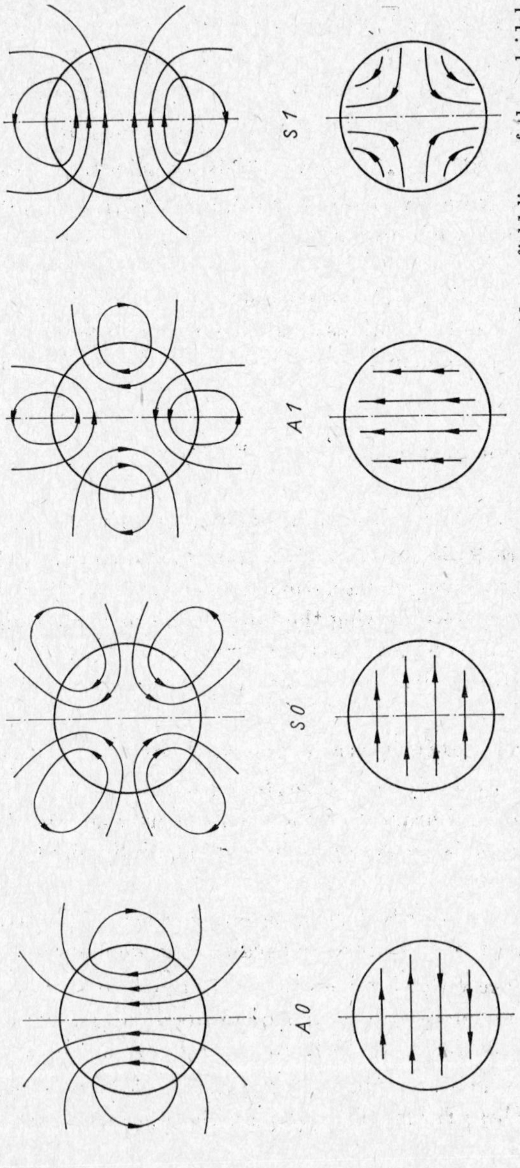

Fig. 15.1. Schematic representation of $\boldsymbol{B}$-fields of the A0, S0, A1, and S1 types. Above field lines of the poloidal parts in meridional planes are given. In the case of the A0 and S0-modes the patterns agree for all such planes, whereas in the case of the A1 and S1-modes the special planes have been chosen which are not crossed by field lines. Below projections of field lines of the toroidal parts on meridional planes are given

real or imaginary part of $\boldsymbol{C}^m \mathrm{e}^{\mathrm{i}m\varphi + pt}$ where $\boldsymbol{C}^m$ now is a complex axisymmetric vector field independent of t, and p is a complex constant.

If a $\overline{\boldsymbol{B}}$-field which is a solution of the basic equations (15.1)...(15.3) does not possess the form of one of the $\overline{\boldsymbol{B}}$-modes introduced here, it may be decomposed into such modes. We may again conclude that each of these modes is also a solution. Investigating dynamo models governed by equations (15.1)...(15.3) we may therefore restrict ourselves to these modes.

In order to study these $\overline{\boldsymbol{B}}$-modes more precisely we put $p = \lambda - \mathrm{i}\Omega$. Then their dependence on φ and t can be expressed by $\cos(m\varphi - \Omega t)\,\mathrm{e}^{\lambda t}$ and $\sin(m\varphi - \Omega t)\,\mathrm{e}^{\lambda t}$. The axisymmetric modes, for which $m = 0$, generally correspond to a growing or damped oscillation; in the case $\Omega = 0$ we have an aperiodic time dependence. Non-axisymmetric modes, for which $m \neq 0$, turn out to be waves migrating in azimuthal direction. In this case Ω depends, of course, on the frame of reference. For a given mode it is always possible to find a frame in which Ω vanishes so that a standing wave appears.

15.4. Specification of the mean velocity field and the turbulent electromotive force

We understand the mean velocity $\overline{\boldsymbol{u}}$ as a sum of two terms,

$$\overline{\boldsymbol{u}} = \overline{\boldsymbol{u}}_\mathrm{t} + \overline{\boldsymbol{u}}_\mathrm{p}, \tag{15.6}$$

where $\overline{\boldsymbol{u}}_\mathrm{t}$ stands for the rotation, and $\overline{\boldsymbol{u}}_\mathrm{p}$ for the meridional circulation. According to the assumptions introduced, both $\overline{\boldsymbol{u}}_\mathrm{t}$ and $\overline{\boldsymbol{u}}_\mathrm{p}$ show symmetry with respect to the axis of rotation and the equatorial plane, i.e. they are invariant under transformations like (15.4a) and (15.5a, b), and are steady. As indicated by our denotation, $\overline{\boldsymbol{u}}_\mathrm{t}$ and $\overline{\boldsymbol{u}}_\mathrm{p}$ are toroidal and poloidal, respectively.

In the following $\overline{\boldsymbol{u}}_\mathrm{t}$ shall be represented in the form

$$\overline{\boldsymbol{u}}_\mathrm{t} = \boldsymbol{\omega} \times \boldsymbol{r}, \quad \boldsymbol{\omega} = \omega \hat{\boldsymbol{\omega}}, \tag{15.7a, b}$$

with ω being an angular velocity and $\hat{\boldsymbol{\omega}}$ the unit vector parallel to the polar axis of the coordinate system. It should be mentioned that in the foregoing chapters the denotations $\boldsymbol{\Omega}$ and Ω have been used instead of $\boldsymbol{\omega}$ and ω. The components of $\overline{\boldsymbol{u}}_\mathrm{p}$ are simply denoted by u_r and u_ϑ. According to (15.3c) we have to require that $u_r = 0$ at $r = R$.

As far as the electromotive force $\mathfrak{E}$ is concerned we may depart from relation (12.7) and specify the coefficients a_{ij} and b_{ijk} according to the special assumptions adopted here. Just as equations (11.1), from which this relation can be deduced, it must be compatible with the rotations and the reflection of the $\boldsymbol{u}$ and $\boldsymbol{B}$-fields as defined by (15.4) and (15.5). Under the rotations and reflection $\mathfrak{E}$ behaves like $\boldsymbol{u}\times\boldsymbol{B}$, and this in turn like $\boldsymbol{u}$; we therefore define $\mathfrak{E}^{\mathrm{rot}}$ and $\mathfrak{E}^{\mathrm{ref}}$ in the same way as $\boldsymbol{u}^{\mathrm{rot}}$ and $\boldsymbol{u}^{\mathrm{ref}}$, respectively. Provided (12.7) holds for given $\mathfrak{E}$, $\boldsymbol{u}$, and $\boldsymbol{B}$, it must continue to hold if they are replaced by $\mathfrak{E}^{\mathrm{rot}}$, $\boldsymbol{u}^{\mathrm{rot}}$, and $\boldsymbol{B}^{\mathrm{rot}}$ or by $\mathfrak{E}^{\mathrm{ref}}$, $\boldsymbol{u}^{\mathrm{ref}}$, and $\boldsymbol{B}^{\mathrm{ref}}$. According to our assumptions the coefficients a_{ij} and b_{ijk}, which depend on $\boldsymbol{u}$ only, cannot be influenced by replacing it by $\boldsymbol{u}^{\mathrm{rot}}$ or $\boldsymbol{u}^{\mathrm{ref}}$. Consequently, the validity of (12.7) must remain unaffected if only $\mathfrak{E}$ and $\overline{\boldsymbol{B}}$ are rotated or reflected. In this way requirements for the structure of the a_{ij} and b_{ijk} result. In addition to this the connection between $\mathfrak{E}$ and $\overline{\boldsymbol{B}}$ and therefore a_{ij} and b_{ijk} have to be independent of time. The most general expression for $\mathfrak{E}$ which satisfies these requirements turns out to be rather extensive. It shall be given at the end of this section.

Let us pay attention to a special case first. We recall the considerations given in section 5.3. and deal with a model in which the electrical conductivity is constant and the mean motion only consists of a rigid body rotation. We suppose an inhomogeneous turbulence the intensity of which only depends on the radius. In the turbulence field no other preferred direction should appear than those given by the intensity gradient and the axis of rotation, i.e. by the unit vectors $\hat{\boldsymbol{r}}$, defined by $\hat{\boldsymbol{r}} = \boldsymbol{r}/r$, and $\hat{\boldsymbol{\omega}}$. We furthermore restrict ourselves to the case where both the intensity gradient and the rotation rate are sufficiently small so that $\mathfrak{E}$ may be considered to be linear with respect to each of these quantities. Then we have

$$\begin{aligned}\mathfrak{E} = &-\beta \,\mathrm{curl}\, \overline{\boldsymbol{B}}\\ &- \gamma\hat{\boldsymbol{r}}\times\overline{\boldsymbol{B}}\\ &- \delta_1(\hat{\boldsymbol{\omega}}\cdot\nabla)\,\overline{\boldsymbol{B}} - (\delta_2-\delta_1)\nabla(\hat{\boldsymbol{\omega}}\cdot\overline{\boldsymbol{B}})\\ &- \alpha_1(\hat{\boldsymbol{r}}\cdot\hat{\boldsymbol{\omega}})\,\overline{\boldsymbol{B}} - \alpha_2((\hat{\boldsymbol{r}}\cdot\overline{\boldsymbol{B}})\,\hat{\boldsymbol{\omega}} + (\hat{\boldsymbol{\omega}}\cdot\overline{\boldsymbol{B}})\,\hat{\boldsymbol{r}})\\ &- \alpha_3((\hat{\boldsymbol{r}}\cdot\overline{\boldsymbol{B}})\,\hat{\boldsymbol{\omega}} - (\hat{\boldsymbol{\omega}}\cdot\overline{\boldsymbol{B}})\,\hat{\boldsymbol{r}}),\end{aligned} \tag{15.8}$$

where some terms may be rewritten according to

$$-\delta_1(\hat{\boldsymbol{\omega}}\cdot\nabla)\,\overline{\boldsymbol{B}} - (\delta_2-\delta_1)\nabla(\hat{\boldsymbol{\omega}}\cdot\overline{\boldsymbol{B}}) = \mu\,\delta_1\hat{\boldsymbol{\omega}}\times\overline{\boldsymbol{j}} - \delta_2\nabla(\hat{\boldsymbol{\omega}}\cdot\overline{\boldsymbol{B}}), \tag{15.9a}$$

$$-\alpha_3((\hat{\boldsymbol{r}}\cdot\overline{\boldsymbol{B}})\,\hat{\boldsymbol{\omega}} - (\hat{\boldsymbol{\omega}}\cdot\overline{\boldsymbol{B}})\,\hat{\boldsymbol{r}}) = \alpha_3(\hat{\boldsymbol{\omega}}\times\hat{\boldsymbol{r}})\times\overline{\boldsymbol{B}}. \tag{15.9b}$$

Provided the turbulence field is fixed the coefficients $\beta, \gamma, \ldots$ depend on r only. Compared with sections 5.4. and 9.3. the notation has been slightly changed by introducing $\delta_1 = \beta_3$ and $\delta_2 = \beta_2 + \beta_3$.

Despite the special suppositions we have referred to when establishing it, relation (15.8) for $\mathfrak{E}$ covers all effects which have been proved to be relevant in the basic turbulent dynamo mechanisms investigated up to now. We shall mainly rely on this relation in the following. As for the coefficients $\beta, \gamma, \ldots$, however, a dependence on both r and ϑ is admitted; of course, symmetry with respect to the equatorial plane $\vartheta = \pi/2$ is required.

Let us briefly recall the meaning of the various contributions to $\mathfrak{E}$ which occur in (15.8). The term with β implies the possibility of introducing a turbulent electrical conductivity, and there are good reasons to assume that $1/\mu\sigma + \beta > 0$ (cf. sections 3.5... 9. and 7.4.). The term with γ describes a transport of magnetic flux similar to that due to a mean motion in radial direction and may also be comprehended by a turbulent magnetic permeability (cf. sections 5.4., 7.2., and 9.4.). The most important feature of the terms with δ_1 and δ_2 is that they cover the $\omega\times\bar{j}$-effect, i.e. the occurrence of an electromotive force proportional to $\hat{\boldsymbol{\omega}}\times\bar{\boldsymbol{j}}$, which is indicated at the right-hand side of (15.9a). This effect is equivalent to the occurrence of a special anisotropic electrical conductivity (cf. sections 5.4. and 9.3.). Finally, the terms with α_1, α_2, and α_3 are connected with the α-effect. The most important of these terms is the one with α_1, which corresponds to the α-effect occurring in a homogeneous isotropic but not mirrorsymmetric turbulence (cf. sections 3.5 ... 10, 5.4., 7.3., and 9.4 ... 6.). In this context we shall speak of the "idealized α-effect" if only this term with α_1 is considered, and of the "real α-effect" if the others are included. The term with α_2 corresponds to an anisotropy of the α-effect. As can be seen from (15.9b) the term with α_3 describes a transport of magnetic flux like that by a rotational motion.

In the following the α-effect will prove to be very important. For many purposes it is useful to introduce the quantity α^+ defined by

$$\alpha^+ = -\alpha_1(\hat{\boldsymbol{r}}\cdot\hat{\boldsymbol{\omega}}) = -\alpha_1\cos\vartheta\,. \tag{15.10}$$

Obviously, the role of α^+ is similar to that of the quantity α in the case of a homogeneous isotropic non-mirrorsymmetric turbulence. It should be emphasized, however, that α^+ does not agree with α as defined by (7.2), i.e. it is not identical with $(1/3)\,a_{ii}$. In the second order correlation approximation the dependence of α on the fluctuating motions may be expressed in the simple form (7.3). The dependence of α^+ is more complex; we refer to (9.51) ... (9.54).

Let us finally return to the most general expression for $\mathfrak{E}$ compatible with the symmetry conditions discussed above. As shown by RÄDLER [16] it may be written in the form

$$\begin{aligned}\mathfrak{E} = &-\boldsymbol{\alpha}\circ\overline{\boldsymbol{B}} - \boldsymbol{\gamma}\times\overline{\boldsymbol{B}}\\ &-\boldsymbol{\beta}\circ\operatorname{curl}\overline{\boldsymbol{B}} - \boldsymbol{\delta}\times\operatorname{curl}\overline{\boldsymbol{B}}\\ &-\boldsymbol{\beta}_r\circ(\hat{\boldsymbol{r}}\circ\nabla\overline{\boldsymbol{B}}) - \boldsymbol{\delta}_r\times(\hat{\boldsymbol{r}}\circ\nabla\overline{\boldsymbol{B}})\\ &-\boldsymbol{\beta}_\omega\circ(\hat{\boldsymbol{\omega}}\circ\nabla\overline{\boldsymbol{B}}) - \boldsymbol{\delta}_\omega\times(\hat{\boldsymbol{\omega}}\circ\nabla\overline{\boldsymbol{B}})\,.\end{aligned} \tag{15.11}$$

This form corresponds to that which results from (15.8) after inserting (15.9). The quantities $\boldsymbol{\alpha}$, $\boldsymbol{\beta}$, $\boldsymbol{\beta}_r$, and $\boldsymbol{\beta}_\omega$ are symmetric tensors, and $\boldsymbol{\gamma}$, $\boldsymbol{\delta}$, $\boldsymbol{\delta}_r$, and $\boldsymbol{\delta}_\omega$

are vectors. Furthermore, $(\hat{\boldsymbol{r}} \circ \nabla \overline{\boldsymbol{B}})$ is a vector defined by $(\hat{\boldsymbol{r}} \circ \nabla \overline{\boldsymbol{B}})_i = \hat{r}_j \, \partial \overline{B}_j / \partial x_i$, and $(\hat{\boldsymbol{\omega}} \circ \nabla \overline{\boldsymbol{B}})$, which is defined analogously, is identical with $\nabla(\hat{\boldsymbol{\omega}} \cdot \overline{\boldsymbol{B}})$. We may represent $\boldsymbol{\alpha} \circ \overline{\boldsymbol{B}}$ and $\boldsymbol{\gamma}$ by

$$\begin{aligned}
\boldsymbol{\alpha} \circ \overline{\boldsymbol{B}} = \alpha^{(0)}(\hat{\boldsymbol{r}} \cdot \hat{\boldsymbol{\omega}})\, \overline{\boldsymbol{B}} &\\
+ \alpha^{(1)}(\hat{\boldsymbol{r}} \cdot \hat{\boldsymbol{\omega}})\, (\hat{\boldsymbol{r}} \cdot \overline{\boldsymbol{B}})\, \hat{\boldsymbol{r}} + \alpha^{(2)}(\hat{\boldsymbol{r}} \cdot \hat{\boldsymbol{\omega}})\, (\hat{\boldsymbol{\omega}} \cdot \overline{\boldsymbol{B}})\, \hat{\boldsymbol{\omega}} &\\
+ \alpha^{(3)}(\hat{\boldsymbol{r}} \cdot \hat{\boldsymbol{\omega}})\, (\boldsymbol{\lambda} \cdot \overline{\boldsymbol{B}})\, \boldsymbol{\lambda} &\\
+ \alpha^{(4)}((\hat{\boldsymbol{\omega}} \cdot \overline{\boldsymbol{B}})\, \hat{\boldsymbol{r}} + (\hat{\boldsymbol{r}} \cdot \overline{\boldsymbol{B}})\, \hat{\boldsymbol{\omega}}) &\qquad (15.12\,\mathrm{a})\\
+ \alpha^{(5)}(\hat{\boldsymbol{r}} \cdot \hat{\boldsymbol{\omega}})\, ((\boldsymbol{\lambda} \cdot \overline{\boldsymbol{B}})\, \hat{\boldsymbol{r}} + (\hat{\boldsymbol{r}} \cdot \overline{\boldsymbol{B}})\, \boldsymbol{\lambda}) &\\
+ \alpha^{(6)}((\boldsymbol{\lambda} \cdot \overline{\boldsymbol{B}})\, \hat{\boldsymbol{\omega}} + (\hat{\boldsymbol{\omega}} \cdot \overline{\boldsymbol{B}})\, \boldsymbol{\lambda}), &
\end{aligned}$$

$$\gamma = \gamma^{(1)}\hat{\boldsymbol{r}} + \gamma^{(2)}(\hat{\boldsymbol{r}} \cdot \hat{\boldsymbol{\omega}})\, \hat{\boldsymbol{\omega}} + \gamma^{(3)}\boldsymbol{\lambda}, \qquad (15.12\,\mathrm{b})$$

and $\boldsymbol{\beta} \circ \operatorname{curl} \overline{\boldsymbol{B}}$ and $\boldsymbol{\delta}$ by

$$\begin{aligned}
\boldsymbol{\beta} \circ \operatorname{curl} \overline{\boldsymbol{B}} = \beta^{(0)} \operatorname{curl} \overline{\boldsymbol{B}} &\\
+ \beta^{(1)}(\hat{\boldsymbol{r}} \cdot \operatorname{curl} \overline{\boldsymbol{B}})\, \hat{\boldsymbol{r}} + \beta^{(2)}(\hat{\boldsymbol{\omega}} \cdot \operatorname{curl} \overline{\boldsymbol{B}})\, \hat{\boldsymbol{\omega}} &\\
+ \beta^{(3)}(\boldsymbol{\lambda} \cdot \operatorname{curl} \overline{\boldsymbol{B}})\, \boldsymbol{\lambda} &\\
+ \beta^{(4)}(\hat{\boldsymbol{r}} \cdot \hat{\boldsymbol{\omega}})\, ((\hat{\boldsymbol{\omega}} \cdot \operatorname{curl} \overline{\boldsymbol{B}})\, \hat{\boldsymbol{r}} + (\hat{\boldsymbol{r}} \cdot \operatorname{curl} \overline{\boldsymbol{B}})\, \hat{\boldsymbol{\omega}}) &\qquad (15.13\,\mathrm{a})\\
+ \beta^{(5)}((\boldsymbol{\lambda} \cdot \operatorname{curl} \overline{\boldsymbol{B}})\, \hat{\boldsymbol{r}} + (\hat{\boldsymbol{r}} \cdot \operatorname{curl} \overline{\boldsymbol{B}})\, \boldsymbol{\lambda}) &\\
+ \beta^{(6)}(\hat{\boldsymbol{r}} \cdot \hat{\boldsymbol{\omega}})\, ((\boldsymbol{\lambda} \cdot \operatorname{curl} \overline{\boldsymbol{B}})\, \hat{\boldsymbol{\omega}} + (\hat{\boldsymbol{\omega}} \cdot \operatorname{curl} \overline{\boldsymbol{B}})\, \boldsymbol{\lambda}), &
\end{aligned}$$

$$\boldsymbol{\delta} = \delta^{(1)}(\hat{\boldsymbol{r}} \cdot \hat{\boldsymbol{\omega}})\, \hat{\boldsymbol{r}} + \delta^{(2)}\hat{\boldsymbol{\omega}} + \delta^{(3)}(\hat{\boldsymbol{r}} \cdot \hat{\boldsymbol{\omega}})\, \boldsymbol{\lambda}, \qquad (15.13\,\mathrm{b})$$

where $\boldsymbol{\lambda}$ stands for $\hat{\boldsymbol{r}} \times \hat{\boldsymbol{\omega}}$. For $\boldsymbol{\beta}_r \circ (\hat{\boldsymbol{r}} \circ \nabla \overline{\boldsymbol{B}})$ and $\boldsymbol{\delta}_r$ we may write relations analogous to (15.12), with coefficients $\beta_r^{(0)}, \beta_r^{(1)}, \ldots \delta_r^{(3)}$, and for $\boldsymbol{\beta}_\omega \circ (\hat{\boldsymbol{\omega}} \circ \nabla \overline{\boldsymbol{B}})$ and $\boldsymbol{\delta}_\omega$ relations analogous to (15.13), with coefficients $\beta_\omega^{(0)}, \beta_\omega^{(1)}, \ldots \delta_\omega^{(3)}$. All coefficients $\alpha^{(0)}, \alpha^{(1)}, \ldots \delta_\omega^{(3)}$ may depend on r and ϑ but must be symmetric with respect to the equatorial plane $\vartheta = \pi/2$. It should be mentioned that, without any loss of generality, some of these coefficients may be fixed arbitrarily. We may choose, e.g.,

$$\alpha^{(3)} = \beta^{(3)} = \beta_r^{(3)} = \beta_\omega^{(3)} = \delta_r^{(3)} = \delta_\omega^{(3)} = 0, \qquad (15.14\,\mathrm{a})$$

$$\beta^{(1)} + \beta^{(2)} + 2\beta^{(4)}(\hat{\boldsymbol{r}} \cdot \hat{\boldsymbol{\omega}})^2 - \delta_r^{(1)} - \delta_r^{(2)}(\hat{\boldsymbol{r}} \cdot \hat{\boldsymbol{\omega}})^2 - \delta_\omega^{(1)}(\hat{\boldsymbol{r}} \cdot \hat{\boldsymbol{\omega}})^2 - \delta_\omega^{(2)} = 0. \qquad (15.14\,\mathrm{b})$$

Incidentally, (15.14b) ensures that $\beta^{(0)}$ coincides with $(1/6)\, \varepsilon_{ijk}\, b_{ijk}$ (cf. section 7.4.).

The general form of $\mathfrak{E}$ as defined by (15.11), (15.12), and (15.13) takes the special form given by (15.8) and (15.9) if we specify the coefficients $\alpha^{(0)}, \alpha^{(1)}, \ldots \delta_\omega^{(3)}$ so that

$$\alpha^{(0)} = \alpha_1, \quad \alpha^{(4)} = \alpha_2, \quad \gamma^{(1)} = \gamma, \quad \gamma^{(3)} = \alpha_3, \qquad (15.15\,\mathrm{a}\ldots\mathrm{d})$$

$$\beta^{(0)} = \beta, \quad \delta^{(2)} = -\delta_1, \quad \beta_\omega^{(0)} = \delta_2 \qquad (15.15\,\mathrm{e}\ldots\mathrm{g})$$

and all the others are zero.

Let us add some remarks on the meaning of the several contributions to $\mathfrak{E}$ occurring in (15.11). The contributions $\boldsymbol{\alpha} \circ \overline{\boldsymbol{B}}$ and $\boldsymbol{\gamma} \times \overline{\boldsymbol{B}}$ are generalizations of the $\alpha_1, \alpha_2, \alpha_3$, and γ-terms in (15.8). Here $\boldsymbol{\alpha} \circ \overline{\boldsymbol{B}}$ describes an anisotropic α-effect, and $\boldsymbol{\gamma} \times \overline{\boldsymbol{B}}$ a transport of magnetic flux in $\boldsymbol{\gamma}$-direction. Likewise the contributions $\boldsymbol{\beta} \circ \operatorname{curl} \overline{\boldsymbol{B}}$ and $\boldsymbol{\delta} \times \operatorname{curl} \overline{\boldsymbol{B}}$ are generalizations of the β and δ_1-terms. The occurrence of $\boldsymbol{\beta} \circ \operatorname{curl} \overline{\boldsymbol{B}}$ is equivalent to that of an anisotropic turbulent conductivity with a symmetric conductivity tensor. As for $\boldsymbol{\delta} \times \operatorname{curl} \overline{\boldsymbol{B}}$ we may speak of a $\delta \times j$-effect analogous to the more special $\omega \times j$-effect. Of course, this $\delta \times j$-effect can also be comprehended by an anisotropic turbulent conductivity with a conductivity tensor no longer being symmetric. The remaining contributions to $\mathfrak{E}$ are more difficult to interpret; in a wider sense they are related to the δ_2-term.

The induction effects of the mean motions described by ω und $\overline{\boldsymbol{u}}_\mathrm{p}$ are for short also referred to as ω-effect and u_p-effect in the following, and the induction effects of the fluctuating motions covered by the single terms of $\mathfrak{E}$ as β-effect, γ-effect etc.

15.5. A further symmetry property of the basic equations

We want to point out a remarkable property of the basic equations (15.1)...(15.3) which shall be utilized in the following. For this purpose we at first consider our models again within the scope of equations (11.1)...(11.4). Whereas before we restricted ourselves to rotations of the $\boldsymbol{u}$ and $\boldsymbol{B}$-fields about the axis of rotation and to their reflections in the equatorial plane we now pay attention to reflections in meridional planes. With regard to the possibility of rotations it is sufficient to consider the reflection in the plane $\varphi = 0$, which transforms $\boldsymbol{u}$ into $\boldsymbol{u}^{\mathrm{ref}}$, and $\boldsymbol{B}$ into $\boldsymbol{B}^{\mathrm{ref}}$, so that

$$u_{r,\vartheta}^{\mathrm{ref}}(r, \vartheta, \varphi) = u_{r,\vartheta}(r, \vartheta, -\varphi), \; u_\varphi^{\mathrm{ref}}(r, \vartheta, \varphi) = -u_\varphi(r, \vartheta, -\varphi), \qquad (15.16\,\mathrm{a,\,b})$$

$$B_{r,\vartheta}^{\mathrm{ref}}(r, \vartheta, \varphi) = -B_{r,\varphi}(r, \vartheta, -\varphi), \; B_\varphi^{\mathrm{ref}}(r, \vartheta, \varphi) = B_\varphi(r, \vartheta, -\varphi). \qquad (15.16\,\mathrm{c,\,d})$$

Again, the validity of equations (11.1)...(11.4) remains unaffected if $\boldsymbol{u}$ and $\boldsymbol{B}$ are replaced by $\boldsymbol{u}^{\mathrm{ref}}$ and $\boldsymbol{B}^{\mathrm{ref}}$, and the same holds for the mean-field equations

(12.1)...(12.5) too. Contrary to the rotations and reflections considered before, the reflection of $\boldsymbol{u}$ in a meridional plane, i.e. the transition from $\boldsymbol{u}$ to $\boldsymbol{u}^{\mathrm{ref}}$, generally changes both $\overline{\boldsymbol{u}}$ and $\mathfrak{E}$. If according to (15.6) $\overline{\boldsymbol{u}}$ is split into $\overline{\boldsymbol{u}}_{\mathrm{t}}$ and $\overline{\boldsymbol{u}}_{\mathrm{p}}$, we have $\overline{\boldsymbol{u}}_{\mathrm{t}}^{\mathrm{ref}} = -\overline{\boldsymbol{u}}_{\mathrm{t}}$ and $\overline{\boldsymbol{u}}_{\mathrm{p}}^{\mathrm{ref}} = \overline{\boldsymbol{u}}_{\mathrm{p}}$. As far as $\mathfrak{E}$ is concerned, for simplicity we restrict ourselves to the special relation (15.8). By similar considerations as in the case of rotations and reflections considered before we find that the validity of this relation may not be disturbed if $\mathfrak{E}$ is replaced by $\mathfrak{E}^{\mathrm{ref}}$, again defined like $\boldsymbol{u}^{\mathrm{ref}}$, likewise $\overline{\boldsymbol{B}}$ by $\overline{\boldsymbol{B}}^{\mathrm{ref}}$, and if, according to the transition from $\boldsymbol{u}$ to $\boldsymbol{u}^{\mathrm{ref}}$, the coefficients $\beta, \gamma, \ldots$ are replaced by others, β^{ref}, γ^{ref}, ... From this requirement we may easily conclude that $\beta^{\mathrm{ref}} = \beta$, $\gamma^{\mathrm{ref}} = \gamma$, but $\beta_1^{\mathrm{ref}} = -\beta_1$, $\beta_2^{\mathrm{ref}} = -\beta_2$, $\alpha_1^{\mathrm{ref}} = -\alpha_1$, $\alpha_2^{\mathrm{ref}} = -\alpha_2$, and $\alpha_3^{\mathrm{ref}} = -\alpha_3$. Bearing in mind that transformation (15.16a, b) changes the $\boldsymbol{u}$-field in a way that right-handed helical motions turn into left-handed ones, and vice versa, the signs occurring here seem to be plausible.

This result allows an interesting conclusion. Let us suppose that a problem has been solved which is described by equations (15.1)...(15.3), (15.6), and (15.8) with some specification of $\overline{\boldsymbol{u}}_{\mathrm{t}}$, $\overline{\boldsymbol{u}}_{\mathrm{p}}$, β, γ, ... In other words we suppose that corresponding $\boldsymbol{B}$-fields are known. Then we are also able to treat another problem which is described by these equations after the transformation

$$\overline{\boldsymbol{u}}_{\mathrm{t}} \to -\overline{\boldsymbol{u}}_{\mathrm{t}}, \quad \overline{\boldsymbol{u}}_{\mathrm{p}} \to \overline{\boldsymbol{u}}_{\mathrm{p}}, \tag{15.17a, b}$$

$$\beta \to \beta, \ \gamma \to \gamma, \tag{15.17c, d}$$

$$\beta_1 \to -\beta_1, \ \beta_2 \to -\beta_2, \tag{15.17e, f}$$

$$\alpha_1 \to -\alpha_1, \ \alpha_2 \to -\alpha_2, \ \alpha_3 \to -\alpha_3. \tag{15.17g, h, i}$$

Its solution is given by the transformation

$$\overline{\boldsymbol{B}} \to \overline{\boldsymbol{B}}^{\mathrm{ref}}, \tag{15.18}$$

with $\overline{\boldsymbol{B}}^{\mathrm{ref}}$ defined in (15.16c, d).

15.6. Reduction of the basic equations

We now recall section 14.2. in which the equations (14.1)...(14.3) governing a very simple model of a turbulent dynamo have been reduced to equations for some scalar functions. Equations (15.1)...(15.3) shall be treated in a similar way. For simplicity, instead of $\overline{\boldsymbol{B}}$ and $\overline{\boldsymbol{u}}$ we write $\boldsymbol{B}$ and $\boldsymbol{u}$ in the following.

As before we split the $\boldsymbol{B}$-field into a toroidal part, $\boldsymbol{B}_{\mathrm{t}}$, and a poloidal one, $\boldsymbol{B}_{\mathrm{p}}$, so that

$$\boldsymbol{B} = \boldsymbol{B}_{\mathrm{t}} + \boldsymbol{B}_{\mathrm{p}}, \tag{15.19}$$

and by reason of (15.1 b) and (15.2 b) we put

$$\boldsymbol{B}_{\mathrm{p}} = \operatorname{curl} \boldsymbol{A}_{\mathrm{t}}, \tag{15.20}$$

with $\boldsymbol{A}_{\mathrm{t}}$ being a purely toroidal vector potential.

At first we restrict ourselves to the case of axisymmetric $\boldsymbol{B}$-fields. Equations (15.1 a) together with (15.19) lead to

$$\frac{1}{\mu} \operatorname{curl} \left(\frac{1}{\sigma} \operatorname{curl} \boldsymbol{B}_{\mathrm{t}} \right) - \operatorname{curl} \left((\boldsymbol{u} \times \boldsymbol{B})_{\mathrm{p}} + \mathfrak{E}_{\mathrm{p}} \right) + \frac{\partial \boldsymbol{B}_{\mathrm{t}}}{\partial t} = 0, \tag{15.21 a}$$

$$\frac{1}{\mu} \operatorname{curl} \left(\frac{1}{\sigma} \operatorname{curl} \boldsymbol{B}_{\mathrm{p}} \right) - \operatorname{curl} \left((\boldsymbol{u} \times \boldsymbol{B})_{\mathrm{t}} + \mathfrak{E}_{\mathrm{t}} \right) + \frac{\partial \boldsymbol{B}_{\mathrm{p}}}{\partial t} = 0, \quad \text{for } r < R. \tag{15.21 b}$$

By virtue of (15.20) the last equation can be integrated to give

$$\frac{1}{\mu\sigma} \Delta \boldsymbol{A}_{\mathrm{t}} + (\boldsymbol{u} \times \boldsymbol{B})_{\mathrm{t}} + \mathfrak{E}_{\mathrm{t}} - \frac{\partial \boldsymbol{A}_{\mathrm{t}}}{\partial t} = 0, \quad \text{for} \quad r < R. \tag{15.22}$$

From equation (15.2 a) we obtain

$$\boldsymbol{B}_{\mathrm{t}} = 0, \ \Delta \boldsymbol{A}_{\mathrm{t}} = 0, \ \text{for } r > R. \tag{15.23 a, b}$$

The conditions (15.3) result in

$$\boldsymbol{B}_{\mathrm{t}} = 0, \ [\operatorname{curl} \boldsymbol{A}_{\mathrm{t}}] = 0, \text{ on } r = R, \tag{15.24 a, b}$$

$$\boldsymbol{A}_{\mathrm{t}} = O(r^{-2}), \text{ as } r \to \infty. \tag{15.24 c}$$

Let us now deal with the decomposition of $\boldsymbol{u} \times \boldsymbol{B}$ and $\mathfrak{E}$ into toroidal and poloidal parts. Splitting $\boldsymbol{u}$ according to (15.6) we obtain

$$(\boldsymbol{u} \times \boldsymbol{B})_{\mathrm{t}} = \boldsymbol{u}_{\mathrm{p}} \times \boldsymbol{B}_{\mathrm{p}}, \tag{15.25 a}$$

$$(\boldsymbol{u} \times \boldsymbol{B})_{\mathrm{p}} = \boldsymbol{u}_{\mathrm{t}} \times \boldsymbol{B}_{\mathrm{p}} + \boldsymbol{u}_{\mathrm{p}} \times \boldsymbol{B}_{\mathrm{t}}. \tag{15.25 b}$$

If $\mathfrak{E}$ is given by (15.8) we have

$$\mathfrak{E}_{\mathrm{t}} = -\beta \operatorname{curl} \boldsymbol{B}_{\mathrm{p}} - \gamma \hat{\boldsymbol{r}} \times \boldsymbol{B}_{\mathrm{p}} - \delta_1 (\hat{\boldsymbol{\omega}} \cdot \nabla) \boldsymbol{B}_{\mathrm{t}} - \alpha_1 (\hat{\boldsymbol{r}} \cdot \hat{\boldsymbol{\omega}}) \boldsymbol{B}_{\mathrm{t}}, \tag{15.26 a}$$

$$\begin{aligned} \mathfrak{E}_{\mathrm{p}} = & -\beta \operatorname{curl} \boldsymbol{B}_{\mathrm{t}} - \gamma \hat{\boldsymbol{r}} \times \boldsymbol{B}_{\mathrm{t}} - \delta_1 (\hat{\boldsymbol{\omega}} \cdot \nabla) \boldsymbol{B}_{\mathrm{p}} - (\delta_2 - \delta_1) \nabla (\hat{\boldsymbol{\omega}} \cdot \boldsymbol{B}_{\mathrm{p}}) \\ & - \alpha_1 (\hat{\boldsymbol{r}} \cdot \hat{\boldsymbol{\omega}}) \boldsymbol{B}_{\mathrm{p}} - \alpha_2 ((\hat{\boldsymbol{r}} \cdot \boldsymbol{B}_{\mathrm{p}}) \hat{\boldsymbol{\omega}} + (\hat{\boldsymbol{\omega}} \cdot \boldsymbol{B}_{\mathrm{p}}) \hat{\boldsymbol{r}}) \\ & - \alpha_3 ((\hat{\boldsymbol{r}} \cdot \boldsymbol{B}_{\mathrm{p}}) \hat{\boldsymbol{\omega}} - (\hat{\boldsymbol{\omega}} \cdot \boldsymbol{B}_{\mathrm{p}}) \hat{\boldsymbol{r}}), \end{aligned} \tag{15.26 b}$$

where $\boldsymbol{B}_{\mathrm{p}}$ can be replaced by $\operatorname{curl} \boldsymbol{A}_{\mathrm{t}}$. Analogously, also other expressions for $\mathfrak{E}$ can be treated.

We now put

$$\boldsymbol{B}_{\mathrm{t}} = B \boldsymbol{e}_{\varphi}, \quad \boldsymbol{A}_{\mathrm{t}} = A \boldsymbol{e}_{\varphi}, \tag{15.27 a, b}$$

where B and A are scalar functions and $\boldsymbol{e}_\varphi$ is again the azimuthal unit vector. In this way we have

$$\boldsymbol{B} = B\boldsymbol{e}_\varphi + (1/r \sin \vartheta) \nabla (r \sin \vartheta A) \times \boldsymbol{e}_\varphi. \tag{15.28}$$

Analogous to (15.27) we introduce

$$(\boldsymbol{u} \times \boldsymbol{B})_{\mathrm{t}} = C^{(uB)} \boldsymbol{e}_\varphi, \quad \operatorname{curl} (\boldsymbol{u} \times \boldsymbol{B})_{\mathrm{p}} = D^{(uB)} \boldsymbol{e}_\varphi, \tag{15.29 a, b}$$

$$\mathfrak{E}_{\mathrm{t}} = C^{(\mathfrak{E})} \boldsymbol{e}_\varphi, \quad \operatorname{curl} \mathfrak{E}_{\mathrm{p}} = D^{(\mathfrak{E})} \boldsymbol{e}_\varphi, \tag{15.29 c, d}$$

with $C^{(uB)}$, $D^{(uB)}$, $C^{(\mathfrak{E})}$, and $D^{(\mathfrak{E})}$ being scalar functions.

In this context it is of interest that $\operatorname{curl} (\boldsymbol{e}_\varphi / r \sin \vartheta) = 0$ and therefore, for any scalar F and any vector $\boldsymbol{U}$,

$$\operatorname{curl} (F\boldsymbol{e}_\varphi / r \sin \vartheta) = \nabla F \times \boldsymbol{e}_\varphi / r \sin \vartheta, \tag{15.30 a}$$

$$\operatorname{div} (\boldsymbol{U} \times \boldsymbol{e}_\varphi / r \sin \vartheta) = \operatorname{curl} \boldsymbol{U} \cdot \boldsymbol{e}_\varphi / r \sin \vartheta. \tag{15.30 b}$$

If F is independent of φ we have

$$\Delta (F\boldsymbol{e}_\varphi) = (\Delta' F)\, \boldsymbol{e}_\varphi, \tag{15.31}$$

with an operator Δ' defined by

$$\Delta' F = (\Delta - 1/r^2 \sin^2 \vartheta)\, F. \tag{15.32}$$

Equations (15.21 a) and (15.22) for $\boldsymbol{B}_{\mathrm{t}}$ and $\boldsymbol{A}_{\mathrm{t}}$ can now be reduced to equations for B and A, namely

$$\frac{1}{\mu\sigma} \left(\Delta' B - \frac{1}{\sigma r \sin \vartheta} \nabla \sigma \cdot \nabla (r \sin \vartheta\, B) \right) + D^{(uB)} + D^{(\mathfrak{E})} - \frac{\partial B}{\partial t} = 0, \tag{15.33 a}$$

$$\frac{1}{\mu\sigma} \Delta' A + C^{(uB)} + C^{(\mathfrak{E})} - \frac{\partial A}{\partial t} = 0, \quad \text{for} \quad r < R. \tag{15.33 b}$$

From (15.23) and (15.24) results

$$B = 0, \ \Delta' A = 0, \text{ for } r > R, \tag{15.34}$$

and

$$B = 0, \ [A] = [\partial A / \partial r] = 0, \text{ on } r = R, \tag{15.35 a, b}$$

$$A = O(r^{-2}), \text{ as } r \to \infty. \tag{15.35 c}$$

These equations have to be supplemented by relations connecting $C^{(uB)}$, $D^{(uB)}$, $C^{(\mathfrak{E})}$, and $D^{(\mathfrak{E})}$ with B and A. Relations of that kind can be derived by means of (15.25), (15.26), and (15.29). It should be noted, however, that the determination of $C^{(uB)}$, $D^{(uB)}$, $C^{(\mathfrak{E})}$, and $D^{(\mathfrak{E})}$ can be carried out even without the decomposition of $\boldsymbol{u}\times\boldsymbol{B}$ and $\mathfrak{E}$ as given by (15.25) and (15.26) because from (15.29) we may conclude

$$C^{(uB)} = (\boldsymbol{u}\times\boldsymbol{B}) \cdot \boldsymbol{e}_\varphi , \tag{15.36a}$$

$$D^{(uB)} = \operatorname{curl} (\boldsymbol{u}\times\boldsymbol{B}) \cdot \boldsymbol{e}_\varphi = r \sin\vartheta \operatorname{div} ((\boldsymbol{u}\times\boldsymbol{B})\times\boldsymbol{e}_\varphi/r \sin\vartheta), \tag{15.36b}$$

$$C^{(\mathfrak{E})} = \mathfrak{E} \cdot \boldsymbol{e}_\varphi , \tag{15.36c}$$

$$D^{(\mathfrak{E})} = \operatorname{curl} \mathfrak{E} \cdot \boldsymbol{e}_\varphi = r \sin\vartheta \operatorname{div} (\mathfrak{E}\times\boldsymbol{e}_\varphi/r \sin\vartheta). \tag{15.36d}$$

Expressing $\boldsymbol{u}$ by ω and $\boldsymbol{u}_{\mathrm{p}}$, and $\boldsymbol{B}$ by B and A, we find

$$C^{(uB)} = -(1/r \sin\vartheta)\, \boldsymbol{u}_{\mathrm{p}} \cdot \nabla(r \sin\vartheta A), \tag{15.37a}$$

$$D^{(uB)} = -r \sin\vartheta \operatorname{div} (\boldsymbol{u}_{\mathrm{p}} B/r \sin\vartheta) + (\nabla\omega \times \nabla(r \sin\vartheta A)) \cdot \boldsymbol{e}_\varphi . \tag{15.37b}$$

With $\mathfrak{E}$ as given by (15.8) we obtain

$$C^{(\mathfrak{E})} = \beta \Delta' A + \frac{\gamma}{r} \frac{\partial}{\partial r} (rA) - \delta_1 \hat{\boldsymbol{\omega}} \cdot \nabla B - \alpha_1(\hat{\boldsymbol{r}} \cdot \hat{\boldsymbol{\omega}})\, B , \tag{15.38a}$$

$$\begin{aligned}
D^{(\mathfrak{E})} = {} & r \sin\vartheta \operatorname{div} \left(\frac{\beta}{(r \sin\vartheta)^2} \nabla(r \sin\vartheta B)\right) + \frac{\hat{\boldsymbol{r}}}{r} \cdot \nabla(\gamma r B) \\
& + r \sin\vartheta\, \hat{\boldsymbol{\omega}} \cdot \nabla \left(\frac{\delta_1}{r \sin\vartheta} \Delta' A\right) + \left(\nabla\delta_2 \times \nabla \left(\frac{\hat{\boldsymbol{\omega}}\times\boldsymbol{e}_\varphi}{r \sin\vartheta} \cdot \nabla(r \sin\vartheta A)\right)\right) \cdot \boldsymbol{e}_\varphi \\
& + r \sin\vartheta \operatorname{div} \left(\frac{\alpha_1(\hat{\boldsymbol{r}} \cdot \hat{\boldsymbol{\omega}})}{(r \sin\vartheta)^2} \nabla(r \sin\vartheta A)\right) \\
& + (\hat{\boldsymbol{r}}\times\boldsymbol{e}_\varphi) \cdot \nabla \left(\frac{\alpha_2}{r \sin\vartheta} (\hat{\boldsymbol{\omega}}\times\boldsymbol{e}_\varphi) \cdot \nabla(r \sin\vartheta A)\right) \\
& + (\hat{\boldsymbol{\omega}}\times\boldsymbol{e}_\varphi) \cdot \nabla \left(\frac{\alpha_2}{r \sin\vartheta} (\hat{\boldsymbol{r}}\times\boldsymbol{e}_\varphi) \cdot \nabla(r \sin\vartheta A)\right) \\
& + \left(\nabla \left(\frac{\alpha_3}{r}\right) \times\nabla(r \sin\vartheta A)\right) \cdot \boldsymbol{e}_\varphi .
\end{aligned} \tag{15.38b}$$

We may conclude from (15.37) that an axisymmetric $\boldsymbol{B}$-field cannot be influenced by a rigid body rotation, i.e. by a rotation with $\omega = \text{const}$. Likewise, (15.38) show that the δ_2 or the α_3-effect does not play any role if $\delta_2 = \text{const}$ or $\alpha_3/r = \text{const}$, respectively.

Let us now take the case of general, i.e. not necessarily axisymmetric $\boldsymbol{B}$-fields. For reasons of simplicity, in this case we assume that the conductivity σ only depends on r. Then equations (15.21)...(15.24) can again be justified. Relations (15.25) and (15.26) are, however, no longer valid.

We now put

$$B_t = -r \times \nabla T, \tag{15.39a}$$

$$A_t = -r \times \nabla S, \tag{15.39b}$$

furthermore

$$(u \times B)_t = -r \times \nabla U^{(uB)}, \tag{15.40a}$$

$$\text{curl}\,(u \times B)_p = -r \times \nabla V^{(uB)}, \tag{15.40b}$$

$$\mathfrak{E}_t = -r \times \nabla U^{(\mathfrak{E})}, \tag{15.40c}$$

$$\text{curl}\,\mathfrak{E}_p = -r \times \nabla V^{(\mathfrak{E})}, \tag{15.40d}$$

and we require

$$\oiint T\,\mathrm{d}\tilde{\omega} = \oiint S\,\mathrm{d}\tilde{\omega} = \ldots = \oiint V^{(\mathfrak{E})}\,\mathrm{d}\tilde{\omega} = 0. \tag{15.41}$$

On these grounds we may reduce equations (15.21a) and (15.22) to

$$\frac{1}{\mu\sigma}\left(\Delta T - \frac{1}{\sigma r}\frac{\mathrm{d}\sigma}{\mathrm{d}r}\frac{\partial}{\partial r}(rT)\right) + V^{(uB)} + V^{(\mathfrak{E})} - \frac{\partial T}{\partial t} = 0, \tag{15.42a}$$

$$\frac{1}{\mu\sigma}\Delta S + U^{(uB)} + U^{(\mathfrak{E})} - \frac{\partial S}{\partial t} = 0, \quad \text{for} \quad r < R, \tag{15.42)b}$$

equations (15.23) to

$$T = 0, \ \Delta S = 0, \quad \text{for} \ \ r > R, \tag{15.43a, b}$$

and conditions (15.24) to

$$T = 0, \ [S] = [\partial S/\partial r] = 0, \ \text{on}\ r = R, \tag{15.44a, b}$$

$$S = O(r^{-2}), \ \text{as}\ r \to \infty. \tag{15.44c}$$

The problem remains to find relations connecting $U^{(uB)}$, $V^{(uB)}$, $U^{(\mathfrak{E})}$, and $V^{(\mathfrak{E})}$ with T and S. Taking the curls of equations (15.40) and multiplying the resulting equations by r we may conclude that

$$\Omega U^{(uB)} = -r \cdot \text{curl}\,(u \times B), \tag{15.45a}$$

$$\Omega V^{(uB)} = -r \cdot \text{curl curl}\,(u \times B), \tag{15.45b}$$

$$\Omega U^{(\mathfrak{E})} = -r \cdot \text{curl}\,\mathfrak{E}, \tag{15.45c}$$

$$\Omega V^{(\mathfrak{E})} = -r \cdot \text{curl curl}\,\mathfrak{E}, \tag{15.45d}$$

where the right-hand sides can also be rewritten by means of (13.34). Expressing $\boldsymbol{u}$ and $\boldsymbol{B}$ by ω, $\boldsymbol{u}_\mathrm{p}$, T, and S we get after some analysis

$$\begin{aligned}\Omega U^{(uB)} = & -\boldsymbol{r}(\nabla(\boldsymbol{r}\cdot\boldsymbol{u}_\mathrm{p})\times\nabla T)\\ & - \omega\Omega\frac{\partial S}{\partial\varphi} - \operatorname{div}(\boldsymbol{u}_\mathrm{p}\,\Omega S) + \nabla(\boldsymbol{r}\cdot\boldsymbol{u}_\mathrm{p})\cdot\left(\boldsymbol{r}\,\Delta S - \nabla\frac{\partial}{\partial r}(rS)\right), \qquad (15.46\mathrm{a})\end{aligned}$$

$$\begin{aligned}\Omega V^{(uB)} = & -\Omega\left(\omega\frac{\partial T}{\partial\varphi} + \boldsymbol{u}_\mathrm{p}\cdot\nabla T\right) + \operatorname{div}(\boldsymbol{r}\times(\nabla(\boldsymbol{r}\cdot\boldsymbol{u}_\mathrm{p})\times\nabla T))\\ & - \Omega\left(\omega\hat{\boldsymbol{\omega}}\cdot\left(\boldsymbol{r}\,\Delta S - \nabla\frac{\partial}{\partial r}(rS)\right)\right) - \operatorname{div}(\boldsymbol{r}\times(\hat{\boldsymbol{\omega}}\times\nabla(\omega\,\Omega S)))\\ & + \operatorname{div}\left(\boldsymbol{r}\times\operatorname{curl}\left((\boldsymbol{r}\times\boldsymbol{u}_\mathrm{p})\,\Delta S + \boldsymbol{u}_\mathrm{p}\times\nabla\frac{\partial}{\partial r}(rS)\right)\right). \qquad (15.46\mathrm{b})\end{aligned}$$

If we choose $\mathfrak{E}$ in the form (15.8) and again express $\boldsymbol{B}$ by T and S, we find

$$\begin{aligned}\Omega U^{(\mathfrak{E})} = & \;\boldsymbol{r}\cdot\left(\nabla\beta\times\nabla\frac{\partial}{\partial r}(rT)\right) - \operatorname{div}(\boldsymbol{r}\times(\boldsymbol{r}\times\beta\nabla\,\Delta S))\\ & + \boldsymbol{r}\cdot(\nabla(\gamma r)\times\nabla T) - \operatorname{div}\left(\boldsymbol{r}\times\left(\boldsymbol{r}\times(\gamma/r)\nabla\frac{\partial}{\partial r}(rS)\right)\right)\\ & - \hat{\boldsymbol{\omega}}\cdot\nabla(\delta_1\Omega T) + \nabla(\delta_1(\boldsymbol{r}\cdot\hat{\boldsymbol{\omega}}))\cdot\left(\boldsymbol{r}\,\Delta T - \nabla\frac{\partial}{\partial r}(rT)\right)\\ & + \boldsymbol{r}\cdot(\nabla(\delta_1(\boldsymbol{r}\cdot\hat{\boldsymbol{\omega}}))\times\nabla\,\Delta S) \qquad (15.47\mathrm{a})\\ & - \boldsymbol{r}\cdot\left(\nabla\delta_2\times\nabla\frac{\partial T}{\partial\varphi}\right) - \boldsymbol{r}\cdot\left(\nabla\delta_2\times\nabla\left(\hat{\boldsymbol{\omega}}\cdot\left(\boldsymbol{r}\,\Delta S - \nabla\frac{\partial}{\partial r}(rS)\right)\right)\right)\\ & + \operatorname{div}(\boldsymbol{r}\times(\boldsymbol{r}\times\alpha_1(\hat{\boldsymbol{r}}\cdot\hat{\boldsymbol{\omega}})\nabla T)) + \boldsymbol{r}\cdot\left(\nabla(\alpha_1(\hat{\boldsymbol{r}}\cdot\hat{\boldsymbol{\omega}}))\times\nabla\frac{\partial}{\partial r}(rS)\right)\\ & - \frac{\alpha_2+\alpha_3}{r}\,\Omega\frac{\partial S}{\partial\varphi},\end{aligned}$$

$$\begin{aligned}\Omega V^{(\mathfrak{E})} = & -\operatorname{div}\left(\boldsymbol{r}\times\left(\boldsymbol{r}\times\nabla(\beta\,\Delta T) + \nabla\beta\times\nabla\frac{\partial}{\partial r}(rT)\right)\right)\\ & - \frac{1}{r}\frac{\partial}{\partial r}(r^2\boldsymbol{r}\cdot(\nabla\beta\times\nabla\,\Delta S))\\ & + \Omega\left(\gamma\frac{\partial T}{\partial r}\right) - \operatorname{div}(\boldsymbol{r}\times(\nabla(\gamma r)\times\nabla T))\\ & - \frac{1}{r}\frac{\partial}{\partial r}\left(r^2\boldsymbol{r}\cdot\left(\nabla(\gamma/r)\times\nabla\frac{\partial}{\partial r}(rS)\right)\right) -\end{aligned}$$

$$
\begin{aligned}
&-\left(\delta_1 - r\frac{\partial \delta_1}{\partial r}\right)\Delta\frac{\partial T}{\partial \varphi} - \nabla\delta_1 \cdot \nabla\frac{\partial}{\partial r}\left(r\frac{\partial T}{\partial \varphi}\right)\\
&+\boldsymbol{r}\cdot\left((\hat{\boldsymbol{\omega}}\cdot\nabla)\left(\nabla\delta_1\times\nabla\frac{\partial}{\partial r}(rT)\right)\right)\\
&+\Omega(\delta_1\hat{\boldsymbol{\omega}}\cdot\nabla\Delta S) - \operatorname{div}(\boldsymbol{r}\times(\nabla(\delta_1(\boldsymbol{r}\cdot\hat{\boldsymbol{\omega}}))\times\nabla\Delta S)) \qquad (15.47\text{b})\\
&+\operatorname{div}\left(\boldsymbol{r}\times\left(\nabla\delta_2\times\nabla\frac{\partial T}{\partial \varphi}\right)\right)\\
&+\operatorname{div}\left(\boldsymbol{r}\times\left(\nabla\delta_2\times\nabla\left(\hat{\boldsymbol{\omega}}\cdot\left(\boldsymbol{r}\Delta S - \nabla\frac{\partial}{\partial r}(rS)\right)\right)\right)\right)\\
&+\frac{1}{r}\frac{\partial}{\partial r}(r^2\boldsymbol{r}\cdot(\nabla(\alpha_1(\hat{\boldsymbol{r}}\cdot\hat{\boldsymbol{\omega}}))\times\nabla T))\\
&-\operatorname{div}(\boldsymbol{r}\times(\boldsymbol{r}\times\alpha_1(\hat{\boldsymbol{r}}\cdot\hat{\boldsymbol{\omega}})\nabla\Delta S))\\
&+\operatorname{div}\left(\boldsymbol{r}\times\left(\nabla(\alpha_1(\hat{\boldsymbol{r}}\cdot\hat{\boldsymbol{\omega}}))\times\left(\boldsymbol{r}\Delta S - \nabla\frac{\partial}{\partial r}(rS)\right)\right)\right)\\
&-\operatorname{div}\left(\boldsymbol{r}\times\left(\hat{\boldsymbol{\omega}}\times\nabla\left(\frac{\alpha_2+\alpha_3}{r}\Omega S\right)\right)\right)\\
&+\Omega\left(\frac{\alpha_2-\alpha_3}{r}\frac{\partial T}{\partial \varphi}\right) + \Omega\left(\frac{\alpha_2-\alpha_3}{r}\hat{\boldsymbol{\omega}}\cdot\left(\boldsymbol{r}\Delta S - \nabla\frac{\partial}{\partial r}(rS)\right)\right).
\end{aligned}
$$

It may be of interest to note that, for any f and F,

$$\operatorname{div}(\boldsymbol{r}\times(\boldsymbol{r}\times f\nabla F)) = -\Omega(fF) - \operatorname{div}(\boldsymbol{r}\times(\boldsymbol{r}\times F\nabla f)). \qquad (15.48)$$

If $\partial f/\partial\varphi = 0$ we have

$$\boldsymbol{r}\cdot(\nabla f\times\nabla F) = \frac{1}{r\sin\vartheta}\frac{\partial f}{\partial\vartheta}\frac{\partial F}{\partial\varphi}. \qquad (15.49)$$

Unfortunately, equations (15.46) and (15.47) cannot readily be integrated. Nevertheless, they show that in contrast to an axisymmetric $\boldsymbol{B}$-field a non-axisymmetric one can be influenced by a mean rotation with $\omega = \text{const}$, and also by the α_3-effect with $\alpha_3/r = \text{const}$. As before, the δ_2-effect does not occur if $\delta_2 = \text{const}$.

Let us return once more to the statement formulated with transformations (15.17) and (15.18). For an axisymmetric $\boldsymbol{B}$-field the transformation (15.18) simply means $\boldsymbol{B}_t \to -\boldsymbol{B}_t$ and $\boldsymbol{B}_p \to \boldsymbol{B}_p$, and the statement mentioned can immediately be confirmed by means of equations (15.21)...(15.26). In the general case, including also non-axisymmetric $\boldsymbol{B}$-fields, the transformation

(15.18) is equivalent to $T(r, \vartheta, \varphi, t) \to -T(r, \vartheta, -\varphi, t)$ and $S(r, \vartheta, \varphi, t) \to S(r, \vartheta, -\varphi, t)$. The statement under consideration may also be derived from equations (15.42)...(15.44), (15.46) and (15.47), albeit in a troublesome manner.

15.7. Possibilities of dynamo mechanisms

In order to know what types of turbulent dynamo mechanisms are possible we shall investigate which parts of the mean motion and of the electromotive force caused by the fluctuating motions are able to generate or to maintain magnetic fields.

The mean velocity field $\boldsymbol{u}$ is, according to (15.6), again understood as a sum of two parts, $\boldsymbol{u}_t$ and $\boldsymbol{u}_p$, describing rotation and meridional circulation. As for the electromotive force $\mathfrak{E}$, until further notice only the terms given in (15.8) are taken into account, and $1/\mu\sigma + \beta$ is considered to be positive.

We consider the $\boldsymbol{B}$-field again as a sum of a toroidal and a poloidal part, $\boldsymbol{B}_t$ and $\boldsymbol{B}_p$. Discussing the possibilities of dynamo mechanisms we should distinguish between different kinds of coupling between $\boldsymbol{B}_t$ and $\boldsymbol{B}_p$.

First we deal with the case in which the $\boldsymbol{B}$-field, and therefore $\boldsymbol{B}_t$ and $\boldsymbol{B}_p$, are axisymmetric. By means of equations (15.21), (15.25), and (15.26) it can easily be followed up how the effects involved in $\boldsymbol{u}$ and $\mathfrak{E}$ influence $\boldsymbol{B}_t$ and $\boldsymbol{B}_p$. A scheme of these influences is given in figure 15.2. One should bear in mind that the mean rotation becomes effective, i.e. an ω-effect occurs, only if ω varies in space. Likewise, the δ_2-effect plays a role only if δ_2 is not constant.

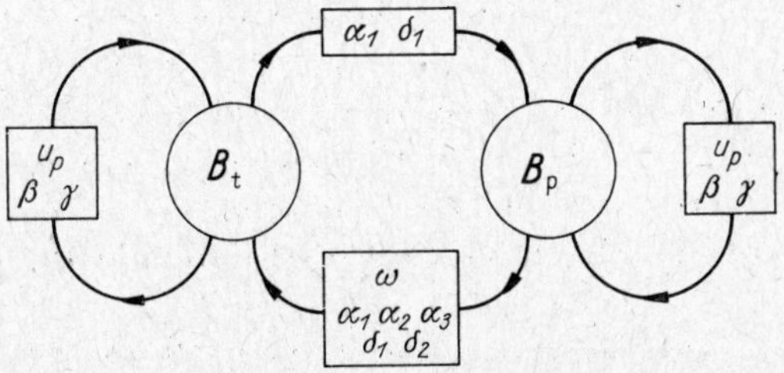

Fig. 15.2. The influences of the induction effects covered by $\boldsymbol{u}$ and $\mathfrak{E}$ on the toroidal and the poloidal parts, $\boldsymbol{B}_t$ and $\boldsymbol{B}_p$, of an axisymmetric $\boldsymbol{B}$-field

To begin with we ask for the possibility of dynamos without any coupling between the $\boldsymbol{B}_t$ and $\boldsymbol{B}_p$-fields. According to the scheme of figure 15.2 a dynamo of that kind could only exist due to the u_p, β, or γ-effect. In order to avoid any coupling between the $\boldsymbol{B}_t$ and $\boldsymbol{B}_p$-fields generally the ω, α_1, α_2, α_3, δ_1 and δ_2-effects have to be excluded. A dynamo as envisaged here would be able to generate a purely toroidal or a purely poloidal $\boldsymbol{B}$-field. We point out that a dynamo with a purely toroidal $\boldsymbol{B}$-field represents an "invisible dynamo" since a field of this kind is always confined in the conducting sphere, i.e. is zero at its surface and in outer space. Besides, in this case only the α_1 and δ_1-effects have to be excluded; the ω, α_2, α_3 and δ_2-effects are not involved anyway.

Since we consider only axisymmetric $\boldsymbol{B}$-fields here, it can be concluded from COWLING's theorem that there are no dynamos due to the u_p, β or γ-effect with steady fields. If $\boldsymbol{u}_\mathrm{d} + \gamma\hat{\boldsymbol{r}}$ is solenoidal, and β is constant, due to BRAGINSKIJ's version of COWLING's theorem the same also holds for non-steady fields. Finally, if $\boldsymbol{u}_\mathrm{p} = 0$ and $\gamma = 0$ any dynamo proves to be impossible by relation (12.8). In conclusion we may state that in the majority of cases the existence of dynamos without coupling between the $\boldsymbol{B}_\mathrm{t}$ and $\boldsymbol{B}_\mathrm{p}$-fields can be excluded. In all other cases it is at least not hinted.

Let us now assume that there is a coupling between $\boldsymbol{B}_\mathrm{t}$ and $\boldsymbol{B}_\mathrm{p}$ which consists, however, only in an influence of $\boldsymbol{B}_\mathrm{t}$ on $\boldsymbol{B}_\mathrm{p}$, or vice versa, but does not allow an interaction between $\boldsymbol{B}_\mathrm{t}$ and $\boldsymbol{B}_\mathrm{p}$. As long as dynamos without coupling between $\boldsymbol{B}_\mathrm{t}$ and $\boldsymbol{B}_\mathrm{p}$ are impossible, we may conclude that there cannot be any dynamo with this simple type of coupling.

In this way we arrive at the presumption that dynamos with axisymmetric $\boldsymbol{B}$-fields can only exist due to the simultaneous influences of both $\boldsymbol{B}_\mathrm{t}$ on $\boldsymbol{B}_\mathrm{p}$ and vice versa, i.e. an interaction of $\boldsymbol{B}_\mathrm{t}$ and $\boldsymbol{B}_\mathrm{p}$. As can be seen from figure 15.2 this interaction requires the α_1 or the δ_1-effect. Thus the presence of the α_1 or the δ_1-effect seems to be a necessary condition for the existence of dynamos with axisymmetric fields.

Dealing with the dynamo action of the α_1-effect, we begin with the case in which no turbulence effects other than the α_1-effect and no mean motion other than a rigid body rotation are taken into account. Even in this simple case a dynamo mechanism indeed turns out to be possible, which is able to maintain steady $\boldsymbol{B}$-fields (cf. section 16.3.). Since both the generation of $\boldsymbol{B}_\mathrm{t}$ from $\boldsymbol{B}_\mathrm{p}$ and that of $\boldsymbol{B}_\mathrm{p}$ from $\boldsymbol{B}_\mathrm{t}$ is due to the α_1-effect, we speak of an α^2-mechanism, or an α^2-process. A typical feature of this mechanism is that $\boldsymbol{B}_\mathrm{t}$ and $\boldsymbol{B}_\mathrm{p}$ are of the same order of magnitude.

Let us have a short look at the energy relation (12.8). In the case of a dynamo the right-hand side has to be non-negative. Since the first term gives a negative contribution and the second one is zero, the third one must provide for a sufficiently strong positive contribution. Hence, a necessary condition for an α^2-mechanism consists in the requirement that $\alpha_1(\hat{\boldsymbol{r}} \cdot \hat{\boldsymbol{\omega}})\,(\boldsymbol{j} \cdot \boldsymbol{B})$ is mainly negative so that the integral of this quantity taken over the volume of the conducting body is negative. This is equivalent to the requirement that $\alpha^+\boldsymbol{B} \cdot \operatorname{curl} \boldsymbol{B}$, with α^+ being defined according to (15.10), is mainly positive in the above sense.

We now modify the assumptions by replacing the α_1-effect by the δ_1-effect. In this case no dynamo has been found. The reasons for this can easily be seen from the relation (12.8). Like the second term of the right-hand side also the third term is always zero; for even the integrand vanishes everywhere. Hence, any $\boldsymbol{B}$-field is bound to decay.

We now consider the possibility of dynamos which act on the basis of a combination of the α_1 or the δ_1-effect with other induction effects.

If the α_1-effect is combined with other induction effects, certain modifications of the α^2-mechanism have to be expected. Only the combination of the α_1-effect with a non-uniform rotation, i.e. the ω-effect, shall be discussed here. As long as the ω-effect is small a dynamo may occur which mainly works on the basis of an α^2-mechanism. A sufficiently strong rotational shear, however, will be much more effective in generating $\boldsymbol{B}_\mathrm{t}$ from $\boldsymbol{B}_\mathrm{p}$ than the α_1-effect. Thus another dynamo mechanism turns out to be possible in which the generation of $\boldsymbol{B}_\mathrm{t}$ from $\boldsymbol{B}_\mathrm{p}$ is mainly by virtue of the rotational shear, i.e. the ω-effect, and the essential role of the α_1-effect is restricted to the generation of $\boldsymbol{B}_\mathrm{p}$ from $\boldsymbol{B}_\mathrm{t}$. By this mechanism both steady and oscillatory $\boldsymbol{B}$-fields can be maintained (cf. section 16.4.). We speak of an $\alpha\omega$-mechanism, or an $\alpha\omega$-process. It clearly implies that the order of $\boldsymbol{B}_\mathrm{t}$ exceeds that of $\boldsymbol{B}_\mathrm{p}$.

Again a look at relation (12.8) is of interest. As long as the non-uniform rotation supplies energy to the mean magnetic field the second term on the right-hand side will give a positive contribution to the expression on this side. With the $\alpha\omega$-mechanism it is essentially this second term which compensates the first one, and the third one is of minor importance.

Whereas the δ_1-effect alone is unable to provide for a dynamo, a combination of this effect with other induction effects may lead to dynamo action. Like the α_1-effect also the δ_1-effect together with a sufficiently strong rotational shear, i.e. the ω-effect, implies the possibility of a dynamo mechanism. Again, $\boldsymbol{B}_\mathrm{t}$ is generated from $\boldsymbol{B}_\mathrm{p}$ mainly by virtue of the ω-effect, but $\boldsymbol{B}_\mathrm{p}$ from $\boldsymbol{B}_\mathrm{t}$ due to the δ_1-effect. In this way again both steady and oscillatory $\boldsymbol{B}$-fields can be maintained (cf. section 16.5.). Analogous to the $\alpha\omega$-mechanism we then speak of a $\delta\omega$-mechanism or a $\delta\omega$-process. Here too, $\boldsymbol{B}_\mathrm{t}$ will be much greater than $\boldsymbol{B}_\mathrm{p}$.

Let us once more return to the relation (12.8). In the case of a $\delta\omega$-mechanism the third term of the right-hand side is zero. Only the non-uniform rotation supplies energy to the mean magnetic field, more precisely to its toroidal part. The δ_1-effect is unable to feed the mean magnetic field; it only provides for a transfer of energy from the toroidal to the poloidal part.

It remains to be scrutinized whether a combination of the δ_1-effect with the δ_2, α_2, or α_3-effect may lead to dynamo action. As far as we can see there are no rigorous arguments against dynamos working on this basis. On the other hand no example of a dynamo of this kind has been constructed up to now. Since the α_3-effect plays the same role as an ω-effect a combination of the δ_1-effect with a sufficiently strong α_3-effect must allow a dynamo process analogous to the $\delta\omega$-process. In this context, however, we have to bear in mind that together with the δ_1 and the α_3-effect generally other induction effects occur too, e.g. the α_1-effect, so that the mechanism envisaged here is in concurrence with other ones, e.g. the α^2-mechanism, and possibly never dominates.

An illustration of the generation of axisymmetric $\boldsymbol{B}$-fields of both A0 and S0 types by the α^2, $\alpha\omega$, and $\delta\omega$-mechanisms is given in figures 15.3 and 15.4. There we have sketched which $\boldsymbol{B}_\mathrm{p}$-fields result from $\boldsymbol{B}_\mathrm{t}$-fields, and vice versa, as a consequence of α_1-effect, δ_1-effect, or non-uniform rotation with radial shear, i.e. ω-effect. For the original fields simple dipole or quadrupole structures are assumed. We restrict ourselves to the cases where α_1, δ_1, and $\mathrm{d}\omega/\mathrm{d}r$ do not change their signs within the conducting sphere. With the resulting fields, again only the dipole and quadrupole parts are regarded; parts corresponding to higher multipoles have been ignored. By means of figures 15.3 and 15.4 we may follow up the different types of interaction between $\boldsymbol{B}_\mathrm{t}$ and $\boldsymbol{B}_\mathrm{p}$-fields, which correspond to the different types of dynamo mechanisms.

It can easily be seen in which way the interaction of $\boldsymbol{B}_\mathrm{t}$ and $\boldsymbol{B}_\mathrm{p}$ due to the α_1-effect may lead to a maintenance of steady $\boldsymbol{B}$-fields, i.e. in which way the α^2-mechanism works. Besides, no possibility is indicated of generating oscillatory fields by means of this mechanism.

Replacing the α_1-effect by the δ_1-effect we arrive at a different situation. The difference consists in the orientation of $\boldsymbol{B}_\mathrm{t}$ and $\boldsymbol{B}_\mathrm{p}$. In agreement with the above considerations according to which the δ_1-effect alone cannot lead to a dynamo, no mechanism becomes apparent which could maintain steady $\boldsymbol{B}$-fields. The fact that also oscillatory $\boldsymbol{B}$-fields have to be excluded is more difficult to explain.

Let us now turn to the combination of the α_1-effect with a strong rotational shear, i.e. a strong ω-effect, which results in an $\alpha\omega$-mechanism. As long as α^+ is positive in the northern hemisphere we see that steady $\boldsymbol{B}$-fields of the A0 or S0 type are to be expected if $\mathrm{d}\omega/\mathrm{d}r$ is positive or negative, respectively. Extending the considerations to both signs of α^+ we may conclude that steady fields of the A0 or S0 type will occur depending on whether the signs of α^+ and $\mathrm{d}\omega/\mathrm{d}r$ in the northern hemisphere coincide or differ. In addition, the possibility of oscillatory fields of S0 or A0 type is indicated for the other sign combinations.

Finally, we deal with the combination of the δ_1-effect and a rotational shear, i.e. a strong ω-effect, which allows a $\delta\omega$-mechanism. For both signs of δ_1 we find that steady $\boldsymbol{B}$-fields, of both the A0 and the S0 types, seem to be possible if the signs of δ_1 and $\mathrm{d}\omega/\mathrm{d}r$ differ, and oscillatory fields if these signs coincide.

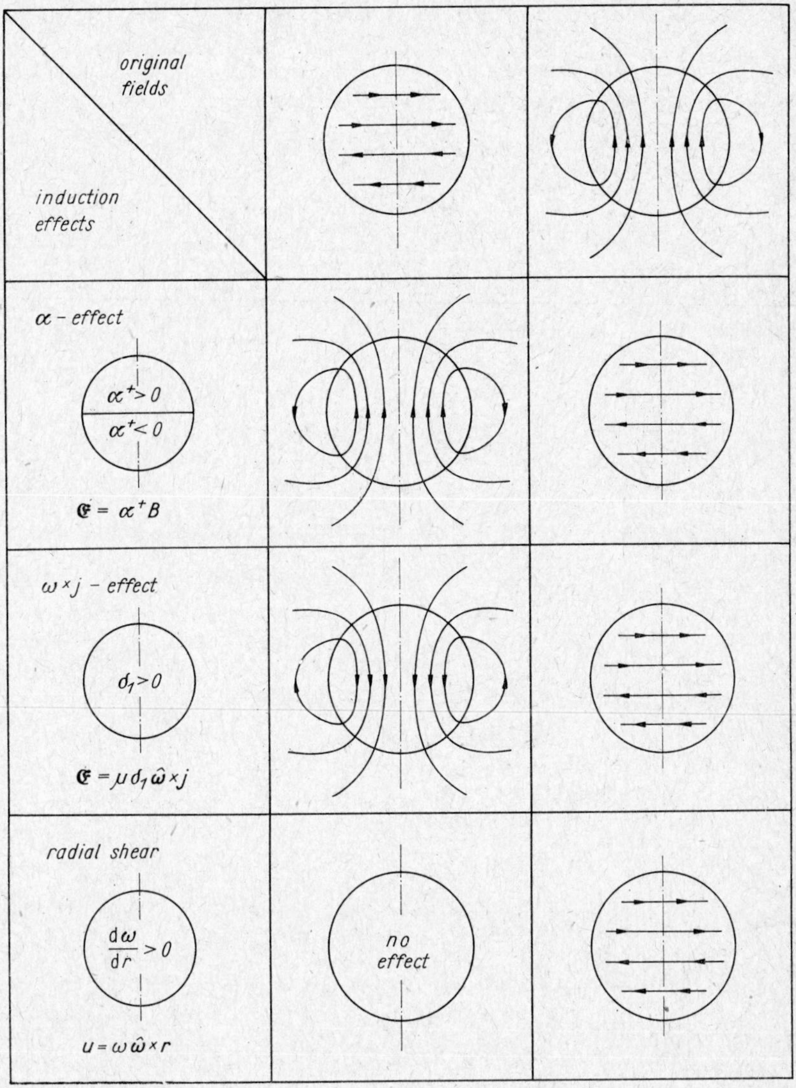

Fig. 15.3. $\boldsymbol{B}_t$-fields of A0 type generated from $\boldsymbol{B}_p$-fields of the same type, and vice versa, by the idealized α-effect, the $\omega \times j$ effect and the rotational shear. The representations of the fields correspond to those in figure 15.1. The signs of α^+ correspond to $\alpha_1 < 0$. If the signs of α^+, δ_1 or $\mathrm{d}\omega/\mathrm{d}r$ are inverted, the directions of the generated fields have to be inverted too

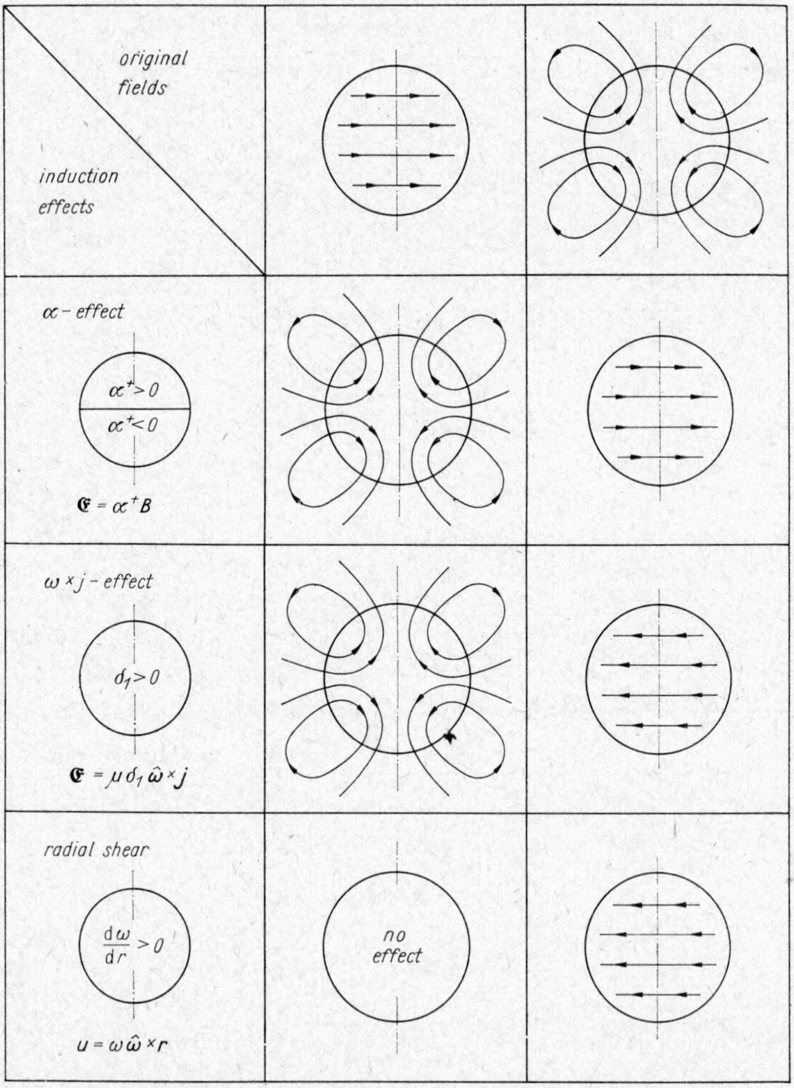

Fig. 15.4. $\boldsymbol{B}_t$-fields of S0 type generated from $\boldsymbol{B}_p$-fields of the same type, and vice versa, by the idealized α-effect, the $\omega \times j$ effect and the rotational shear. The explanations given with figure 15.3 apply here too

We want to emphasize that these considerations carried out by means of figures 15.3 and 15.4 hold, of course, only for the mentioned special assumptions on the signs of α_1, δ_1 and $d\omega/dr$. Furthermore, we have ignored all parts of the $\boldsymbol{B}$-fields which correspond to higher multipoles. It is already for this that our statements are just presumptions. They might give some orientation. Detailed investigations show, however, that some of these statements, especially those concerning the conditions for oscillatory fields, have to be modified.

We now remove the restriction to axisymmetric $\boldsymbol{B}$-fields, i.e. we admit again arbitrary $\boldsymbol{B}$-fields. Then the influences of $\boldsymbol{u}$ and $\mathfrak{E}$ on $\boldsymbol{B}_t$ and $\boldsymbol{B}_p$ can no longer be concluded from the equations (15.21), (15.25), and (15.26) but require a more complicated investigation on the basis of (15.42), (15.46), and (15.47). A scheme of these influences is given in figure 15.5.

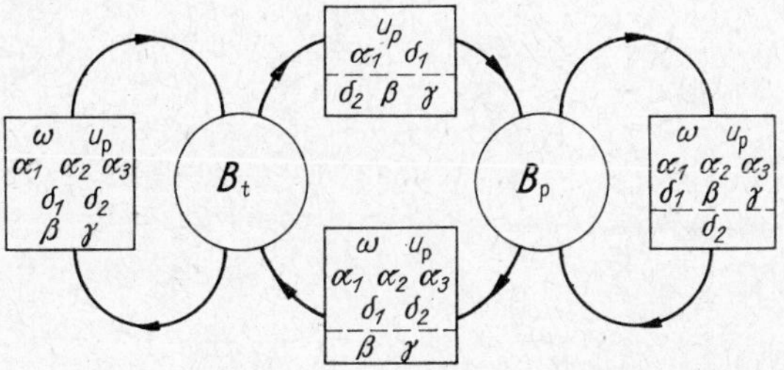

Fig. 15.5. The influences of the induction effects covered by $\boldsymbol{u}$ and $\mathfrak{E}$ on the toroidal and poloidal parts, $\boldsymbol{B}_t$ and $\boldsymbol{B}_p$, of an arbitrary, i.e. not necessarily axisymmetric $\boldsymbol{B}$-field. The effects given below the dotted lines only occur if the corresponding coefficients, i.e. β, γ, and δ_2, depend on latitude

Compared with the case of axisymmetric $\boldsymbol{B}$-fields, the situation looks more complex and more effort would have to be put into a detailed discussion of the possibilities of dynamo mechanisms. Therefore we shall only deal with some special points.

As above, we first ask for possibilities of dynamos without any coupling between $\boldsymbol{B}_t$ and $\boldsymbol{B}_p$, or with a coupling which, however, does not allow an interaction between $\boldsymbol{B}_t$ and $\boldsymbol{B}_p$. Again, the majority of these possibilities can be excluded. As for the remaining ones, there is no suggestion that dynamos exist.

Let us now take the case of interaction between $\boldsymbol{B}_t$ and $\boldsymbol{B}_p$. According to figure 15.5, this interaction occurs due to the u_p, α_1 or δ_1-effects or, if β, γ or δ_2 depend on latitude, also by the β, γ or δ_2-effects.

We have good reasons to assume that there are spherical dynamos operating due to a meridional circulation, i.e. the u_p-effect. Dynamos of this kind would be only a modification of that proposed by GAILITIS [2]. In this case fluctuating motions do not play any role. Of course, only $\boldsymbol{B}$-fields without any axisymmetric part may be generated.

Dynamos due to the α_1-effect, i.e. due to an α^2-mechanism have been discussed before. We now add that the α_1-effect is not only able to generate axisymmetric but also non-axisymmetric $\boldsymbol{B}$-fields (cf. section 16.3.). The arguments excluding dynamos due to the δ_1-effect also hold for non-axisymmetric $\boldsymbol{B}$-fields.

Again, combinations of the α_1 or δ_1-effect with non-uniform rotation, i.e. the ω-effect, should be considered, which lead to the $\alpha\omega$ or the $\delta\omega$-mechanism. We must, however, bear in mind that these mechanisms presuppose the generation of a sufficiently strong toroidal field from a poloidal field by means of the rotational shear. Whereas this is possible for axisymmetric fields, no comparable process is known for non-axisymmetric fields. This suggests the presumption that the $\alpha\omega$ and $\delta\omega$-mechanisms are not able to generate or maintain non-axisymmetric fields.

The possibility of dynamos due to the β-effect can, of course, again be excluded on grounds of relation (12.8). As for the γ and δ_2-effects, a dynamo process is at least not hinted.

The illustrations given for axisymmetric $\boldsymbol{B}$-fields in figures 15.3 and 15.4 shall now be extended to non-axisymmetric $\boldsymbol{B}$-fields of both A1 and S1 types in figures 15.6 and 15.7. The general explanations are the same as before; the rotational shear, however, is not considered. Again it can be seen that the interaction of $\boldsymbol{B}_\mathrm{t}$ and $\boldsymbol{B}_\mathrm{p}$ due to the α_1-effect may provide for the maintenance of $\boldsymbol{B}$-fields, and it is also demonstrated that a dynamo is impossible due to the δ_1-effect alone.

It should be mentioned that in these considerations one important point has been ignored. In the case of a non-axisymmetric $\boldsymbol{B}_\mathrm{p}$-field the α-effect does not only result in a generation of a $\boldsymbol{B}_\mathrm{t}$-field but also in a variation of the $\boldsymbol{B}_\mathrm{p}$-field. This variation consists mainly in a rotation about the axis of rotation of the conducting body. Therefore the non-axisymmetric $\boldsymbol{B}$-fields maintained by an α^2-mechanism have the form of waves travelling round the equator.

Let us sum up the results of these considerations on the possibilities of dynamo mechanisms. It should be emphasized that in these considerations the general form (15.6) of the mean velocity field $\boldsymbol{u}$ was taken into account but only the special form (15.8) of the electromotive force $\mathfrak{E}$. Apart from a dynamo mechanism which is only due to meridional circulation, three others, the α^2, $\alpha\omega$, and $\delta\omega$-mechanisms, have been found which presuppose induction effects by fluctuating motions. Whereas the α^2-process is able to generate and maintain both axisymmetric and non-axisymmetric $\boldsymbol{B}$-fields, the $\alpha\omega$ and $\delta\omega$-processes preferably lead to axisymmetric fields. On the other hand only in the case

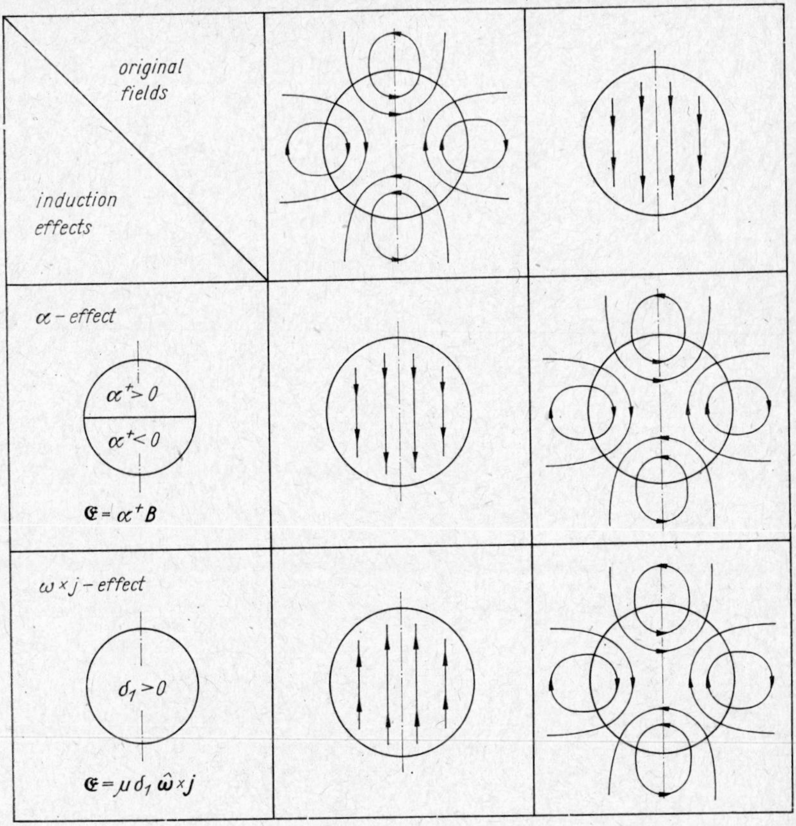

Fig. 15.6. $\boldsymbol{B}_t$-fields of the A1 type generated from $\boldsymbol{B}_p$-fields of the same type, and vice versa, by the idealized α-effect and the $\omega \times j$-effect. The representations of the fields correspond to those in figure 15.1. The signs of α^+ correspond to $\alpha_1 < 0$. If the signs of α^+ or δ_1 are inverted, the directions of the generated fields have to be inverted too

of the last two processes one has also to expect axisymmetric fields with an oscillatory time dependence.

If we replace the special form (15.8) of the electromotive force $\mathfrak{E}$ by the general form (15.11), a variety of further possibilities for dynamo mechanisms occur. Some of them turn out to be only modifications of the mechanisms mentioned before. Still to be thoroughly investigated remain processes resulting from, e.g., combinations of the $\beta^{(5)}$, $\beta^{(6)}$, or $\delta^{(1)}$-effects with the ω-effect.

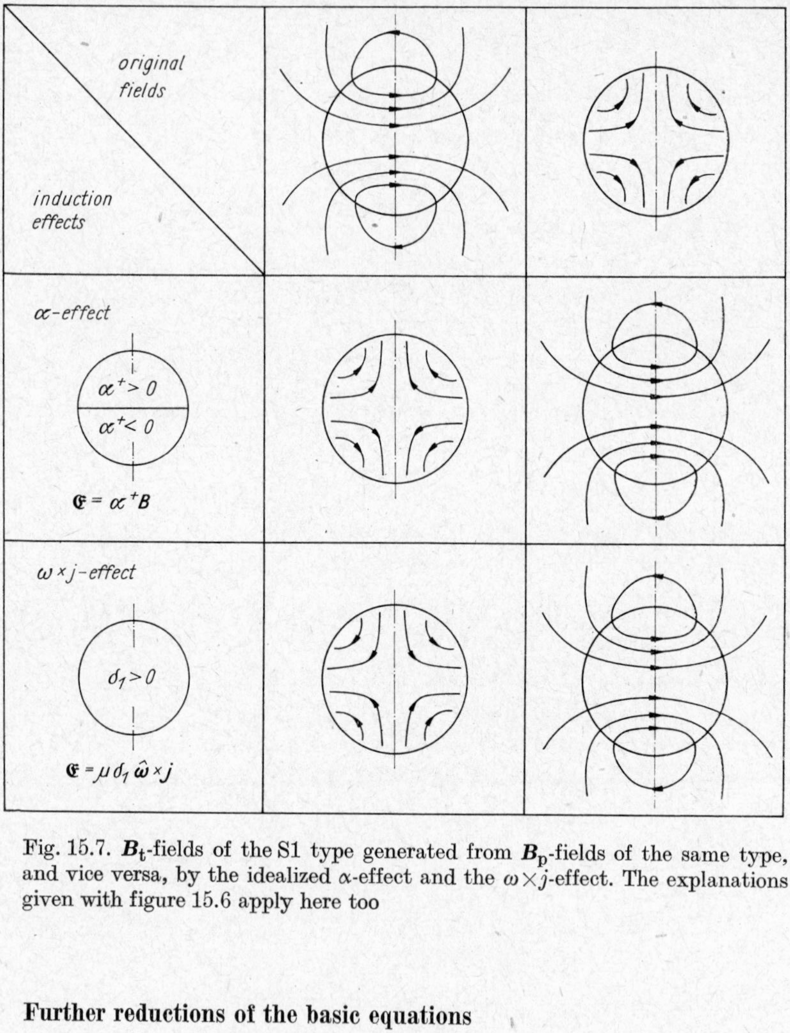

Fig. 15.7. $\boldsymbol{B}_t$-fields of the S1 type generated from $\boldsymbol{B}_p$-fields of the same type, and vice versa, by the idealized α-effect and the $\omega \times j$-effect. The explanations given with figure 15.6 apply here too

15.8. Further reductions of the basic equations

In the foregoing we have reduced the equations governing the $\boldsymbol{B}$-field to equations for certain scalar functions. When looking only for axisymmetric $\boldsymbol{B}$-fields we may restrict ourselves to equations (15.33)...(15.35), (15.37), and (15.38) for B and A. The occurrence of the operator Δ' suggests B and A be represented by series of associated Legendre polynomials $P_n^1(\cos\vartheta)$. On this basis the first investigations of spherical models of turbulent dynamos have been carried out, cf. e.g. Steenbeck and Krause [6, 7]. In the general case, where also non-axisymmetric **B**-fields are included, we have to deal with the equations (15.42)...(15.44), (15.46), and (15.47) for T and S. It proves useful to develop T and S in series of spherical harmonics $Y_n^m(\vartheta, \varphi)$. We shall derive those relations most relevant for the treatment of dynamo models in this way.

At first some dimensionless quantities are introduced. For the conductivity we write

$$\sigma = \sigma_0 \tilde{\sigma}, \tag{15.50}$$

where σ_0 is a constant and $\tilde{\sigma}$ a dimensionless function; if σ is constant we put $\sigma = \sigma_0$, i.e. $\tilde{\sigma} = 1$. Furthermore, we define dimensionless coordinates by

$$x = r/R, \quad \tau = t/\mu\sigma_0 R^2, \tag{15.51a, b}$$

and an operator, D_n, by

$$D_n f = \frac{1}{x^2}\left(\frac{\mathrm{d}}{\mathrm{d}x}\left(x^2 \frac{\mathrm{d}f}{\mathrm{d}x}\right) - n(n+1)\, f\right), \tag{15.52}$$

where f is any function and n an integer.

As explained in section 15.3. we may decompose any **B**-field which is a solution of the basic equations into **B**-modes of the Am and Sm types, which are again solutions of them. We ask only for these special modes. For clarity we speak of Al and Sl modes at this stage, i.e. denote the parameter m introduced above by l.

For each of these **B**-modes the functions T and S, and also $U^{(uB)}$, $V^{(uB)}$, $U^{(\mathfrak{E})}$, and $V^{(\mathfrak{E})}$ are represented by series of spherical harmonics. We write

$$T = T^+, \quad S = RS^+, \tag{15.53a, b}$$

$$U^{(uB,\mathfrak{E})} = (\mu\sigma_0 R)^{-1} U^{(uB,\mathfrak{E})+}, \quad V^{(uB,\mathfrak{E})} = (\mu\sigma_0 R^2)^{-1} V^{(uB,\mathfrak{E})}, \tag{15.53c, d}$$

where T^+, S^+, $U^{(uB,\mathfrak{E})+}$, and $V^{(uB,\mathfrak{E})+}$ are defined as

$$F^+ = \Big(\sum_{n,m} F_n^m(x)\; Y_n^m(\vartheta, \varphi)\Big) \exp\{(\lambda - \mathrm{i}\Omega)\,\tau\}. \tag{15.54}$$

The factors R, $(\mu\sigma_0 R)^{-1}$, and $(\mu\sigma_0 R^2)^{-1}$ ensure that T_n^m, S_n^m, $U_n^{(uB,\mathfrak{E})m}$, and $V_n^{(uB,\mathfrak{E})m}$ have the same dimensions. As for the Y_n^m we refer to (13.24)...(13.27). Both λ and Ω are real quantities independent of n and m. The summation is at first over all n and m with $n \geqq 0$ and $|m| \leqq n$. Because of (15.41), however, the terms with $n = 0$ have to be omitted. Furthermore, for an Al mode the T_n^m and $V_n^{(uB,\mathfrak{E})m}$ are non-zero only if $n = l + 2k$ and $m = \pm l$, the S_n^m and $U_n^{(uB,\mathfrak{E})m}$ only if $n = l + 2k + 1$ and $m = \pm l$ where k is a non-negative integer. For an Sl mode we have the same situation; we then have only to exchange the two relations for n with each other.

We are now able to reduce equations (15.42) for T and S to

$$D_n T_n^m - \frac{1}{x}\frac{\mathrm{d}\ln\tilde{\sigma}}{\mathrm{d}x}\frac{\mathrm{d}}{\mathrm{d}x}(xT_n^m) + V_n^{(uB)m} + V_n^{(\mathfrak{E})m} - (\lambda - \mathrm{i}\Omega)\, T_n^m = 0\,, \quad (15.55\mathrm{a})$$

$$D_n S_n^m + U_n^{(uB)m} + U_n^{(\mathfrak{E})m} - (\lambda - \mathrm{i}\Omega)\, S_n^m = 0, \quad \text{for} \quad 0 \leqq x \leqq 1. \quad (15.55\mathrm{b})$$

From (15.43) and (15.44 c) follows

$$T_n^m = 0, \quad S_n^m = a_n^m x^{-(n+1)}, \quad \text{for} \quad x \geqq 1, \quad (15.56\mathrm{a, b})$$

where the a_n^m are constants. Finally, (15.44 a, b) result in

$$T_n^m = 0, \quad \mathrm{d}S_n^m/\mathrm{d}x + (n+1)\, S_n^m = 0, \quad \text{at} \quad x = 1. \quad (15.57\mathrm{a, b})$$

From the relations (15.45) for $U_n^{(uB)m}$, $V_n^{(uB)m}$, $U_n^{(\mathfrak{E})m}$, and $V_n^{(\mathfrak{E})m}$ we conclude

$$U_n^{(uB)m} = \mu\sigma_0 R B_n^m \oiint (\boldsymbol{r} \cdot \mathrm{curl}\,(\boldsymbol{u}\times\boldsymbol{B}))\, Y_n^{-m}\, \mathrm{d}\tilde{\omega}\,, \quad (15.58\mathrm{a})$$

$$V_n^{(uB)m} = \mu\sigma_0 R^2 B_n^m \oiint (\boldsymbol{r} \cdot \mathrm{curl}\,\mathrm{curl}\,(\boldsymbol{u}\times\boldsymbol{B}))\, Y_n^{-m}\, \mathrm{d}\tilde{\omega}\,, \quad (15.58\mathrm{b})$$

$$U_n^{(\mathfrak{E})m} = \mu\sigma_0 R B_n^m \oiint (\boldsymbol{r} \cdot \mathrm{curl}\,\boldsymbol{\mathfrak{E}})\, Y_n^{-m}\, \mathrm{d}\tilde{\omega}\,, \quad (15.58\mathrm{c})$$

$$V_n^{(\mathfrak{E})m} = \mu\sigma_0 R^2 B_n^m \oiint (\boldsymbol{r} \cdot \mathrm{curl}\,\mathrm{curl}\,\boldsymbol{\mathfrak{E}})\, Y_n^{-m}\, \mathrm{d}\tilde{\omega}\,, \quad (15.58\mathrm{d})$$

with

$$B_n^m = (n(n+1)\, A_n^m)^{-1}. \quad (15.59)$$

The integrals are defined as in (13.2), and A_n^m is given by (13.27). Incidentally, the integrals can be rewritten according to (13.35). In $\boldsymbol{u}\times\boldsymbol{B}$, and $\boldsymbol{\mathfrak{E}}$ as well, $\boldsymbol{B}$ can be expressed by T and S. If this is done the integrands differ from the right-hand sides of relations (15.46) and (15.47) only by the factor Y_n^{-m}. For T and S in turn the series mentioned above should be inserted. Then the $U_n^{(uB)m}$, $V_n^{(uB)m}$, $U_n^{(\mathfrak{E})m}$, and $V_n^{(\mathfrak{E})m}$ can be calculated as functions of T_n^m, S_n^m, and their derivatives. For special forms of $\boldsymbol{u}$ and $\boldsymbol{\mathfrak{E}}$ explicit but lengthy expressions for $U_n^{(uB,\mathfrak{E})m}$ and $V_n^{(uB,\mathfrak{E})m}$ have been given in a paper by Rädler [9].

We may interpret the investigations of the kinematic dynamo models considered in this chapter again in terms of stability theory. Then $\lambda = 0$ corresponds to marginal stability. In this sense we shall speak of marginal values of the parameters determining λ, of marginal $\boldsymbol{B}$-modes etc. in the following.

CHAPTER 16

SPHERICAL MODELS OF TURBULENT DYNAMOS AS SUGGESTED BY COSMICAL BODIES. RESULTS OF COMPUTATIONS

16.1. General definitions

In the foregoing chapter we dealt with some general considerations of those spherical models of turbulent dynamos which are of interest for cosmical bodies. Quite a number of numerical investigations have been carried out for models working on the basis of α^2, $\alpha\omega$, and $\delta\omega$-mechanisms. Here we shall give a survey of the essential features of the results obtained.

We shall adhere to the assumptions introduced in the foregoing chapter. Unless defined otherwise the conductivity σ shall be constant. As for the electromotive force $\mathfrak{E}$ only the contributions given by the expression (15.8) are taken into account. For the coefficients appearing there we put

$$\alpha_1 = \alpha_0\tilde{\alpha}_1, \quad \alpha_2 = \alpha_0\tilde{\alpha}_2, \quad \alpha_3 = \alpha_0\tilde{\alpha}_3, \tag{16.1a, b c}$$

$$\delta_1 = \delta_0\,\tilde{\delta}_1, \quad \delta_2 = \delta_0\,\tilde{\delta}_2, \tag{16.1d, e}$$

$$\beta = \beta_0\tilde{\beta}, \qquad \gamma = \gamma_0\tilde{\gamma}, \tag{16.1f, g}$$

where α_0, δ_0, β_0, and γ_0 are constants and $\tilde{\alpha}_1$, $\tilde{\alpha}_2$, ... are dimensionless quantities which may generally depend on x and ϑ. As for the mean motion given by $\boldsymbol{u}$ we recall (15.6) and (15.7) and add

$$\omega = \omega_0 + \Delta\omega\, w, \tag{16.2a}$$

$$u_r = u_0 v_r, \quad u_\vartheta = u_0 v_\vartheta, \tag{16.2b, c}$$

with ω_0, $\Delta\omega$, and u_0 being constants, and w, v_r, and v_ϑ dimensionless functions of x and ϑ.

We furthermore define the parameters

$$C_\alpha = \mu\sigma_0\alpha_0 R, \qquad C_\delta = \mu\sigma_0\,\delta_0, \tag{16.3a, b}$$

$$s = \mu\sigma_0\beta_0, \qquad g = \mu\sigma_0\gamma_0 R, \tag{16.3c, d}$$

$$C_\omega = \mu\sigma_0\,\Delta\omega\, R^2, \quad C_u = \mu\sigma_0 u_0 R, \tag{16.3e, f}$$

which are also dimensionless. They are normalized measures for induction effects involved in $\mathfrak{E}$ and $\boldsymbol{u}$.

In relation (15.54) we have already introduced the dimensionless frequency, Ω, of a $\boldsymbol{B}$ mode. For an axisymmetric mode we choose $\Omega \geqq 0$ in the following. In the case of a non-axisymmetric mode we relate Ω to a frame of reference in which the equator is at rest, and we fix the sign such that $\Omega > 0$ or $\Omega < 0$ corresponds to eastward or westward migration, respectively.

When describing results of numerical investigations we refrain from giving information on the numerical methods applied and refer to the papers cited.

16.2. Definitions for special types of models

In the following, various models shall be considered which are determined by various specifications of the electromotive force $\mathfrak{E}$ and the mean velocity $\boldsymbol{u}$, i.e. of the functions $\alpha_1, \alpha_2, \ldots, \omega, u_r$, and u_ϑ. On several occasions special examples are taken from investigations by Rädler [16, 17]. The corresponding special choice of these functions shall be explained first.

In order to motivate the specification of $\mathfrak{E}$, we recall the corresponding results obtained in the second order correlation approximation (cf. sections 9.3., 9.4., and 9.5.) and assume that the dependence of $\alpha_1, \alpha_2, \ldots$ on the correlation tensor of the turbulent velocity field $\boldsymbol{u}'$ may simply be reduced to a dependence on $\overline{\boldsymbol{u}'^2}$, λ_{cor}, and τ_{cor}. Furthermore, $\overline{\boldsymbol{u}'^2}$ may depend on radius but λ_{cor} and τ_{cor} are considered as constant. We then put

$$\tilde{\alpha}_1 = \mathrm{d}\chi/\mathrm{d}x, \quad \tilde{\alpha}_2 = \zeta_\alpha/\tilde{\alpha}_1, \quad \tilde{\alpha}_3 = 0, \tag{16.4a, b, c}$$

$$\tilde{\delta}_1 = \chi, \quad \tilde{\delta}_2 = \zeta_\delta\, \tilde{\delta}_1, \tag{16.4d, e}$$

$$\tilde{\beta} = \chi, \qquad \tilde{\gamma} = \mathrm{d}\chi/\mathrm{d}x, \tag{16.4f, g}$$

and, in addition,

$$\overline{\boldsymbol{u}'^2} = u_0'^2 \chi. \tag{16.5}$$

The function χ, which may depend on x only, represents a normalized distribution of the turbulence intensity; u_0' means a characteristic value of the turbulent velocity. The parameters ζ_α and ζ_δ are supposed to be independent of x. As explained above, the α_3-effect is equivalent to a special non-uniform rotation and so the assumption $\tilde{\alpha}_3 = 0$ is no serious restriction.

The function χ is specified to be

$$\chi = \begin{cases} 0 & \text{for} \quad x \leqq x_{\chi 1} - d_{\chi 1} \\ \dfrac{1}{16}(8 + 15\xi - 10\xi^3 + 3\xi^5) & \xi = (x - x_{\chi 1})/d_{\chi 1} \\ & \text{for} \quad x_{\chi 1} - d_{\chi 1} \leqq x \leqq x_{\chi 1} + d_{\chi 1} \\ 1 & \text{for} \quad x_{\chi 1} + d_{\chi 1} \leqq x \leqq x_{\chi 2} - d_{\chi 2} \\ \dfrac{1}{16}(8 - 15\xi + 10\xi^3 - 3\xi^5) & \xi = (x - x_{\chi 2})/d_{\chi 2} \\ & \text{for} \quad x_{\chi 2} - d_{\chi 2} \leqq x \leqq x_{\chi 2} + d_{\chi 2} \\ 0 & \text{for} \quad x_{\chi 2} + d_{\chi 2} \leqq x . \end{cases} \tag{16.6}$$

In the following we distinguish between two different cases, *a* and *b*, characterized by $\chi(0) \neq 0$ and $\chi(0) = 0$, respectively. In the case *a* the turbulence also covers the centre of the sphere, in the case *b* it is restricted to some outer layers of the sphere. The corresponding χ-profiles are depicted in figure 16.1. In view of later considerations, in figure 16.2 the distribution of the sign of α^+, defined by (15.10), for different types of χ-profiles is given.

We may, of course, digress from the above motivation given for relations (16.4) and use them without interpreting χ according to (16.5), i.e. as normalized turbulence intensity. Thus the results of computations for which (16.4) and (16.6) have been used allow applications to a wide range of suppositions on $\mathfrak{E}$.

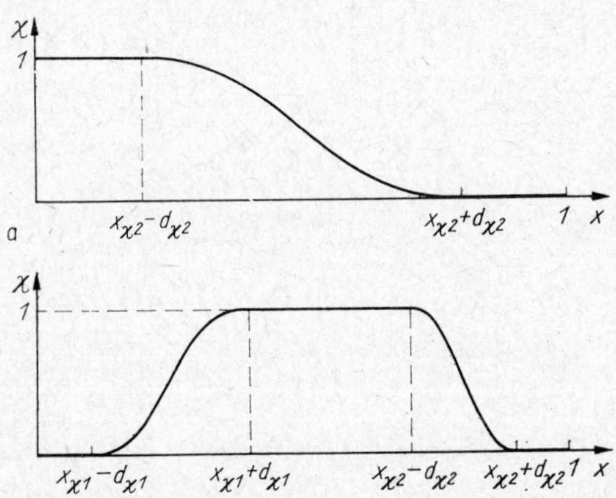

Fig. 16.1. The profiles of the normalized turbulence intensity χ, defined by (16.6).
Case a: The turbulence covers the centre of the sphere. Case b: The turbulence is restricted to some outer layers of the sphere

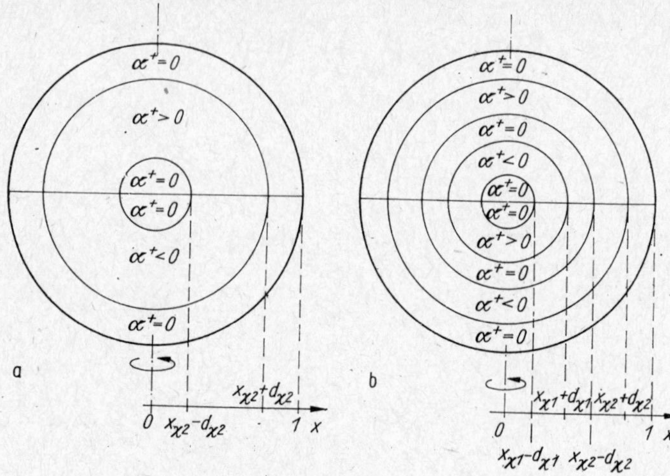

Fig. 16.2. The distribution of the signs of the quantity α^+, defined by (15.10), for the profiles of the turbulence intensity of types *a* and *b* as depicted in figure 16.1. The signs given here correspond to $\alpha_0 > 0$, i.e. $C_\alpha > 0$, and have to be inverted if $\alpha_0 < 0$, i.e. $C_\alpha < 0$

Let us now turn to the mean velocity $\boldsymbol{u}$. We assume that the angular velocity ω of the rotation depends on radius only and put

$$w = \begin{cases} 0 & \text{for} \quad x \leqq x_w - d_w \\ \dfrac{1}{4}(2 + 3\xi - \xi^3) & \xi = (x - x_w)/d_w \\ & \text{for} \quad x_w - d_w \leqq x \leqq x_w + d_w \\ 1 & \text{for} \quad x_w + d_w \leqq x . \end{cases} \tag{16.7}$$

As for the meridional circulation we suppose $\boldsymbol{u}_p$ to be solenoidal so that

$$v_r = \frac{1}{x \sin\vartheta} \frac{\partial}{\partial\vartheta} (\sin\vartheta\psi), \quad v_\vartheta = -\frac{1}{x} \frac{\partial}{\partial x} (x\psi), \tag{16.8a. b}$$

with ψ being a stream function specified to be

$$\psi = -f \sin\vartheta \cos\vartheta, \tag{16.9a}$$

$$f = \begin{cases} 0 & \text{for} \quad x \leqq x_u \\ \xi^3 & \xi = 4(x - x_u)(1 - x)/(1 - x_u)^2 \\ & \text{for} \quad x_u \leqq x \leqq 1 . \end{cases} \tag{16.9b}$$

Profiles of ω and streamlines $\psi = \text{const}$ are given in figures 16.3 and 16.4.

According to these specifications we have $C_\omega > 0$ for a differential rotation with $\mathrm{d}\omega/\mathrm{d}r > 0$, and $C_\omega < 0$ for $\mathrm{d}\omega/\mathrm{d}r < 0$. If $C_u > 0$ the meridional circulation runs in the outer layers from the equator to the poles, and if $C_u < 0$ from the poles to the equator. $C_\omega = \pm 2\pi$ means that the outer parts carry out one revolution more, or one less, than the inner ones within the time $\mu\sigma_0 R^2$. Analogously, $C_u = \pm 1$ corresponds to about one revolution in maximum velocity region within that time $\mu\sigma_0 R^2$.

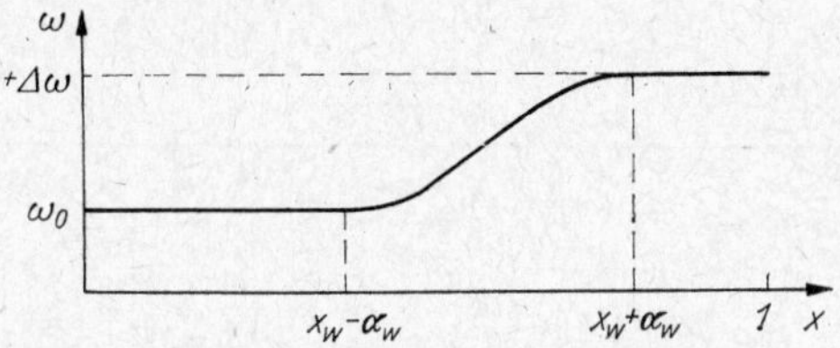

Fig. 16.3. The profile of the angular velocity ω of the non-uniform rotation, defined by (16.2 a) and (16.7). This ascending curve corresponds to $\Delta\omega > 0$, i.e. $C_\omega > 0$, a descending one occurs if $\Delta\omega < 0$, i.e. $C_\omega < 0$

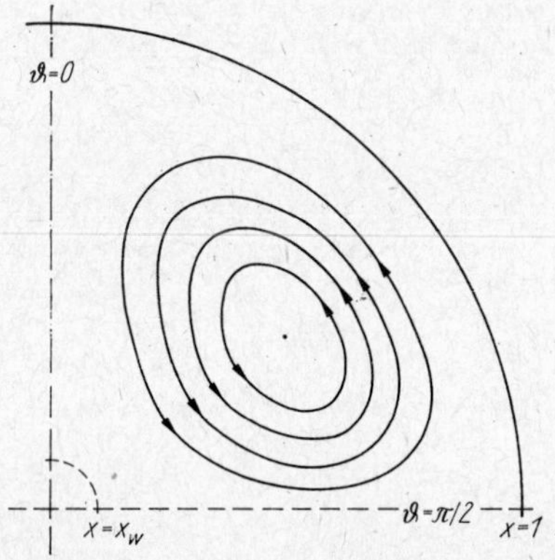

Fig. 16.4. The streamlines $\psi = \mathrm{const}$ of the meridional circulation, defined by (16.2b, c), (16.8) and (16.9). The arrows correspond to $u_0 > 0$, i.e. $C_u > 0$, and have to be inverted if $u_0 < 0$, i.e. $C_u < 0$

16.3. Models with α^2-mechanism

We direct our attention to models in which the magnetic field is generated by means of the α^2-mechanism. As explained above, in this case the generation of the toroidal from the poloidal field, and vice versa, is due to the α-effect. This presupposes that C_α is non-zero. We may restrict our discussion to the case $C_\alpha > 0$. Starting from results for $C_\alpha > 0$ we easily arrive at corresponding results for $C_\alpha < 0$ by virtue of the statement formulated with (15.17) and (15.18).

In the most simple case apart from the idealized α-effect no other induction effects are included. As long as this holds we speak of a pure α^2-mechanism.

The first models with magnetic field regeneration due to an α^2-mechanism have been proposed and computed by Steenbeck and Krause [3, 7] (1966, 1969). These models have been constructed with the intention to explain the origin of the Earth's magnetic field. Only a pure α^2-mechanism was taken into account. For the main part of the investigation referred to here, the coefficient $\tilde{\alpha}_1$ was specified according to

$$\tilde{\alpha}_1 = \mathrm{d}\chi/\mathrm{d}x, \quad \chi = \begin{cases} 0 & \text{for} \quad x \leqq x_\alpha \\ \dfrac{27}{4} \dfrac{(x - x_\alpha)^2 (1 - x)}{(1 - x_\alpha)^3} & \\ & \text{for} \quad x_\alpha \leqq x \leqq 1, \end{cases} \tag{16.10a, b}$$

with various values of x_α. Only axisymmetric $\boldsymbol{B}$-fields, i.e. A0 and S0-modes were considered, and these turned out to be steady. Marginal values of C_α in dependence on x_α are given in table 16.1. Field patterns for a special case are depicted in figures 16.5 and 16.6.

Table 16.1. Model with a pure α^2-mechanism; $\tilde{\alpha}_1$ given by (16.10). Marginal values of C_α for the A0 and S0-modes in dependence on x_α. These modes are steady, i.e. $\Omega = 0$

x_α	A0	S0
0.0	3.734	3.732
0.1	3.656	3.654
0.3	3.467	3.471
0.5	3.164	3.174
0.7	2.712	2.726
0.8	2.629	2.638

The results obtained by Steenbeck and Krause have been confirmed and supplemented by calculations by Roberts [4], which are also based on the assumption of a pure α^2-mechanism and are also restricted to axisymmetric fields. Roberts and Stix [2] studied models with an α^2-mechanism modified by the presence of a differential rotation. They also included the case of non-axisymmetric fields. Finally, in investigations by Rädler [17] in addition to the idealized α-effect other induction effects have also been taken into account, and again the cases of both axisymmetric and non-axisymmetric modes have been studied.

Let us explain and illustrate the essential findings of models with α^2-mechanisms on the basis of special results gathered from the investigations by RÄDLER [17]. For these investigations the specifications of section 16.2. hold.

At first we again consider a pure α^2-mechanism and suppose that $\tilde{\alpha}_1$ is given by (16.4a) and (16.6). In addition to the axisymmetric also non-axisymmetric $\boldsymbol{B}$-modes turn out to be possible. For special χ-profiles of types a and b marginal values of C_α and corresponding values of Ω are presented in tables 16.2 and 16.3 and in figure 16.7.

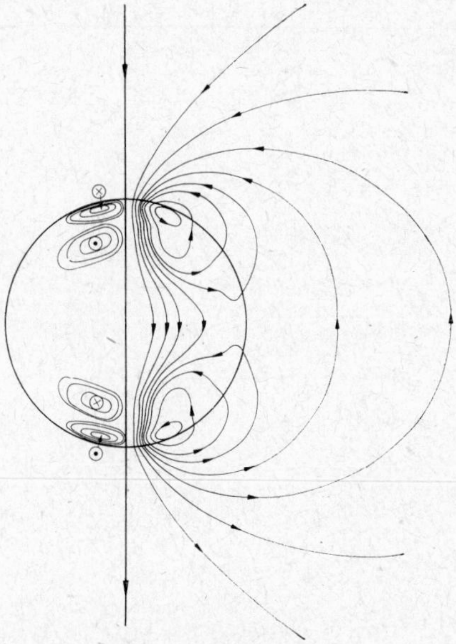

Fig. 16.5. Model with a pure α^2-mechanism; $\tilde{\alpha}_1$ given by (16.10), $x_\alpha = 0.5$. Field pattern for the A0-mode belonging to $C_\alpha = 3.164$. The left-hand side shows lines of constant magnitude of the toroidal field, the right-hand side field lines of the poloidal field

Table 16.2. Model with a pure α^2-mechanism; $\tilde{\alpha}_1$ given by (16.4a) and (16.6); χ-profile of type a, $x_{\chi 1} + d_{\chi 1} \leqq 0$, $x_{\chi 2} = 0.5$, $d_{\chi 2} = 0.4$. Marginal values of C_α and corresponding values of Ω

	A0	S0	A1	S1	A2	S2
C_α	4.66	4.75	4.86	4.75	6.31	6.27
Ω	0	0	−0.20	1.95	−0.16	0.81

Table 16.3. Model with a pure α^2-mechanism; $\tilde{\alpha}_1$ given by (16.4a) and (16.6); χ-profile of type *b*, $x_{\chi 1} = 0.5$, $x_{\chi 2} = 0.9$, $d_{\chi 1} = d_{\chi 2} = 0.1$. Marginal values of C_α and corresponding values of Ω

	A0	S0	A1	S1	A2	S2
C_α	2.68	2.70	2.78	2.77	3.12	3.12
Ω	0	0	−1.39	−2.65	−0.24	−0.85

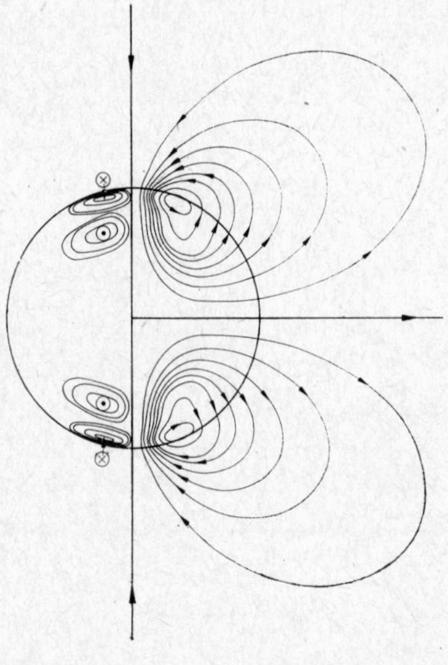

Fig. 16.6. Model with a pure α^2-mechanism; $\tilde{\alpha}_1$ given by (16.10), $x_\alpha = 0.5$. Field pattern of the S0-mode belonging to $C_\alpha = 3.174$. As for the representation of field lines the explanations given with figure 16.5 apply

In all investigated cases the A0, S0, A1, and S1-modes show marginal values of C_α which are rather close together. Mostly the A0-mode has the smallest of these marginal values. Therefore all these modes can be excited with almost equal ease; generally the A0-mode is slightly preferred. The A2 and S2-modes have higher marginal values of C_α and are less easily excited. These different properties of different modes may be understood as a consequence of the

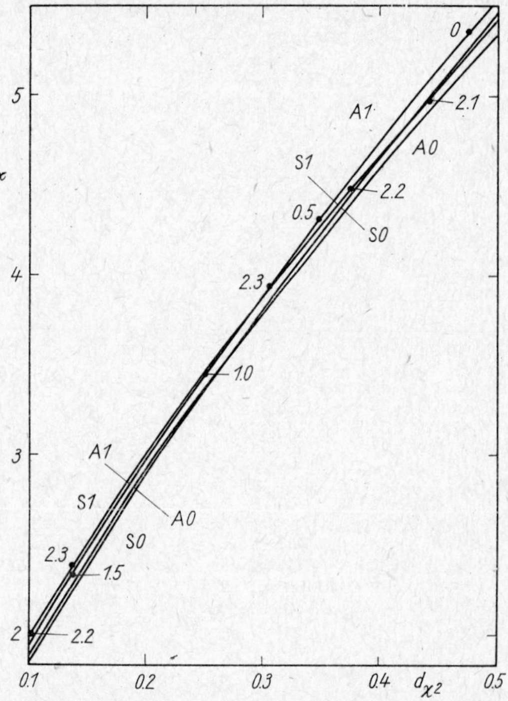

Fig. 16.7. Model with a pure α^2-mechanism; $\tilde{\alpha}_1$ given by (16.4a) and (16.6); χ-profile of type a, $x_{\chi 1} + d_{\chi 1} \leqq 0$, $x_{\chi 2} = 0.5$. Marginal values of C_α in dependence on $d_{\chi 2}$. For the A0 and S0-modes holds $\Omega = 0$, for the A1 and S1-modes Ω is given by the figures at the curves

fact that fields with less complicated geometrical structures show less dissipative effects than those with more complicated structures.

In the case of χ-profiles of type a with non-axisymmetric modes the eastward migration dominates. This result becomes plausible if we consider the distribution of the electromotive force due to the α-effect for simple field structures. It can easily be followed up that these structures indeed should change in time in the sense of an eastward migration. For χ-profiles of type b, however, a westward migration appears. Since such profiles lead to more complex distributions of the electromotive force this result is more difficult to explain.

Now in addition to the idealized α-effect other induction effects are taken into account too. More precisely, in each case only one of the other effects is included.

Studying at first the transition from the idealized to the real α-effect we allow ζ_α to be non-zero. There are some reasons to believe that ζ_α is generally negative; in special cases $\zeta_\alpha = -1/4$ has been found (cf. section 9.6.). For all hitherto investigated cases, with χ-profiles of types a and b, the marginal values of C_α increase if ζ_α decreases. Again, the A0, S0, A1, and S1-modes were observed to have marginal values of C_α lying closely together. A typical example is given in figure 16.8. In this case for sufficiently small ζ_α it is no longer the A0-mode but the S1-mode which has the smallest marginal value of C_α, i.e. which is excited most easily.

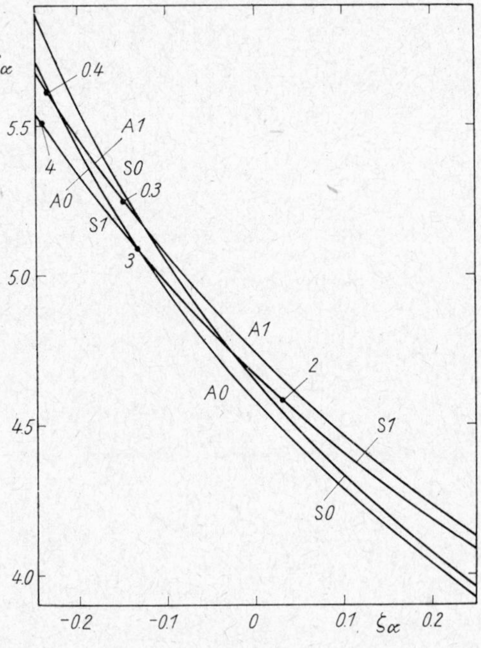

Fig. 16.8. Model including α_1 and α_2-effect; $\tilde{\alpha}_1$ given by (16.4a) and (16.6); χ-profile of type a, $x_{\chi 1} + d_{\chi 1} \leqq 0$, $x_{\chi 2} = 0.5$, $d_{\chi 2} = 0.4$. Marginal values of C_α in dependence on ζ_α. As for Ω the explanations given with figure 16.7 apply

Let us now replace the original by the turbulent electrical conductivity, i.e. include the β-effect so that the parameter s is no longer zero. Contrary to the original conductivity the turbulent conductivity generally depends on radius. Assuming that the turbulent conductivity is always smaller than the original one we have $s \geqq 0$. It is to be expected that the marginal values of C_α grow with s. Indeed in all investigations so far, again involving χ-profiles of types a and b, the marginal values of C_α show such dependence on s. A typical example is given in figure 16.9.

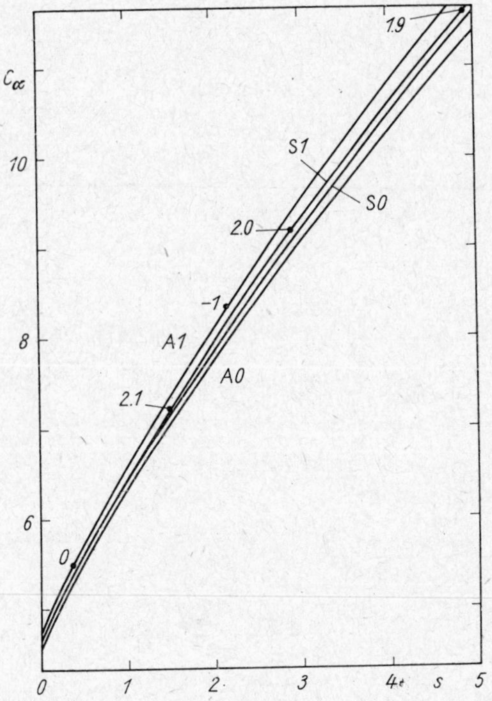

Fig. 16.9. Model including α_1 and β-effect; $\tilde{\alpha}_1$ and $\tilde{\beta}$ given by (16.4) and (16.6); χ-profile of type a, $x_{\chi 1} + d_{\chi 1} \leqq 0$, $x_{\chi 2} = 0.5$, $d_{\chi 2} = 0.4$. Marginal values of C_α in dependence on s. As for Ω the explanations given with figure 16.7 apply

Now the transport of magnetic flux due to the γ-effect is considered, which is connected with the parameter g. As long as the γ-effect is due to a gradient of turbulence intensity we have $g \geqq 0$. Then this transport of magnetic flux can also be described in terms of an enhanced magnetic permeability. This fact suggests that the marginal values of C_α grow with g. Again all investigations prove that this is indeed the case. An example is given in figure 16.10.

As far as the influence of the δ_1 and δ_2-effects is concerned, unfortunately only rather incomplete results are available.

We now consider the influence of a differential rotation, i.e. allow C_ω to be non-zero. The marginal values of C_α for different modes turn out to depend on C_ω in a different way. In all investigated cases for sufficiently high values of $|C_\omega|$ always one of the axisymmetric modes is clearly favoured from all other modes. With χ-profiles of type a the A0-mode is favoured for $C_\omega > 0$, and

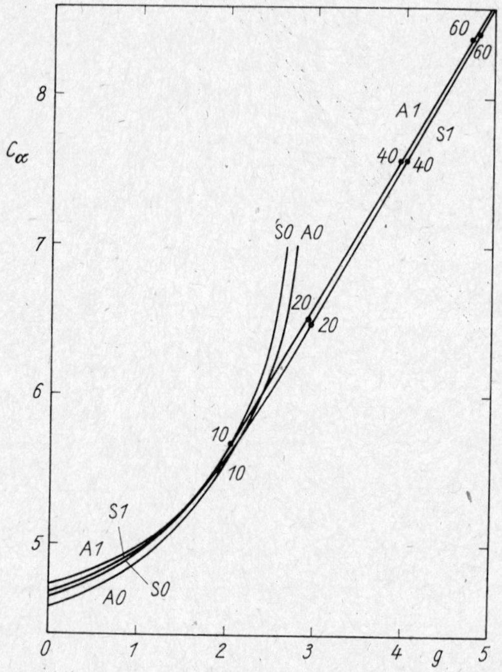

Fig. 16.10. Model including α_1 and γ-effect; $\tilde{\alpha}_1$ and $\tilde{\gamma}$ given by (16.4) and (16.6); χ-profile of type a, $x_{\chi 1} + d_{\chi 1} \leqq 0$, $x_{\chi 2} = 0.5$, $d_{\chi 2} = 0.4$. Marginal values of C_α in dependence on g. As for Ω the explanations given with figure 16.7 apply

the S0-mode for $C_\omega < 0$. An example is given in figure 16.11. The different behaviour of the A0 and S0-modes is plausible. As can easily be followed up in figures 15.3 and 15.4, in the case of an A0-mode a differential rotation with $C_\omega > 0$ is able to assist the α-effect in generating the toroidal from the poloidal field, and the same holds for an S0 mode and $C_\omega < 0$. In the region of small values of $|C_\omega|$ we have a rather complex situation. The results strongly depend on the χ and ω-profiles. It should be noted that Roberts and Stix [2] observed a preferrence for the S1 mode for special values of $|C_\omega|$.

Finally, a meridional circulation is taken into account, i.e. C_u is allowed to be non-zero. There are striking similarities between the dependencies of the marginal values of C_α on C_ω and on C_u. In all investigated cases for sufficiently high values of $|C_u|$ the axisymmetric modes are favoured compared with the non-axisymmetric ones. Among the axisymmetric modes again the A0 or S0-mode is more easily to be excited for $C_u < 0$ or $C_u > 0$, respectively.

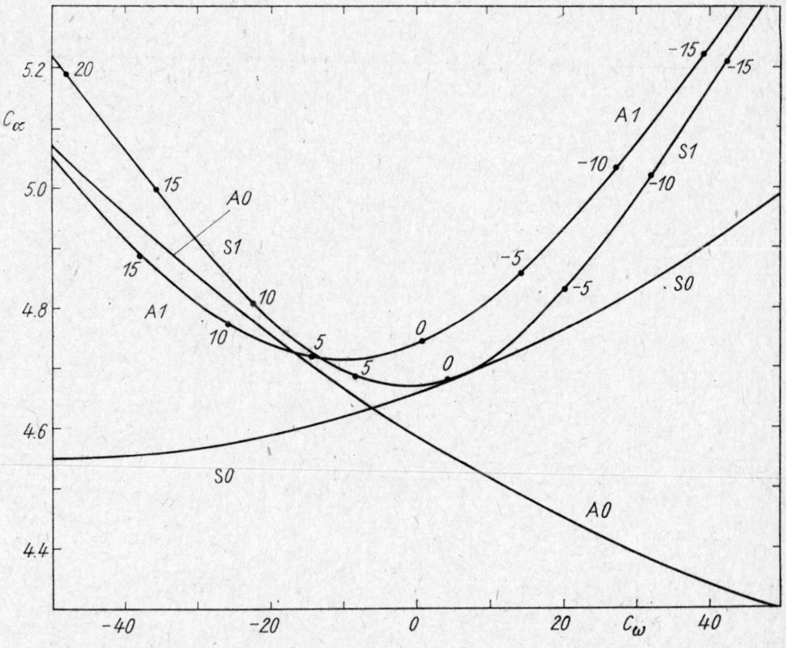

Fig. 16.11. Model including α_1 and ω-effect; $\tilde{\alpha}_1$ given by (16.4a) and (16.6); χ-profile of type a, $x_{\chi1} + d_{\chi1} \leqq 0$, $x_{\chi2} = 0.5$, $d_{\chi2} = 0.4$; w given by (16.7), $x_w = 0.5$, $d_w = 0.4$. Marginal values of C_α in dependence on C_ω. As for Ω the explanations given with figure 16.7 apply

An example is shown in figure 16.12. Contrary to the differential rotation, in the case of axisymmetric modes the meridional circulation does not provide for any coupling between the toroidal and poloidal fields. However, it is, obviously, able to influence axisymmetric field structures in a way that the interaction due to the α-effect is enhanced.

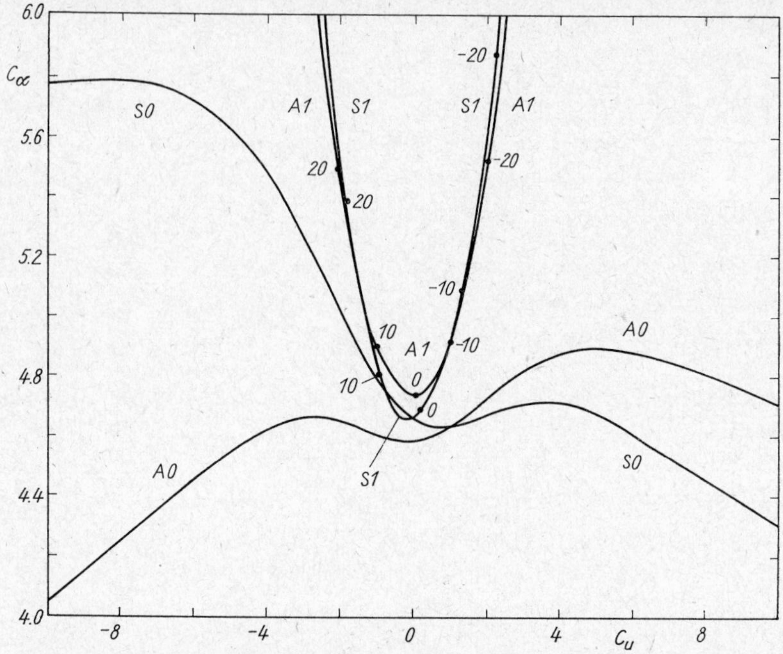

Fig. 16.12. Model including α_1 and u_p-effect; $\tilde{\alpha}_1$ given by (16.4a) and (16.6); χ-profile of type a, $x_{\chi 1} + d_{\chi 1} \leqq 0$, $x_{\chi 2} = 0.5$, $d_{\chi 2} = 0.4$; v_r and v_ϑ given by (16.8) and (16.9), $x_u = 0.1$. Marginal values of C_α in dependence on C_u. As for Ω the explanations given with figure 16.7 apply

16.4. Models with $\alpha\omega$-mechanism

We shall now deal with models allowing the maintenance of magnetic fields on the basis of an $\alpha\omega$-mechanism. In this case the generation of the poloidal from the toroidal field is due to the α-effect, more precisely the α_1-effect, and for the generation of the toroidal from the poloidal field the main role is played by the rotational shear, i.e. the ω-effect. Therefore, both C_α and C_ω have to be non-zero. Again, we shall largely restrict our considerations to $C_\alpha > 0$. They can easily be extended to $C_\alpha < 0$ by the help of the statement formulated with (15.17) and (15.18).

Let us at first assume that beside the idealized α-effect and the ω-effect no other induction effects are present, and that the generation of the toroidal field is exclusively due to the ω-effect, i.e. without any assistance of the α-effect. As long as these assumptions are fulfilled we speak of a pure $\alpha\omega$-mechanism.

The conditions for the generation of axisymmetric $\boldsymbol{B}$-fields by a pure $\alpha\omega$-mechanism show a remarkably simple dependence on C_α and C_ω, that is one on $C_\alpha C_\omega$ only. In order to see this it is useful to specify equations (15.21a) and (15.22) for $\boldsymbol{B}_\mathrm{t}$ and $\boldsymbol{A}_\mathrm{t}$ according to the above suppositions and to rewrite them, after introducing C_α and C_ω, into equations for $\boldsymbol{B}_\mathrm{t}$ and $\boldsymbol{A}_\mathrm{t}^+$ where $\boldsymbol{A}_\mathrm{t}^+ = C_\alpha^{-1}\boldsymbol{A}_\mathrm{t}$. Since in (15.21a) the term with α_1 is to be omitted the resulting equations contain C_α and C_ω only in the form $C_\alpha C_\omega$. Contrary to the conditions for the generation of axisymmetric $\boldsymbol{B}$-fields their structures do vary with C_α or C_ω even if $C_\alpha C_\omega$ remains unchanged. As can be seen from the foregoing considerations the ratio of $\boldsymbol{B}_\mathrm{t}$ to $\boldsymbol{B}_\mathrm{p}$ behaves like $(C_\omega/C_\alpha)^{1/2}$ if $C_\alpha C_\omega$ is fixed. It should be emphasized that these statements do not apply to non-axisymmetric $\boldsymbol{B}$-fields.

The first models in which magnetic fields are produced by an $\alpha\omega$-mechanism have been elaborated by Steenbeck and Krause [3, 6] (1966, 1969). These models are intended to reflect the situation at the Sun. Only a pure $\alpha\omega$-mechanism is taken into account. Calculations have been carried out with, e.g.,

$$\tilde{\alpha}_1 = -\frac{1}{2}\left(1 + \Phi((x - x_\alpha)/d_\alpha)\right), \tag{16.11a}$$

$$w = \frac{1}{2}\left(1 - \Phi((x - x_w)/d_w)\right), \tag{16.11b}$$

where Φ denotes the error function, for various x_α, d_α, x_w, and d_w. The authors looked only for axisymmetric $\boldsymbol{B}$-modes. They found such modes with an oscillatory time dependence. As an example we take the special case determined by $x_\alpha = 0.9$, $x_w = 0.7$, and $d_\alpha = d_\omega = 0.075$. In this case there is an oscillatory A0-mode with the marginal value $C_\alpha C_w = 2.07 \cdot 10^4$, and $\Omega = 31.8$. The oscillatory S0-mode is less easy to excite. As for the A0-mode a sequence of field patterns of different phases of the oscillation is depicted in figure 16.13.

The results by Steenbeck and Krause have been supported by an investigation by Deinzer and Stix [1]. Their model also acts on the basis of a pure $\alpha\omega$-mechanism but $\tilde{\alpha}_1$ is essentially reduced to a delta function and w to a step function. Again only axisymmetric $\boldsymbol{B}$-modes were considered. The authors found such modes with an oscillatory dependence on time. Stix [1] modified this model so that the α-effect takes part in the generation of the toroidal field too, and studied also non-axisymmetric $\boldsymbol{B}$-modes. Further investigation of models with an $\alpha\omega$-mechanism allowing oscillatory $\boldsymbol{B}$-modes has been carried out by Roberts and Stix [2], Roberts [4], Köhler [1], Stix [4, 6], Deinzer, v. Kusserow and Stix [1], Jepps [1], Ivanova and Ruzmaikin [1, 2, 3]. In some of these investigations in addition to the α-effect also other turbulent induction effects have been included, and in addition to the differential rotation also a meridional circulation.

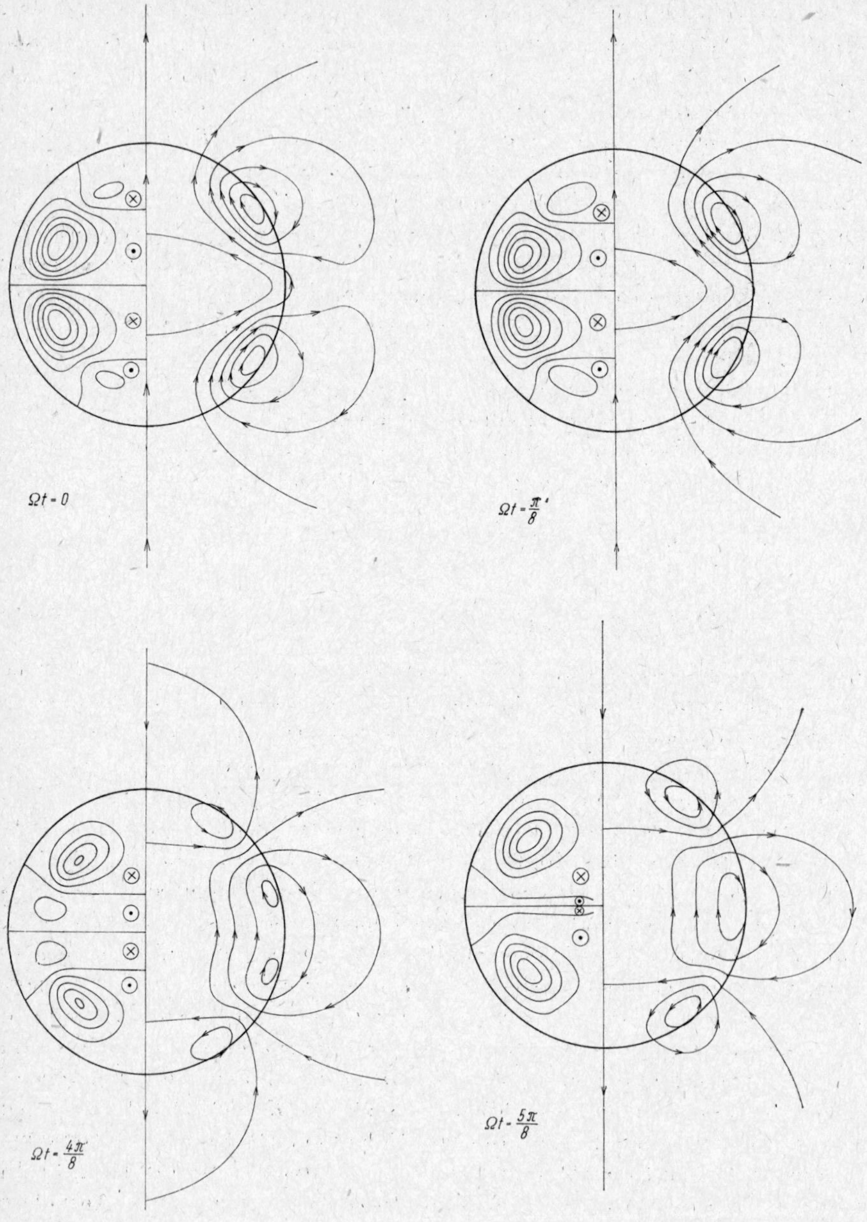

Fig. 16.13. Model with a pure $\alpha\omega$-mechanism; $\tilde{\alpha}_1$ and w given by (16.11), $x_\alpha = 0.9$, $x_w = 0.7$, $d_\alpha = d_w = 0.075$. Field patterns for the oscillatory A0-mode belonging to

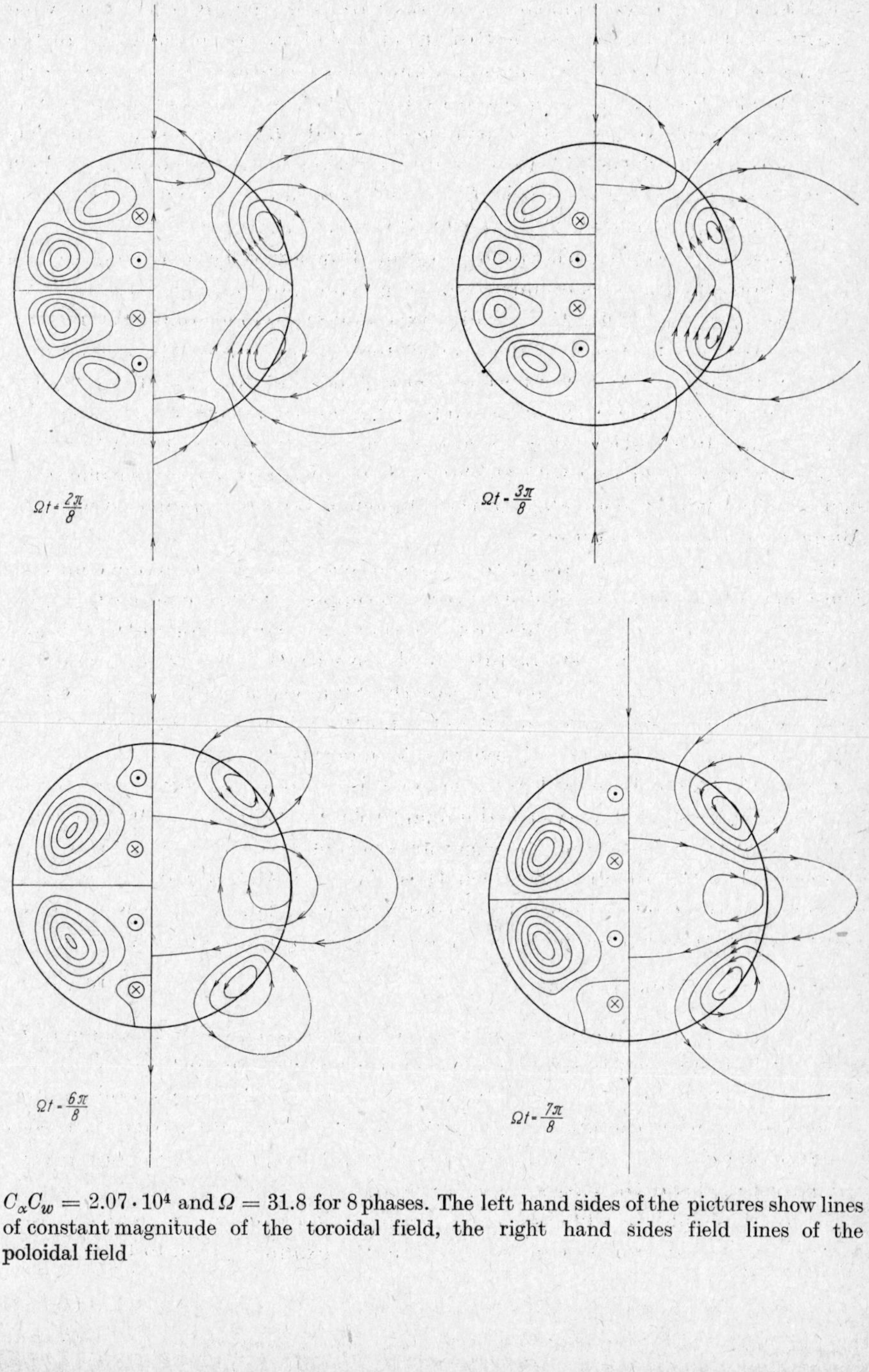

$C_\alpha C_w = 2.07 \cdot 10^4$ and $\Omega = 31.8$ for 8 phases. The left hand sides of the pictures show lines of constant magnitude of the toroidal field, the right hand sides field lines of the poloidal field

Up to now we have paid attention only to oscillatory $\boldsymbol{B}$-fields. Following BRAGINSKIJ [3] it was proved by ROBERTS [4] that an $\alpha\omega$-mechanism in the presence of a meridional circulation admits the existence of axisymmetric steady modes too. The investigations by LEVY [1] have shown also a pure $\alpha\omega$-mechanism is able to generate axisymmetric steady $\boldsymbol{B}$-modes. This result has been confirmed by STIX [4]. Further investigation on $\alpha\omega$-mechanisms which can generate axisymmetric steady $\boldsymbol{B}$-modes has been carried out by DEINZER, v. KUSSEROW and STIX [1, 2, 3] and by RÄDLER [17].

The fact that models with $\alpha\omega$-mechanism allow a variety of $\boldsymbol{B}$-modes raises the question which modes dominate under given conditions.

Let us first deal with the steady axisymmetric modes only. We restrict ourselves to cases in which α^+ does not change its sign inside the northern or southern hemisphere and, furthermore, ω depends only on r, and $\mathrm{d}\omega/\mathrm{d}r$ does not change its sign. As is to be expected from the considerations with figures 15.3 and 15.4 in all investigated cases among the steady modes, the A0-mode is clearly favoured compared with the S0-mode if in the northern hemisphere the signs of α^+ and $\mathrm{d}\omega/\mathrm{d}r$ coincide, and the S0-mode is favoured compared with the A0-mode, if these signs differ.

The situation with the oscillatory axisymmetric modes is more complex. The maintenance of an oscillatory mode presupposes rather complicated processes. In particular, special phase relations for the regeneration of the toroidal and the poloidal field have to be fulfilled. Even with the above simple assumptions on α^+ and $\mathrm{d}\omega/\mathrm{d}r$ it is not always the S0-mode among the oscillatory modes which is easiest to excite if in the northern hemisphere the signs of α^+ and $\mathrm{d}\omega/\mathrm{d}r$ coincide, and not always the A0-mode if these signs differ.

One would like to know at least whether it is one of the steady or one of the oscillatory modes which plays the dominant role. In the majority of the models investigated the α-effect and the shear are concentrated in certain thin layers. It seems that the distance between these layers essentially determines the character of the mode which is the easiest to excite. DEINZER, v. KUSSEROW, and STIX [3] studied a model with a pure $\alpha\omega$-mechanism and with

$$\tilde{\alpha}_1 = \delta(x - x_\alpha), \quad \mathrm{d}w/\mathrm{d}x = \delta(x - x_w), \tag{16.12 a,b}$$

where x_α and x_w are constants. Only axisymmetric modes were considered. In this model for sufficiently strong spatial separation of the induction effects, i.e. for sufficiently large $|x_\alpha - x_w|$ always one of the steady modes dominantes. In agreement with the above explanation it is the A0-mode if $C_\alpha C_\omega > 0$, and the S0-mode if $C_\alpha C_w < 0$. For certain small values of $|x_\alpha - x_\omega|$ one of the oscillatory modes is the easiest to be excited.

A remark should be added on the axisymmetric oscillatory $\boldsymbol{B}$-modes caused by the $\alpha\omega$-mechanism. As can be seen from the example given in figure 16.13

not only the magnitude of the field but also its geometrical structure varies in time. A typical feature of these modes is a migration of the toroidal field belts. This migration, which plays an important role for solar models, may be interpreted in terms of the dynamo waves discovered by PARKER [7].

As explained in section 15.7. the $\alpha\omega$-mechanism is presumably not able to generate or maintain non-axisymmetric **B**-fields. Indeed, at least in all cases of models with a pure $\alpha\omega$-mechanism, the existence of non-axisymmetric **B**-modes has not yet been proved. At first sight a model by KRAUSE [4], which corresponds to the suppositions of a pure $\alpha\omega$-mechanism, seems to allow of non-axisymmetric **B**-modes. The suppositions used there, however, cannot be justified for these modes.

In order to illustrate some of the foregoing statements let us explain some special results of investigations by RÄDLER [17], for which again the specifications of section 16.2. hold.

We consider models which involve the α-effect, more precisely the α_1-effect, and a rotational shear, i.e. the ω-effect. As long as only C_α is non-zero but C_ω vanishes we will have an α^2-mechanism, and if C_ω grows we may arrive at an

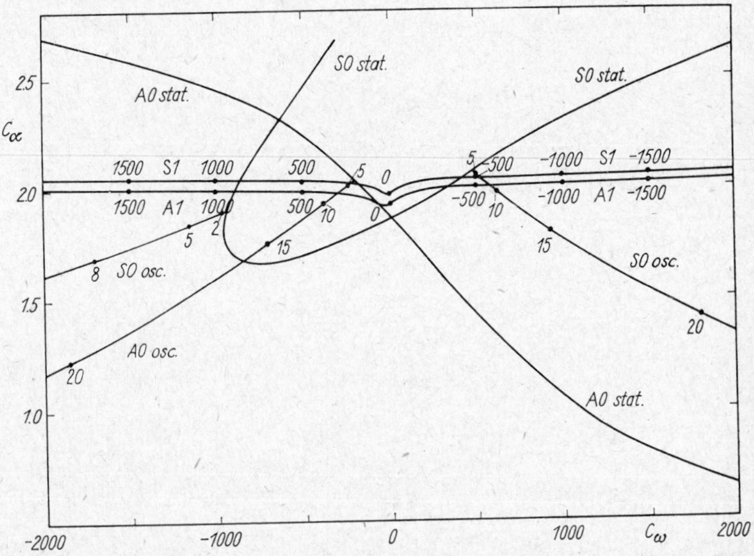

Fig. 16.14. Model including the α_1 and ω-effect; $\tilde{\alpha}_1$ given by (16.4a) and (16.6); χ-profile of type a, $x_{\chi 1} + d_{\chi 1} \leqq 0$, $x_{\chi 2} = 0.5$, $d_{\chi 2} = 0.1$; w given by (16.7), $x_w = 0.8$, $d_w = 0.1$. Marginal values of C_α in dependence on C_ω. For the steady modes holds $\Omega = 0$, for all other modes Ω is given by the figures at the curves

$\alpha\omega$-mechanism. It is instructive to study the transition from the $\boldsymbol{B}$-modes generated by an α^2-mechanism to those due to an $\alpha\omega$-mechanism.

We pay attention to a special model in which the α_1 and the ω-effect are restricted to rather thin separate spherical layers; the α-effect is concentrated in the inner, the ω-effect in the outer layer. In figure 16.14 marginal values of C_α and corresponding values of Ω are given for various $\boldsymbol{B}$-modes.

Let us focus on the case $C_\omega > 0$. Then, apart from very small C_ω, the steady A0-mode is the one to be excited most easily. For small C_ω it is due to an α^2-mechanism. For large C_ω, however, it must be ascribed to an $\alpha\omega$-mechanism. In this case the marginal value of $C_\alpha C_\omega$ may not be influenced by a further increase of C_ω. As can be seen from figure 16.15 this requires at least $C_\omega \gtrapprox 1000$. Only then the steady A0-mode is generated by a more or less pure $\alpha\omega$-process, i.e. the α-effect is by far less effective in the generation of the poloidal field than the ω-effect.

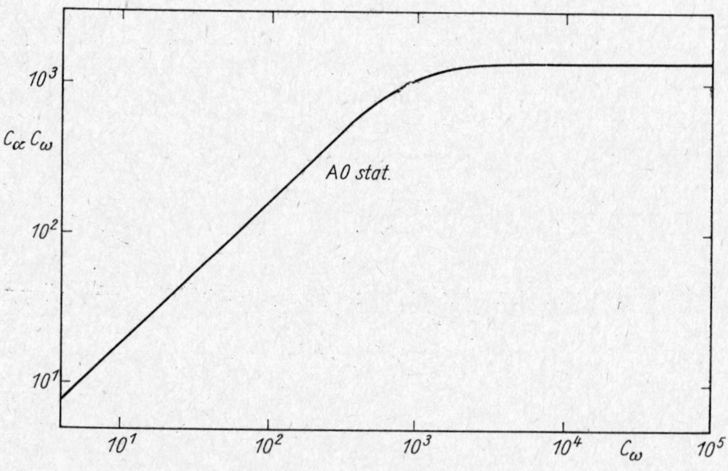

Fig. 16.15. Model including α_1 and ω-effect; $\tilde{\alpha}_1$ given by (16.4a) and (16.6); χ-profile of type a, $x_{\chi 1} + d_{\chi 1} \leqq 0$, $x_{\chi 2} = 0.5$, $d_{\chi 2} = 0.1$; w given by (16.7), $x_w = 0.8$, $d_w = 0.1$. Marginal values of $C_\alpha C_\omega$ in dependence on C_ω for the steady A0-mode

The steady S0-mode, however, shows quite different features for $C_\omega > 0$. It must be interpreted as a consequence of an α^2-mechanism which is to some extent modified by the presence of the ω-effect, but there is no indication of an $\alpha\omega$-mechanism. Likewise, the A1 and S1-modes can scarcely be related to an $\alpha\omega$-mechanism.

Contrary to the hitherto discussed modes which exist for both small and large C_ω, the oscillatory A0 and S0-modes have been found for sufficiently large C_ω only. They must be understood as a consequence of an $\alpha\omega$-mechanism. The marginal values of $C_\alpha C_\omega$ tend to a constant as C_ω grows.

The dependence of the marginal values of $C_\alpha C_\omega$ on the distance between the layer with α-effect and that with ω-effect has been studied for the steady A0-mode and the oscillatory S0-mode. Figure 16.16 shows, for $C_\omega = 2000$, that these values decrease if the distance is diminished. Moreover, the difference of these values for the two modes becomes smaller.

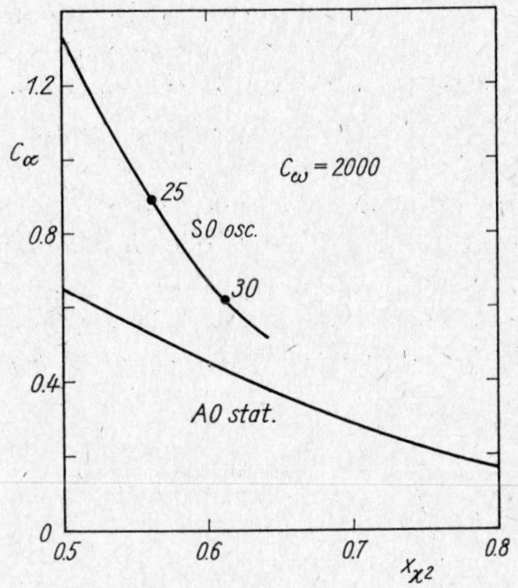

Fig. 16.16. Model including α_1 and ω-effect; $\tilde{\alpha}_1$ given by (16.4a) and (16.6); χ-profile of type a, $x_{\chi 1} + d_{\chi 1} \leqq 0$, $d_{\chi 2} = 0.1$; w given by (16.7), $x_w = 0.8$, $d_w = 0.1$. Marginal values of C_α, at $C_\omega = 2000$, in dependence on $x_{\chi 2}$ for the steady A0-mode and the oscillatory S0-mode. As for Ω the explanations given with figure 16.14 apply

For $C_\omega < 0$ a more complex situation occurs, which should not be discussed in detail. We point out that only the oscillatory A0 and S0-modes result from an $\alpha\omega$-mechanism.

For a model of this kind in which the α-effect is concentrated in the outer layer, and the ω-effect in the inner one, very similar results have been obtained.

Still to be discussed is the way in which the $\alpha\omega$-mechanism will be influenced if in addition to the α_1-effect and the ω-effect other induction effects occur.

The transition from the idealized to the real α-effect is of less importance. Provided that only axisymmetric fields are considered, this transition can only affect the contribution of the α-effect to the generation of the toroidal from the poloidal field, and this contribution is small anyway, or even negligible, compared with that of the rotational shear.

As far as the modification of electrical conductivity due to the β-effect and the turbulent transport of magnetic flux due to the γ-effect are concerned we have a similar situation as in the case of an α^2-mechanism. It is to be expected that the marginal values of C_α and also the corresponding values of Ω will grow with s and g. This is confirmed by results by Köhler [1] and by Ivanova and Ruzmaikin [1].

Studying the influence of the $\omega \times j$-effect we must bear in mind that this effect, even without any assistance of the α-effect, together with a rotational shear may give rise to a dynamo mechanism, that is the $\delta\omega$-mechanism. If in addition to the α_1-effect the δ_1-effect is also present, a combination of an $\alpha\omega$ and a $\delta\omega$-mechanism will occur. The two mechanisms can act in the same or the opposite sense. We again assume that α^+ does not change its sign inside the northern or the southern hemisphere, and that δ_1 does not change its sign at all. Let us first consider the case where both signs in the northern hemisphere coincide. As we may conclude from figures 15.3 and 15.4 the action of the $\alpha\omega$-mechanism for the A0-modes is weakened by the presence of the $\delta\omega$-mechanism,

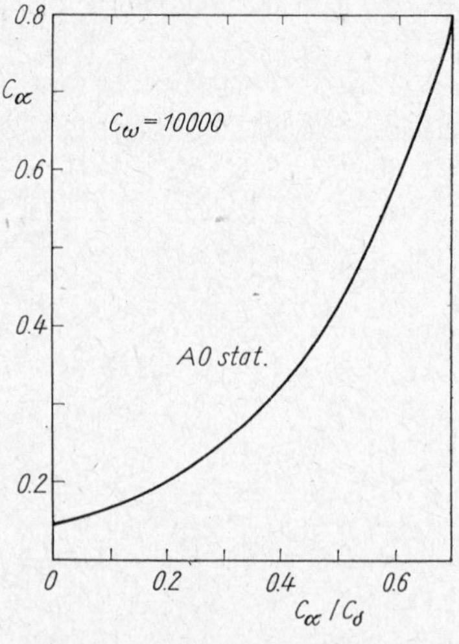

Fig. 16.17. Model including α_1, δ_1, and ω-effect; $\tilde{\alpha}_1$ and $\tilde{\delta}_1$ given by (16.4a), (16.4d) and (16.6); χ-profile of type a, $x_{\chi 1} + d_{\chi 1} \leqq 0$, $x_{\chi 2} = 0.5$, $d_{\chi 2} = 0.1$; w given by (16.7), $x_w = 0.8$, $d_w = 0.2$. Marginal values of C_α, at $C_\omega = 10000$, in dependence on C_δ/C_α for the steady A0-mode

and supported for the S0-modes. If the signs of α^+ and δ_1 in the northern hemisphere differ, the A0-modes are favoured, and the S0-modes are restrained. A special result supporting these statements is shown in figure 16.17.

The influence of meridional circulations is more complex. As can be seen from investigations by ROBERTS and STIX [2] and ROBERTS [4], meridional circulations are able to favour the steady axisymmetric $\boldsymbol{B}$-modes over the oscillatory ones.

Models with $\alpha\omega$-mechanism in which the back-reaction of the magnetic field on the α-effect is taken into account have been studied by JEPPS [1] and by IVANOVA and RUZMAIKIN [3].

16.5. Models with $\delta\omega$-mechanism

After having discussed the maintenance of magnetic fields by virtue of an $\alpha\omega$-mechanism we finally turn to the $\delta\omega$-mechanism. In this case the generation of the poloidal from the toroidal field is no longer a consequence of the α-effect but of the $\omega \times j$-effect whereas the generation of the toroidal from the poloidal field is again due to rotational shear. Hence C_δ and C_ω have to be nonzero. We restrict ourselves to $C_\delta > 0$; with respect to $C_\delta < 0$ we again refer to the statement formulated with (15.17) and (15.18).

There is some analogy between the $\alpha\omega$ and the $\delta\omega$-mechanisms. If besides the $\omega \times j$-effect, i.e. the δ_1-effect, and the rotational shear, i.e. the ω-effect, no other induction effects are included and, furthermore, the generation of the toroidal field is exclusively due to ω-effect we speak of a pure $\delta\omega$-mechanism. The conditions for the generation of axisymmetric $\boldsymbol{B}$-fields by a pure $\delta\omega$-mechanism, which depend on C_δ and C_ω, may be formulated as a requirement for $C_\delta C_\omega$ only.

The first model in which a $\delta\omega$-mechanism is responsible for the generation of a magnetic field has been proposed and calculated by RÄDLER [7, 8] (1969, 1970). A pure $\delta\omega$-mechanism with

$$\tilde{\delta}_1 = 1, \quad w = x \tag{16.13a,b}$$

was taken into account, and only axisymmetric steady modes were considered. For the S0-mode, which turned out to be most readily excited, a marginal value $C_\delta C_\omega = -166.5$ was found. The field pattern of this mode is given in figure 16.18. The essential results of this investigation have been confirmed and supplemented by ROBERTS [4]. Further results have been given by RÄDLER [17].

As RÄDLER [15, 17] has shown, a pure $\delta\omega$-mechanism is also able to generate oscillatory axisymmetric fields. This finding has been confirmed by STIX [6].

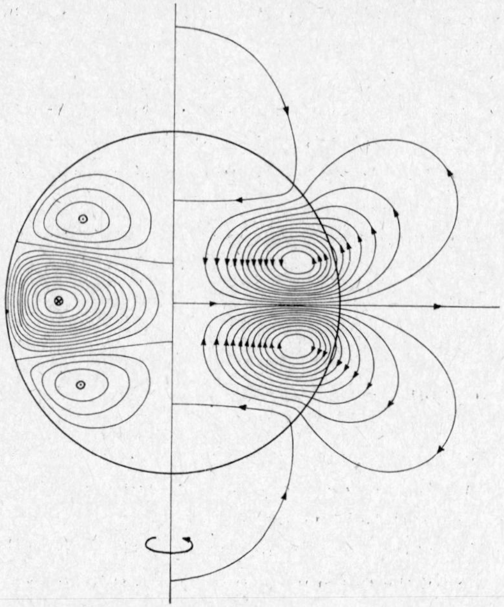

Fig. 16.18. Model with a pure $\delta\omega$-mechanism; $\tilde{\delta}_1$ and w given by (16.13). Field pattern of the S0-mode belonging to $C_\delta C_\omega = -166.5$. The left-hand side shows lines of constant magnitude of the toroidal field, the right-hand side field lines of the poloidal field

As explained in section 15.7. a pure $\delta\omega$-mechanism is probably unable to maintain non-axisymmetric $\boldsymbol{B}$-modes.

Again we are confronted with the question which modes dominate under given conditions. Unfortunately no satisfying answer can be given at present. We restrict ourselves to a pure $\delta\omega$-process with ω depending only on r and suppose that δ_1 and $\mathrm{d}\omega/\mathrm{d}r$ do not change their signs. From the considerations on the basis of figures 15.3 and 15.4 we may conjecture that steady modes only occur if the signs of δ_1 and $\mathrm{d}\omega/\mathrm{d}r$ differ. In all hitherto investigated cases this could be confirmed. Depending on the special δ_1 and ω-profiles in some cases the steady A0-mode, in other cases the steady S0-mode, proves to be dominant. Oscillatory modes, however, have been found for all combinations of the signs of δ_1 and $\mathrm{d}\omega/\mathrm{d}r$.

Let us illustrate the foregoing explanations by some results for special models with pure $\delta\omega$-mechanism by Rädler [17]. Again, the specifications of section 16.2. apply. Only axisymmetric $\boldsymbol{B}$-modes have been investigated.

There is a number of models with different specifications of δ_1 and ω for which steady A0 and S0-modes have been found. For various cases marginal

Table 16.4. Model with a pure $\delta\omega$-mechanism; $\tilde{\delta}_1$ given by (16.4d) and (16.6); w given by (16.7). Marginal values of $C_\delta C_\omega$ for steady A0 and S0-modes for various δ_1 and ω-profiles

δ_1-profile	ω-profile	A0	S0
$x_{\chi 1} + d_{\chi 1} \leqq 0$ $x_{\chi 2} - d_{\chi 2} \geqq 1$	$x_w = 0.5$ $d_w = 0.5$	-115.4	-110.6
$x_{\chi 1} = 0.75 \quad d_{\chi 1} = 0.25$ $x_{\chi 2} - d_{\chi 2} \geqq 1$	$x_w = 0.75$ $d_w = 0.25$	-166.4	-185.6
$x_{\chi 1} + d_{\chi 1} \leqq 0$ $x_{\chi 2} = 0.5 \quad d_{\chi 2} = 0.4$	$x_w = 0.5$ $d_w = 0.4$	-250.2	-291.5

values of $C_\delta C_\omega$ are listed in table 16.4. Depending on the special δ_1 and ω-profiles either the A0 or S0-mode may dominate.

Results for a model allowing oscillatory A0 and S0-modes are given in table 16.5. In this case the S0 or the A0-mode is preferred if $C_\delta C_\omega > 0$ or $C_\delta C_\omega < 0$, respectively. No steady modes have been found which could be excited more easily than the oscillatory modes.

Table 16.5. Model with a pure $\delta\omega$-mechanism; $\tilde{\delta}_1$ given by (16.4d) and (16.6); χ-profile of type b, $x_{\chi 1} = 0.8$, $d_{\chi 1} = 0.2$, $x_{\chi 2} - d_{\chi 2} \geqq 1$; w given by (16.7), $x_w = 0.5$, $d_w = 0.2$. Marginal values of $C_\delta C_\omega$ and corresponding values of Ω for oscillatory A0 and S0-modes

	A0	A0	S0	S0
$C_\delta C_\omega$	-2349	19268	-2217	4914
Ω	16.6	73.4	6.5	35.4

It remains to be discussed in which way the $\delta\omega$-mechanism is modified if in addition to the δ_1-effect and the ω-effect further induction effects are included.

The δ_1-effect is generally accompanied by the δ_2-effect. As explained above the δ_2-effect can influence a dynamo process only if δ_2 is not constant. But even in this case it is of less importance for a $\delta\omega$-mechanism. As long as δ_2 depends only on r the δ_2-effect influences only the generation of the toroidal field, the main contribution to which is provided by the ω-effect.

As for the turbulent electrical conductivity and the turbulent transport of magnetic flux the situation is presumably similar to that in the case of an α^2-mechanism. There are, however, no special investigations relating to this.

If the α-effect is included, we have again a combination of an $\alpha\omega$ and a $\delta\omega$-mechanism. In this respect we refer to the corresponding considerations of the foregoing section.

The influence of a meridional circulation on a $\delta\omega$-mechanism has not been investigated up to now.

CHAPTER 17

APPLICATIONS TO COSMICAL OBJECTS

17.1. Observational facts on the magnetic fields of the Earth, the Moon and the planets

Best and longest known among the cosmical magnetic fields is that of the Earth, the geomagnetic field. Contrary to the fields of other cosmical objects it is, at least at the surface of the Earth, accessible to direct measurements. Information on this field in outer space has been obtained from spacecraft studies.

Roughly speaking, the magnetic field in the closer regions around the Earth shows the structure of a dipole field with some symmetry with respect to the axis of rotation. The flux density at the equator is about $3 \cdot 10^{-1}$ G. A more detailed analysis reveals that the field cannot be completely described by a dipole located in the centre. The dipole plays only the role of the leading term in a multipole expansion. At the surface the dipole part exceeds the remaining part by a factor in the order of 10. The most striking deviation of the field from the symmetry with respect to the axis of rotation consists in an inclination of the dipole axis by an angle of 11°. Also the non-dipole part is not completely symmetric.

At first sight, the magnetic field of the Earth may be assumed to be steady yet there are some remarkable variations in time. We shall refrain here from the spectrum of the weak local or regional variations with time scales from minutes up to years; the causes of these variations are located outside the Earth's body. We refer, however, to the secular variations. The dipole axis shows a westward precession in the order of 0.05° per year, and the non-axisymmetric field structures related to higher multipoles a westward drift of about 0.2° per year. This corresponds to one revolution in $7 \cdot 10^3$ or $2 \cdot 10^3$ years, respectively. Paleomagnetic studies provided some information on the geomagnetic field in earlier epochs. Most of the time a dipole-like field of today's order of magnitude has existed. A remarkable phenomenon is that reversals of the field occur, i.e. sudden changes of its polarity. The length of the intervals with constant polarity varies from 10^5 to more than 10^7 years and seems to be statistically distributed; a reversal ranges over less than 10^5 years.

Quite unexpected, the Moon too showed to possess some magnetism, which was detected by spacecraft missions. The flux densities observed at the surface are, however, only in the order of 10^{-4} or 10^{-3} G. A striking feature of the field is its rather irregular structure. If a dipole component exists at all it is very small compared with the other components.

Some planets too have magnetic fields. As for Jupiter, the existence of a field has already been concluded from its radio emission. Details of this field and some information about fields of other planets are known due to spacecraft missions.

Jupiter's field is stronger than that of the Earth. It also possesses a dominant dipole component. The flux density at the equator is about 4 G, and the dipole axis is inclined against the rotational axis by an angle of about 10°.

On Mercury too a field with a dominant dipole component has been found. The flux density at the equator is about $4 \cdot 10^{-3}$ G, and the dipole axis is probably inclined by about 7°.

It should be noted that Mars possesses a magnetic field, the flux density of which at the surface is, however, only in the order of 10^{-3} G.

As far as Venus is concerned, no magnetic field with flux densities greater than $5 \cdot 10^{-5}$ G has been detected. The other planets could not yet be successfully investigated with respect to the presence of magnetic fields.

17.2. Dynamo theory of the Earth's magnetic field

A first step towards a theory of the Earth's magnetic field consists in the understanding of a mechanism which is able to maintain a steady dipole-like field. It is generally believed that the Earth's magnetic field is a consequence of a dynamo process which essentially takes place inside the core. The core shows an electrical conductivity like that of normal metals, and in all probability at least some parts of the core show internal motions. The mantle and the outer space, which show much smaller electrical conductivities, are of less importance for the process mentioned.

Treating the Earth's magnetic field as steady we refer to the time intervals between reversals, which are generally longer, sometimes much longer than 10^5 years, i.e. than the free decay time (cf. section 11.3.). The reversals require a special explanation.

An interesting and important step towards the understanding of the origin and the behaviour of the Earth's magnetic field is the theory of the nearly symmetric dynamo established by Braginskij [1, 2, 3] (1964) (cf. section 11.5.). As explained above, this theory deals with mean fields which are defined to be the axisymmetric parts of the original fields (cf. section 12.2.). It has been shown within this frame that a steady dipole-like field may exist as a consequence of non-uniform rotation, meridional circulation and some special non-axisymmetric motions. The mechanism of field generation is similar to the $\alpha\omega$-mechanism in the theory of the turbulent dynamo. The toroidal field is produced from the poloidal field by non-uniform rotation and the poloidal field in turn from the toroidal one due to non-axisymmetric motions. The toroidal field is much stronger than the poloidal field. Of course, the theory of the nearly symmetric dynamo can only give an explanation for the axisymmetric part of the geomagnetic field. As for the non-axisymmetric parts and their westward drift BRAGINSKIJ [6] proposed an interpretation in terms of the so-called MAC waves, i.e. special magnetohyrodynamic waves in presence of buoyancy and Coriolis forces (M stands for magnetic, A for Archimedean, and C for Coriolis forces).

Further promising suggestions for an explanation of the origin and some essential features of the Earth's magnetic field have arisen from the theory of the turbulent dynamo. First models were proposed by STEENBECK and KRAUSE [3, 7] (1966, 1969). We shall discuss some of these suggestions in the following.

We consider spherical dynamo models as described in section 15.1. and adopt all assumptions mentioned there. For reasons discussed in section 12.2. we define the mean fields on the basis of a time average with an averaging interval of some decades. We identify the spherical electrically conducting body with the Earth's core and suppose the conductivity to be constant. In agreement with the generally accepted data for the Earth's core, which were already used in section 11.3., we put $R = 3 \cdot 10^6$ m, $\mu = \mu_0 = 4\pi \cdot 10^{-7}$ VsA^{-1} m^{-1}, and $\sigma = \sigma_0 = 3 \cdot 10^5\ \Omega^{-1}$ m^{-1}. The Earth's mantle, and all outer space as well, is then treated as electrically insulating.

As far as the mean motion is concerned, the rotation of the core has to be taken into account. The rotation of the core essentially coincides with that of the mantle, i.e. we have one revolution per day, or $\omega_0 = 7.3 \cdot 10^{-5}$ s^{-1}. Opinions differ on whether or not a small relative motion of core and mantle is present. If it does not exist we put $\Delta\omega = 0$ so that $C_\omega = 0$. There are, however, some reasons to assume that the rotation rate of the main parts of the core slightly differs from that of the mantle. Then the outer layer of the core inevitably show a shear. Some concepts proceed from the assumption that, e.g., the dipole part of the magnetic field participates in the motion of the inner core, and that the westward drift immediately indicates the rotation of the inner core against the mantle. This would mean that $\Delta\omega = 3.10^{-11}$ s^{-1} and $C_\omega \approx 100$. There are hardly reasons to assume a meridional circulation inside the core. Therefore we at first put $u_0 = 0$ and, consequently, $C_u = 0$.

Let us now deal with the fluctuating motions. There are good reasons to assume that the inner core, the radius of which is about half the radius of the whole core, is in a solid state. Consequently, motions of that kind only occur in the remaining part of the core, the outer core. They may be a consequence of an unstable stratification of the core, of the precessional motion of the Earth or of other causes. Unfortunately, only little information is available on the structure, the magnitude and the length and time scales of these motions.

Let us give a rough estimate of the parameters C_α, C_δ, s, and g on the basis of the results of the second order correlation approximation. We first consider the quantity $\mu\sigma_0\lambda_{\text{cor}}^2/\tau_{\text{cor}}$. It seems reasonable to assume that $\lambda_{\text{cor}} > 5 \cdot 10^3$ m and $\tau_{\text{cor}} < 1$ d. In doing so we find that $\mu\sigma_0\lambda_{\text{cor}}^2/\tau_{\text{cor}}$ exceeds 10^2 and, therefore, the relations for the high conductivity limit may be used.

As for C_α we then rely on (16.3a), replace α_0 by $|\alpha|$ given by (9.60) and furthermore $\overline{\boldsymbol{u}'^2}$ by $u_0'^2$, Ω by ω_0, and λ_0 by R. Thus we see that $|C_\alpha|$ is of the order of $\mu\sigma_0 u_0'^2 \omega_0 \tau_{\text{cor}}^2$. If we put, e.g., $u_0' = 10^{-2}$ ms^{-1} and $\tau_{\text{cor}} = 5 \cdot 10^4$ s this expression takes the value of 20. Of course, the data used have a high degree of uncertainty. Nevertheless it can be seen that values of $|C_\alpha|$ exceeding unity may be possible.

A similar estimate can be carried out for C_δ. It shows that $|C_\delta/C_\alpha|$ is of the order of $(\mu\sigma\lambda_{\text{cor}}^2/\tau_{\text{cor}})^{-1}$. According to the above results this expression is smaller than 10^{-2}. Thus $|C_\delta|$ may be assumed to be much smaller than $|C_\alpha|$.

Finally, s and g prove to be of the same order of magnitude, which is given by $\mu\sigma_0 u_0'^2 \tau_{\text{cor}}$. With the data used above this expression takes the value of 2. Hence, s and g may be assumed to have the order of unity.

Let us now deal with some general features of models of turbulent dynamos suitable for an explanation of the origin and some of the basic properties of the Earth's magnetic field.

We first consider the case in which the mean motion consists simply in a rigid body rotation so that $C_\omega = C_u = 0$. Then only an α^2-mechanism will be at work. We discuss this possibility, which was first proposed by Steenbeck and Krause [6] (1969), on the basis of the special results presented in section 16.3.

In the simplest models of that kind apart from the idealized α-effect no other turbulence effect is included, i.e. a pure α^2-mechanism is taken into account. Comparing marginal values of C_α for the above examples with the estimate for C_α given before we see that there are good reasons to believe that the condition for the generation of magnetic fields is fulfilled. The improvement of these models by adding other turbulence effects generally increases the marginal values of $|C_\alpha|$, i.e. makes this condition more difficult to fulfil. But there is at least no strong argument which excludes the generation of magnetic fields.

In almost all investigated cases the marginal values of $|C_\alpha|$ for the A0-mode are slightly smaller than those for the other modes. Therefore a magnetic field is to be expected which to a large extent coincides with the A0-mode, so that the leading multipole of the poloidal part is a dipole parallel to the axis of rotation.

The marginal values of C_α for the A0, S0, A1, and S1-modes are, however, always close together. It seems reasonable to speculate that the magnetic field consists not only of a part similar to the A0-mode but also contains smaller parts similar to the S0, A1, and S1-modes. The coexistence of parts belonging to the A0 and S1-modes corresponds to an inclined dipole.

The non-axisymmetric modes, in particular the A1 and S1-modes are not steady but migrate round the equator. Therefore the non-axisymmetric parts of the magnetic field inevitably show drifts of that kind. In the special cases considered above the direction of migration depends on the turbulence distribution. Since the inner core is presumed to be solid, the distributions of type *a* seem to be less realistic than those of type *b*. For distributions of type *a* we have eastward drifts, for those of type *b*, however, westward drifts. The revolution times are given by $2\pi\mu\sigma_0 R^2/\Omega$. With the data used above we have $(9.2/\Omega) \cdot 10^5$ years. They are rather high compared with those of the drifts observed. We should bear in mind, however, that Ω considerably decreases if parameters like s and g grow.

As we see from these considerations, models of turbulent dynamos working on the basis of an α^2-mechanism offer an interesting possibility to explain not only the existence of the Earth's magnetic field and of its dipole-like structure but also the occurrence of non-axisymmetric parts of this field and its westward drift.

Up to now the mean motion has been considered as a rigid body rotation. The presence of a non-uniform rotation or a meridional circulation provides a completely different situation. Even for rather small values of $|C_\omega|$ and $|C_u|$ the marginal values of C_α for the A0, S0, A1, and S1-modes are no longer close together. Then the above explanation of the occurrence of non-axisymmetric parts of the magnetic field and of its westward drift can not be sustained anymore.

Let us now proceed on the assumption that the Earth's core shows a non-uniform rotation for which $|C_\omega|$ exceeds the order of unity. In this case $\alpha\omega$ or $\delta\omega$-mechanisms have to be envisaged.

We first discuss models which allow an $\alpha\omega$-mechanism. Compared with the case $C_\omega = 0$, in which only an α^2-mechanism is possible, in the region of such values of C_ω which are sufficient for an $\alpha\omega$-mechanism the marginal values of $|C_\alpha|$ for the dominant mode are generally smaller. Thus we may suppose that the conditions for the generation of magnetic fields are fulfilled. In view of the above estimation for C_ω we add that for $C_\omega = 100$ by no means a pure $\alpha\omega$-mechanism is to be expected. On the contrary, the α-effect may still play a considerable role in the generation of the toroidal field.

With respect to the Earth such models have to be selected in which under realistic conditions the smallest of the marginal values of $|C_\alpha|$ belongs to the steady A0-mode, so that this mode is the one to be excited most easily. As long as conditions of this kind can be justified we again find an explanation of the Earth's magnetic field, more precisely of its axisymmetric part.

A remarkable difference of the α^2 and the $\alpha\omega$-mechanism consists in the ratio of the magnitudes of the toroidal and the poloidal field. Whereas in the case of an α^2-mechanism the orders of both magnitudes agree, in the case of an $\alpha\omega$-mechanism the magnitude of the toroidal field exceeds that of the poloidal one considerably. This difference is, of course, of interest with respect to the dynamics of the Earth's core.

Contrary to the situation with the α^2-mechanism in the case of an $\alpha\omega$-mechanism the marginal values of C_α for different modes are generally quite different. Therefore the above straightforward interpretation of the deviations of the Earth's magnetic field from the axisymmetry and of its drift does not apply here.

As explained above there is some analogy between the nearly symmetric dynamo models by BRAGINSKIJ and the turbulent dynamo models with an $\alpha\omega$-mechanism. In both cases we get at first only an explanation of the axisymmetric part of the Earth's magnetic field, although this restriction is due to different causes. Departing from the conception of a strong toroidal part of the field BRAGINSKIJ [6] proposed an interpretation of the occurrence and the behaviour of the non-axisymmetric part in terms of MAC waves. We want to point out that this interpretation also applies in connection with models with an $\alpha\omega$-mechanism.

Finally, the possibility of a $\delta\omega$-mechanism remains to be discussed. According to the above estimation $|C_\delta|$ is much smaller than $|C_\alpha|$. Therefore the $\delta\omega$-mechanism will presumably never assert itself against the $\alpha\omega$-mechanism.

In conclusion we may state that the theory of the turbulent dynamo offers a promising basis for an explanation of the origin and of some fundamental features of the Earth's magnetic field. The main obstacle in the detailed elaboration of suitable models consists in the lack of information on the motions inside the core. As soon as more precise conceptions on these motions are available, more precise models appear to be possible which will reflect more special properties of the Earth's magnetic field.

17.3. Observational facts on magnetic fields at the Sun

The Sun shows a great number of magnetic phenomena. Our knowledge on the magnetic fields originates from the Zeeman displacement of spectral lines.

For a long time it has been well known that the phenomena of solar activity like sunspots, flares, and protuberances are accompanied by local magnetic fields with flux densities of up to a few 10^3 G. There are good reasons to interpret the local magnetic fields as consequences of a general field under the solar surface which has the same order of magnitude and varies in time with the period of the solar cycle, i.e. 22 years.

Sunspots commonly occur in pairs roughly aligned along a line of constant latitude. The spots of a pair have different magnetic polarities. Thinking of the rotation of the solar body we speak of a preceeding and a following spot of a pair. During a half-period of the solar cycle the preceeding spots in one hemisphere have the same polarity, and those of the other hemisphere the opposite polarity. In the next half-period all polarities are reversed. The distribution of sunspots in latitude varies with the phase of the cycle. At the beginning of a half-period the spots preferably occur at moderate latitudes but later on at latitudes near the equator. This behaviour may be represented by MAUNDER's butterfly diagram as reproduced in figure 17.1.

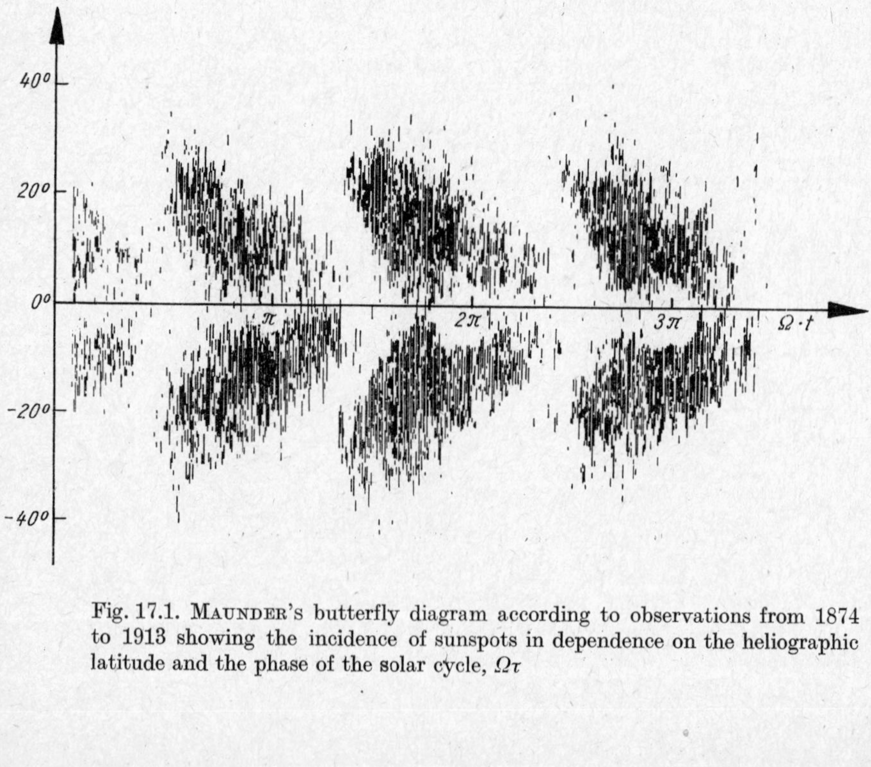

Fig. 17.1. MAUNDER's butterfly diagram according to observations from 1874 to 1913 showing the incidence of sunspots in dependence on the heliographic latitude and the phase of the solar cycle, $\Omega\tau$

From the observational material about sunspots some conclusions may be drawn on the structure and the behaviour of the general magnetic field mentioned. It must be assumed to be a toroidal field symmetric with respect to the axis of rotation and antisymmetric with respect to the equatorial plane. It mainly consists of two oppositely orientated belts, one in each hemisphere. At the beginning of each half-period these belts occur at moderate latitudes. While the field grows and again diminishes, these belts migrate towards the equator. In the next half-period this is repeated with opposite field directions. The extrema and zeros of this field correspond to the maxima and minima of solar activity.

Observations of the corona show that in addition to this toroidal field a poloidal field of dipole-type exists with some symmetry with respect to the axis of rotation. Its flux densities at the surface are, however, only of the order of 1 G. It proves to be also an alternating field with the period of the solar cycle. It reaches its extrema roughly at the minimum phases and disappears at the maximum phases of solar activity.

It must be emphasized that we have drawn a somewhat simplified picture of the solar magnetic field. It should be noted that also some sectorial structures of the fields, i.e. deviations from the axisymmetry, have been observed, and that the oscillations are not purely harmonic.

17.4. Dynamo theory of the solar cycle

The key to an understanding of the magnetic phenomena observed at the Sun seems to be the understanding of the origin of a magnetic field with the structure and the time behaviour as described before. It is hardly conceivable that an oscillatory field with a period of 22 years also covers the deep interior of the Sun. Because of the skin effect it may only be present in some outer layers. It is widely believed that the general magnetic field of the Sun is a consequence of a dynamo process which is essentially restricted to the convection zone the thickness of which is about 15% of the solar radius.

Pioneering work in the theory of the solar dynamo has been done by Parker [2] (1955) even before the concept of mean-field magnetohydrodynamics was available. Following an idea of Babcock [1] he assumed that the toroidal field is generated from the poloidal field by virtue of differential rotation. He showed that the poloidal field in turn can be created from the toroidal field by the action of a cyclonic turbulence. In mean-field magnetohydrodynamics this process is described by the α-effect.

It was the theory of the turbulent dynamo which was most successful in explaining the origin and the behaviour of the Sun's general magnetic field and, therefore, in understanding the solar cycle. Within this frame first models were given in investigations by STEENBECK and KRAUSE [3, 6] (1966, 1969). We shall discuss here some general features of turbulent dynamo models for the Sun.

We consider spherical dynamo models with some symmetric structure as described in section 15.1. For reasons discussed in section 12.2. we suppose a definition of the mean quantities on the basis of a time average with an averaging interval of one or two years. We identify the radius R of the models with the solar radius, i.e. $R = 7 \cdot 10^8$ m, and for the magnetic permeability we put again $\mu = \mu_0 = 4\pi \cdot 10^{-7}\,\mathrm{VsA^{-1}\,m^{-1}}$. As for the electrical conductivity σ at least a radial dependence should be admitted. The conductivity of the convection zone is in the order of $3 \cdot 10^3\,\Omega^{-1}\,\mathrm{m}^{-1}$. As explained in section 3.8. the turbulent conductivity σ_T of the convection zone is smaller by a factor of about 10^{-4}. In the following we refer to that version of the relevant equations where σ is replaced by σ_T, i.e. we do not use (12.1) but (12.2). In this sense we choose $\sigma_0 = 3 \cdot 10^{-1}\,\Omega^{-1}\,\mathrm{m}^{-1}$.

The skin depth, d, for an alternating electromagnetic field with a period T is given by $d = (T/\mu\sigma)^{1/2}$. Using $T = 22$ a, $\mu = \mu_0$, and identifying σ with σ_0, we obtain $d = 4 \cdot 10^7$ m, i.e. $d = 0.06R$. Thus the assumption that the dynamo process essentially takes place in the convection zone seems to be justified.

As far as the mean motion is concerned the rotation of the Sun has to be taken into account. As is well known, the Sun shows a differential rotation. The angular velocity at the surface, which can be immediately derived from the observations, depends on latitude, i.e. $\partial\omega/\partial\vartheta \neq 0$. The revolution time for the equatorial regions is about 25 days. Accordingly we put $\omega_0 = 2.9 \cdot 10^{-6}\,\mathrm{s}^{-1}$. The time for the polar regions, however, is about 27 days, which corresponds to $\Delta\omega = 2.2 \cdot 10^{-7}\,\mathrm{s}^{-1}$. With respect to the dynamo theory it is very important to know in what way the angular velocity depends on depth. Unfortunately this dependence is inaccessible to observations. The opinions differ not only with respect to the magnitude but also with respect to the sign of $\partial\omega/\partial r$. It seems reasonable to assume that the variations of ω with r are not smaller than those with ϑ. With the data given here we may therefore conclude that $|C_\omega| > 4 \cdot 10^4$. There is no observational evidence for a meridional circulation within the convection zone. We therefore put $u_0 = 0$ and, consequently, $C_u = 0$.

Turning now to the fluctuating motions we have to deal essentially with those motions of the convection zone which are connected with the granules and supergranules. These motions can be characterized by typical velocities, u_0', typical lengths, λ_{cor}, and typical times, τ_{cor}. For the granules we put $u_0' = 2 \cdot 10^3\,\mathrm{ms}^{-1}$, $\lambda_{cor} = 2 \cdot 10^6$ m, and $\tau_{cor} = 3 \cdot 10^2$ s, and for the supergranules

$u_0' = 4 \cdot 10^2 \text{ ms}^{-1}$, $\lambda_{\text{cor}} = 3 \cdot 10^7$ m, and $\tau_{\text{cor}} = 5 \cdot 10^4$ s. Not only the typical lengths mentioned but the scale height, h, of the mass density will be of interest. We may put $h = 10^5$ m.

With regard to the parameters C_α, C_δ, s, and g we at first calculate the quantity $\mu\sigma\lambda_{\text{cor}}^2/\tau_{\text{cor}}$. With the above data we get for both granules and supergranules $\mu\sigma\lambda_{\text{cor}}^2/\tau_{\text{cor}} > 10^7$. Consequently, relations developed for the high conductivity limit may be used.

For a very rough estimate of C_α we again rely on (16.3a) and replace α_0 by $|\alpha|$ given in (9.60), $\overline{\boldsymbol{u}'^2}$ by $u_0'^2$, Ω by ω_0, and λ_0 by h. Thus we find that $|C_\alpha|$ should be of the order of $\mu\sigma_0 u_0'^2 \omega_0 \tau_{\text{cor}}^2 R/h$. Depending on whether the data for granules or supergranules are used the last expression takes the values 10 or 10^2, respectively. Thus we have good reasons to assume that $|C_\alpha|$ at least exceeds the order of unity.

From a similar estimate for C_δ it may be concluded that $|C_\delta/C_\alpha|$ is of the order of $(\mu\sigma\lambda_{\text{cor}}^2/\tau_{\text{cor}})^{-1}\, h/R$. With the above data this expression is definitely smaller than 10^{-10}. Hence, $|C_\delta|$ is much smaller than $|C_\alpha|$.

Finally, s and g prove to be of the same order of magnitude. Since σ_0 was chosen to be of the order of σ_{T}, we may conclude that s, and therefore g too, must be in the order of unity.

We have to ask now which type of dynamo mechanism is responsible for the solar magnetic field. The α^2-mechanism can be ruled out. There is no hint that it is able to maintain axisymmetric fields with an oscillatory dependence on time. Thus the $\alpha\omega$ and $\delta\omega$-mechanisms have to be considered.

In all models elaborated up to now an $\alpha\omega$-mechanism has been supposed. Comparing the results given in section 16.4. with the above estimates of C_α and C_ω we see that this mechanism is able to maintain magnetic fields which show the wanted symmetries with respect to the axis of rotation and the equatorial plane and also an oscillatory dependence on time. We recall the example originating from Steenbeck and Krause [6] illustrated by figure 16.13. The period of oscillation is given by $2\pi\mu\sigma_0 R^2/\Omega$, i.e., with the above data, $(4.8 \cdot 10^4/\Omega)$ years, and so for this example we find a period of 80 years, which is at least of the required order of magnitude.

We are now faced with the question whether models of this kind also reflect the equatorward migration of the toroidal field belts during each half-cycle as indicated by Maunder's butterfly diagram. For special models such diagrams have been calculated on the basis of the assumption that sunspots only occur in those regions where the magnitude of the toroidal field in a certain depth exceeds a critical value. For the model illustrated in figure 16.13 the calculated butterfly diagram is given in figure 17.2. It shows an impressive similarity with that derived from observations.

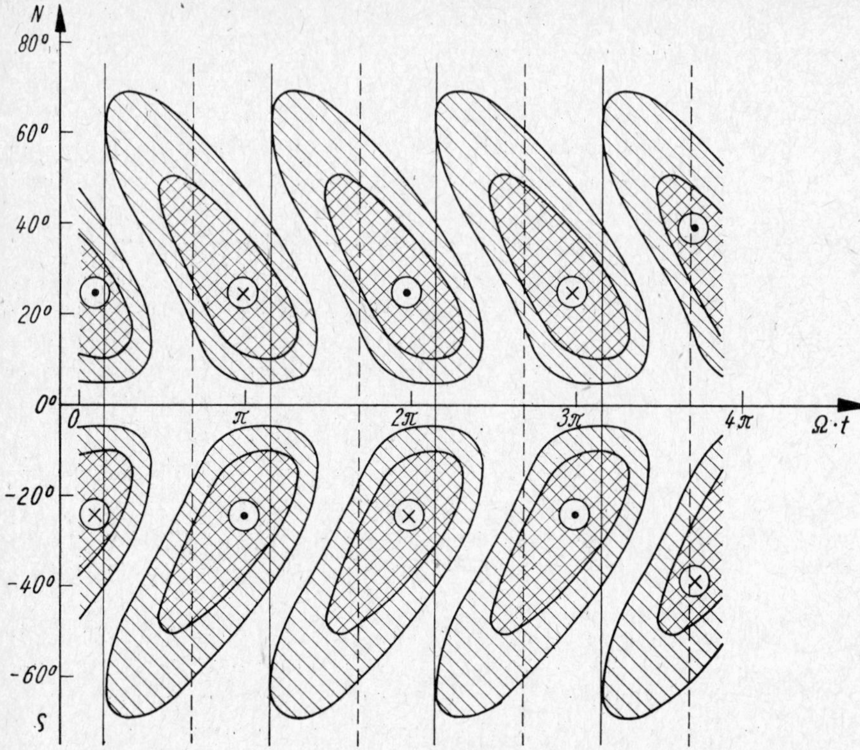

Fig. 17.2. Butterfly diagram calculated for a model with a pure $\alpha\omega$-mechanism mentioned in section 16.4. and illustrated by figure 16.13. The hatched area corresponds to $B_t > \frac{1}{3} B_{t\,max}$, the cross-hatched area to $B_t > \frac{2}{3} B_{t\,max}$, where B_t is the magnitude of the toroidal field and $B_{t\,max}$ its maximum value for all space and time

Many efforts have been made in order to find models which reflect as many features of the observational material as possible. We again refer to the investigations by ROBERTS and STIX [2], ROBERTS [4], YOSHIMURA [2, 4, 5], KÖHLER [1], STIX [4, 6], DEINZER, v. KUSSEROW and STIX [1], JEPPS [1], and IVANOVA and RUZMAIKIN [1, 2, 3]. A survey of the essential findings concerning solar dynamo models has been given by STIX [7].

Only one of these findings shall be explained here. It is apparently a general rule that the requirement of an equatorward migration of the toroidal field belts is only fulfilled if the signs of α^+ and $\partial\omega/\partial r$ in the northern hemisphere coincide; in this context $\partial\omega/\partial\vartheta$ plays a minor role.

According to the considerations of section 9.1., illustrated by figure 9.1, there are preferably left-handed helical motions in the northern hemisphere. This suggests that α^+ in this hemisphere is positive. Besides, the same conclusion may be drawn from the requirement that the phase relation between the toroidal and the poloidal field agrees with the observations; see STIX [7].

In this way we arrive at the requirement that $\partial\omega/\partial r$ is mainly positive. We remark that this result of the dynamo theory is in conflict with some of the theoretical investigations on the differential rotation of the convection zone according to which $\partial\omega/\partial r$ is mainly negative.

Like the $\alpha\omega$-mechanism also the $\delta\omega$-mechanism is able to maintain oscillatory magnetic fields. According to the above considerations, however, $|C_\alpha|$ is much greater than $|C_\delta|$. For this reason it is not very likely that the $\delta\omega$-mechanism is responsible for the solar magnetic field. Besides, according to an investigation by STIX [6] this mechanism hardly provides an equatorward migration of the toroidal field belts.

17.5. Observational facts on magnetic fields of stellar objects

At several stars too magnetic fields have been observed. In almost all cases the existence of these fields has been concluded from the Zeeman displacement of the spectral lines. Compared with the situation at the Sun, the corresponding investigations of stars are more difficult. Of course, only a relatively weak radiation flux can be received. In addition to this, the radiation flux always represents an integral over the radiation originating from different places of the stellar surface with different magnetic flux densities, motions and other properties so that the influence of the Zeeman effect on the spectral lines is rather complicated and can be obscured by the Doppler and other effects. That is why even under the best conditions only global fields of the order of 10^2 G can be detetced. If the Sun were so far away that it appeared like a star, there would be almost no chance of finding any magnetic field. The existence of very strong magnetic fields of stars with flux densities of more than 10^7 G can also be concluded from the polarization in the continuous spectrum.

Most of the stars on which magnetic fields were detected are, according to their spectral type, to be classified as Ap-stars. There are about one hundred such stars for which magnetic fields could be measured. The flux densities are of the orders of 10^3 or 10^4 G. But by no means all stars of this spectral type show discernable magnetic fields. The majority of the fields detected vary periodically in time with periods of the order of some days. The most probable explanation of this variation consists in the concept of a non-axisymmetric rotator. It is assumed that the magnetic field considerably deviates from the symmetry with respect to the axis of rotation; in a very simple example the field may correspond to a dipole which lies in the equatorial plane. Then, due to the rotation of the star, a non-corotating observer would register periodic variations of the field. A remarkable property of these stars with magnetic fields is the fact that some elements occur in anomalous abundances, especially rare earths. In some cases the abundances vary in time, then always having the same period as the magnetic field. It is probably the magnetic field that somehow causes a non-uniform distribution of those elements at the surface. Due to the rotation and this non-uniform distribution variations of abundances occur.

Among the white dwarfs too some objects with strong magnetic fields have been found. The flux densities are of the order of some 10^6 up to 10^8 G. As for the geometrical structure and the time dependence of these fields, no reliable information has been available up to now.

It should be added that for pulsars very strong magnetic fields are assumed because otherwise some observed phenomena like deceleration and the synchrotron radiation could not be explained. Flux densities of up to 10^{12} or 10^{13} G and field structures which correspond to an oblique rotator have been envisaged.

17.6. Remarks on dynamo mechanisms on magnetic stars

As explained in section 11.1. magnetic fields as observed on magnetic stars may exist for several reasons. The assumption that they are maintained by dynamo mechanisms is by no means imperative. As long as no observational evidence is given for time variations other than those due to rotation, the magnetic fields of Ap-stars can be considered to be primordial. Nevertheless, we shall deal here only with the possibility of magnetic fields generated by internal motions.

It seems reasonable to assume that the gravitational energy of magnetic stars is large compared with their magnetic and rotational energies, of which the latter is large compared with the first. As a consequence we consider a magnetic star to be a spherical body and assume that internal motions exist which show some symmetry with respect to the axis of rotation and the equatorial plane. More precisely, models as described in section 15.1. are envisaged. We have to look for dynamo mechanisms which are able to maintain non-axisymmetric magnetic fields.

Let us first consider the case where the motions show no fluctuating part. Then a dynamo might only be possible by virtue of meridional circulations; of course, it may be modified by the presence of differential rotation. As can be concluded from Cowling's theorem the magnetic field generated by a dynamo of that kind cannot contain any axisymmetric part. In particular, if in the multipole expansion a dipole occurs it must lie in the equatorial plane.

If fluctuating motions are present, turbulent dynamo mechanisms are to be envisaged, i.e. the α^2, $\alpha\omega$, and $\delta\omega$-mechanisms. According to the results presented in the foregoing, however, non-axisymmetric fields are only to be expected in the case of α^2-mechanisms. Thus we have to require that the rotational shear is not too strong. More precisely, $|C_\omega|$ may by no means exceed the order of 10^3. We recall that for the solar differential rotation $|C_\omega|$ is hardly smaller than $5 \cdot 10^4$. Therefore the turbulent dynamo mechanisms under consideration can only provide non-axisymmetric fields in the case of rather small deviations from a non-uniform rotation.

Since the magnetic fields on Ap-stars are rather strong, their influence on the motions may play an important role. Like these fields the internal motions too will then considerably deviate from symmetry with respect to the axis of rotation. However it cannot be expected that the symmetry with respect to the equatorial plane is broken too. Apart from the special case in which the excitation conditions for different magnetic field modes coincide, fields showing symmetry with respect to the equatorial plane will be generated. This symmetry will not be destroyed by the interaction of magnetic fields and motions.

These considerations suggest that observational data of magnetic stars should be interpreted in terms of field configurations which show symmetry with respect to the equatorial plane; cf. KRAUSE [4] and KRAUSE and OETKEN [1]. Thus the 'symmetric rotator' constitutes an alternative to the 'oblique rotator', which has been preferred until now. The incompleteness of accessible observational data is the reason that for the time being neither model can be made solely compulsory. A detailed analysis of the most thoroughly investigated Ap-stars was carried out by OETKEN [1, 2] relying on the assumption of equatorial symmetry as well for the magnetic fields as for the element inhomogeneities. In this way a close connection between the geometrical structures of both these quantities becomes visible which promises an access to a better understanding of the mechanisms producing the anomalous structures of these cosmical objects.

BIBLIOGRAPHY

For the papers denoted by an asterisk, which were written in German, a translation into English has been given by ROBERTS *and* STIX [1].

ACUNA, M. H., and N. F. NESS:
[1] The main magnetic field of Jupiter. J. Geophys. Res **81** (1976) 2917–2922

ALEMANY, A., R. MOREAU, P. L. SULEM and U. FRISCH:
[1] Influence of an external magnetic field on homogeneous mhd turbulence. J. Mécanique **18** (1979) 277–313

ANDRÉ, J. D., and M. LESIEUR:
[1] Evolution of high Reynolds number isotropic three-dimensional turbulence; influence of helicity. J. Fluid Mech. **81** (1977) 187–208

ANGEL, J. R. P.:
[1] Strong magnetic fields in White Dwarfs. Ann. New York Acad. Sci. **257** (1975) 80–81

BABCOCK, H. D.:
[1] The Sun's polar magnetic field. Astrophys. J. **130** (1959) 364–380

BABCOCK, H. W.:
[1] Zeeman effect in stellar spectra. Astrophys. J. **105** (1947) 105–119
[2] The topology of the Sun's magnetic field and the 22-year cycle. Astrophys. J. **133** (1961) 572–587

BABCOCK, H. W., and H. D. BABCOCK:
[1] Mapping the magnetic fields of the Sun. In "The Sun", ed. G. R. KUIPER, Univ. of Chicago Press 1953, 70–710
[2] The Sun's magnetic field 1952–1954. Astrophys. J. **121** (1955) 349–366

BACKUS, G. E.:
[1] The axisymmetric self-excited fluid dynamo. Astrophys. J. **125** (1957) 500–524
[2] A class of self-sustaining dissipative spherical dynamos. Ann. Phys. **4** (1958) 372–447

BACKUS, G. E., and S. CHANDRASEKHAR:
[1] On Cowling's theorem on the impossibility of self-maintained axisymmetric homogeneous dynamos. Proc. Nat. Acad. Sci. **42** (1956) 105–109

BATCHELOR, G. K.:
[1] On the spontaneous magnetic field in a conducting liquid in turbulent motion. Proc. Roy. Soc. **A 201** (1950) 405–416
[2] The theory of homogeneous turbulence. Cambridge Univ. Press 1953

BEVIR, M. K.:
[1] Possibility of electromagnetic self-excitation in liquid metal flows in fast reactors. J. Brit. Nuclear Soc. **4** (1973) 455–458

Biermann, L.:
[1] Bemerkungen über das Rotationsgesetz in irdischen und stellaren Instabilitätszonen. Z. Astrophys. **28** (1951) 304–309

Bochner, S.:
[1] Monotone Funktionen, Stieltjessche Integrale und harmonische Analyse. Math. Ann. **108** (1933) 378–410

Braginskij, S. I.: (Брагинский, С. И.):
[1] О самовозбуждении магнитного поля при движении хорошо проводящей жидкости. ЖЭТФ **47** (1964) 1084–1098 (Self excitation of a magnetic field during the motion of a highly conducting fluid. Sov. Phys.-JETP **20** (1964) 726–735)
[2] К теории гидромагнитного динамо. ЖЭТФ **47** (1964) 2178–2193 (Theory of the hydromagnetic dynamo. Sov. Phys.-JETP **20** (1964) 1462–1471)
[3] Кинематические модели гидромагнитного динамо земли. Геомагн. и Аэрономия **4** (1964) 732–747 (Kinematic models of the Earth's hydromagnetic dynamo. Geomagn. Aeron. **4** (1964) 572–583)
[4] Магнитогидродинамика земного ядра. Геомагн. и Аэрономия **4** (1964) 898–916 (Magnetohydrodynamics of the Earth's core. Geomagn. Aeron. **4** (1964) 698–711)
[5] Об основах теории гидромагнитного динамо земли. Геомагн. и Аэрономия **7** (1967) 401–410 (Principles of the theory of the Earth's hydromagnetic dynamo. Geomagn. Aeron. **7** (1967) 323–329)
[6] Магнитные вольны в ядре земли. Геомагн. и Аэрономия **7** (1967) 1050–1060 (Magnetic waves in the Earth's core. Geomagn. Aeron. **7** (1967) 851–859)
[7] Магнитогидродинамические крутильные колебания в земном ядре и вариации длины суток. Геомагн. и Аэрономия **10** (1970) 1–12 (Torsional magnetohydrodynamic vibrations in the Earth's core and variations in day length. Geomagn. Aeron. **10** (1970) 1–8)
[8] О спектре колебаний гидромагнитного динамо земли. Геомагн. и Аэрономия **10** (1970) 221–233 (Oscillation spectrum of the hydromagnetic dynamo of the Earth. Geomagn. Aeron. **10** (1970) 172–181)
[9] Почти аксиально-симметричная модель гидромагнитного динамо земли. 1. Геомагн. и Аэрономия **15** (1975) 149–156
[10] On the nearly axially-symmetrical model of the hydromagnetic dynamo of the Earth. Phys. Earth and Plan. Int. **11** (1976) 191–199

Bräuer, H. J.:
[1] Some properties of the Green's tensor of the induction equation. ZAMM **53** (1973) 719–722
[2] The non-linear dynamo problem: Small oscillatory solutions in a strongly simplified model. Astron, Nachr. **300** (1979) 43-49

Bräuer, H. J., and F. Krause:
[1] Remark on the mean energy of the fluctuating magnetic field in mean-field magnetohydrodynamics. Astron. Nachr. **294** (1973) 179–182
[2] The mean energy of a decaying magnetic field in mhd-turbulence. Astron. Nachr. **295** (1974) 223–228

Brissaud, A., U. Frisch, J. Leorat, M. Lesieur and A. Mazure:
[1] Helicity cascades in fully developed isotropic turbulence. Phys. Fluids **16** (1973) 1366–1367

BULLARD, E. C., and H. GELLMAN:
[1] Homogeneous dynamos and terrestrial magnetism. Phil. Trans. Roy. Soc. **A 247** (1954) 213—278

BULLARD, E. C., and D. GUBBINS:
[1] Generation of magnetic fields by fluid motions of global scale. Geophys. Fluid Dyn. **8** (1977) 43—56

BUMBA, V.:
[1] Development of spot group areas in dependence on the local magnetic field. Bull. Astron. Inst. Czech. **14** (1963) 91—97

BUSSE, F. H.:
[1] Steady fluid flow in a precessing spheroidal shell. J. Fluid Mech. **33** (1968) 739—751
[2] Thermal instabilities in rapidly rotating systems. J. Fluid Mech. **44** (1970) 441—460
[3] Generation of magnetic fields by convection. J. Fluid Mech. **57** (1973) 529—544
[4] A necessary condition for the geodynamo. J. Geophys. Res. **80** (1975) 278—280
[5] A model of the geodynamo. Geophys. J. Roy. Astr. Soc. **42** (1975) 437—459
[6] Generation of planetary magnetism by convection. Phys. Earth Planet. Int. **12** (1976) 350—358
[7] Mathematical problems of dynamo theory. In "Applications of bifurcation theory", Academic Press Inc. New York-San Francisco-London 1977, 175—202
[8] An example of non-linear dynamo action. J. Geophys. **43** (1977) 441—452

CHILDRESS, S.:
[1] Théorie magnétohydrodynamique de l'effet dynamo. Report from Département Méchanique de la Faculté des Sciences Paris 1969

COWLING, T. G.:
[1] The magnetic fields of sunspots. Mon. Not. Roy. Astr. Soc. **94** (1934) 39—48
[2] On the Sun's general magnetic field. Mon. Not. Roy. Astr. Soc. **105** (1945) 166—174
[3] Solar electrodynamics. In "The Sun", ed. G. P. KUIPER, Univ. Chicago Press 1953, 532—591
[4] The axisymmetric dynamo. Mon. Not. Roy. Astr. Soc. **140** (1968) 547—548
[5] Sunspots and the solar cycle. Nature **255** (1975) 189—190

CRAMER, H.:
[1] On the theory of stationary random processes. Annals of Mathematics **41** (1940) 215—230

CSADA, I. K.:
[1] On the magnetic effect of turbulence in ionized gases. Acta Phys. Hung. **1** (1951) 235—246

DEINZER, W.:
[1] Zur Theorie des solaren Magnetfeldes. Mitt. Astron. Gesellsch. **30** (1971) 67—74

DEINZER, W., and M. STIX:
[1] On the eigenvalues of Krause-Steenbeck's solar dynamo. Astron. & Astrophys. **12** (1971) 111—119

DEINZER, W., H. U. VON KUSSEROW and M. STIX:
[1] Gibt es auch stationäre $\alpha\omega$-Dynamos? Mitt. Astron. Gesellsch. **34** (1973) 155—158
[2] Steady and oscillatory $\alpha\omega$-dynamos. Astron. & Astrophys. **36** (1974) 69—78

DEISSLER, R. G.:
[1] Magneto-fluid dynamic turbulence with a uniform imposed magnetic field. Phys. Fluids **6** (1963) 1250–1259

DOLGINOV, A. Z. (Долгинов, А. З.):
[1] Приливные течения в компонентах двойных звёзд с осью вращения наклонной к орбите и генерация магнитного поля. Астрон. Ж. **51** (1974) 388–394
[2] Генерация магнитного поля и ускорение заряженных частиц в двойных звёздных системах. Изв. АН. Серия Физическая **39** (1975) 354–358
[3] Генерация магнитного поля в двойных звёздах и звёздах, окруженных оболочкой. Магнитные Ар-Звёзды, Издательство ,,Элм“, Baku 1975, 147–150

DOLGINOV, SH. SH. (Долгинов, Ш. Ш.):
[1] Магнетизм планет и проблема механизма динамо. Препринт № 15, ИЗМИРАН, Moscow 1976 (Planetary magnetism and dynamo mechanism problem. Preprint Nr. 15a, IZMIRAN Moscow 1976)

DROBYSHEVSKIJ, E. M., and V. S. YUFEREV:
[1] Topological pumping of magnetic flux by three-dimensional convection. J. Fluid Mech. **65** (1974) 33–44

ELSASSER, W. M.:
[1] Induction effects in terrestrial magnetism. Phys. Rev. **69** (1946) 106–116
[2] Hydromagnetic dynamo theory. Rev. Mod. Phys. **28** (1956) 135–163
[3] The terrestrial dynamo. Proc. Nat. Acad. Sci. **43** (1957) 14–24

FRAZER, M. C.:
[1] The dynamo problem and the geomagnetic field. Contemp. Phys. **14** (1973) 213–228

FRENKEL, YA. I. (Френкель, Я. И.):
[1] О происхождении земного магнетизма. ДАН **49** (1945) 98–101

FRISCH, U., J. LÉORAT, A. MAZURE and A. POUQUET:
[1] Possibility of an inverse cascade of magnetic helicity in magnetohydrodynamic turbulence. J. Fluid Mech. **68** (1975) 769–778

FYFE, D., and D. MONTGOMERY:
[1] High-beta turbulence in two-dimensional magnetohydrodynamics. J. Plasma Physics **16** (1976) 181–191

GAILITIS, A.: (Гаилитис, А.):
[1] Условия самовозбуждения лабораторной модели геомагнитного динамо. Магнитная Гидродинамика **1967**, **3**, 45–54
[2] Самовозбуждение магнитного поля парой кольцевых вихрей. Магнитная Гидродинамика **1970**, **1**, 19–22 (Self-excitation of a magnetic field by a pair of annular vortices. Magnetohydrodynamics **6** (1970) 14–17)
[3] К теории динамо Герценберга. Магнитная Гидродинамика **1973**, **4**, 12–16
[4] О генерации магнитного поля зеркало-симметричной турбулентностью. Магнитная Гидродинамика **1974**, **2**, 31–35

GAILITIS, A., and YA. FREIBERG (Гаилитис, А., и Я. Фрейберг):
[1] Самовозбуждение магнитного поля парой колбцевих вихрей. Магнитная Гидродинамика **1974**, **1**, 37–42

[2] К теории винтого МГД-линамо. Магнитная Гидродинамика **1976**, **2**, 3—6
[3] Расчёт динамо-неустойчивости винтого потока. АН ЛатвССР Инст. Физ. Рига, ,,Зинатне'' (1977) 1—48

GAILITIS, A., and O. LIELAUSIS (ГАИЛИТИС, А., и О. ЛИЕЛАУСИС):
[1] Теория МГД-динамо и лабораторный эксперимент. In: ,,Магнитные Ар-Звёзды'', Издателбство ,,Элм'', Baku 1975, 140—146

GAILITIS, A., O. LIELAUSIS and YA. FREIBERG: (ГАИЛИТИС, А., О. ЛИЕЛАУСИС и Я. ФРЕЙБЕРГ):
[1] О путах наблюдения генерации магнитного поля в потоках жидкого натрия. Препринт № 1 (1977), АН ЛатвССР Инст. Физ. Рига, 1—35

GIBSON, R. D.:
[1] The Herzenberg dynamo. I. Q. J. Mech. Appl. Math. **21** (1968) 243—255
[2] The Herzenberg dynamo. II. Q. J. Mech. Appl. Math. **21** (1968) 257—287

GIBSON, R. D., and P. H. ROBERTS:
[1] Some comments on the theory of homogeneous dynamos. In: Magnetism and the Cosmos, eds. W. R. HINDMARSH, F. J. LOWES, P. H. ROBERTS and S. K. RUNCORN, Oliver & Boyd Ltd., Edinburgh 1967, 108—120
[2] The Bullard-Gellman dynamo. In: The Application of Modern Physics to the Earth and Planetary Interiors, ed. S. K. RUNCORN, Wiley Interscience 1969, 577—601

GILLILAND, J. M.:
[1] Mean-field electrodynamics and dynamo theories of planetary magnetic fields. Thesis, University of Alberta (Edmonton, Canada) 1973

GOLITSYN, G. S. (Голицин, Г. С.):
[1] Флуктуации магнитного поля и плотности тока в турбулентном потоке слабо проводящей жидкости. ДАН **132** (1960) 315—318 (Fluctuations of the magnetic field and current density in a turbulent flow of a weakly conducting fluid. Sov. Phys. Dokl. **5** (1960), 536—539)

GREENSPAN, H.:
[1] On α-dynamos. Studies in Appl. Math. **13** (1974) 35—43

GUBBINS, D.:
[1] Kinematic dynamos and geomagnetism. Nature Phys. Sci. **238** (1972) 119-122
[2] Numerical solutions of the kinematic dynamo problem. Phil. Trans. Roy. Soc. **A 274** (1973) 493—521
[3] Theories of the geomagnetic and the solar dynamo. Revs. Geophys. Space Phys. **12** (1974) 137—154
[4] Dynamo action of isotropically driven motions of a rotating fluid. Studies in Appl. Math. **8** (1974) 157-164
[5] Observational constraints on the generation process of the Earth's magnetic field. Geophys. J. Roy. Astr. Soc. **47** (1976) 19-39

GUREVICH, L. S., and A. I. LEBEDINSKIJ (ГУРЕВИЧ, Л. Э., и А. И. ЛЕБЕДИНСКИЙ):
[1] Магнитное поле солнечных пятен. ДАН **49** (1945) 92—94

HELMIS, G.:
[1] Untersuchungen zur Elektrodynamik turbulent bewegter leitender Medien unter Berücksichtigung des Halleffektes. Mon.ber. dtsch. Akad. Wiss. Berlin **10** (1968) 280—291
[2] Zur Elektroydnamik turbulent bewegter leitender Medien unter Berücksichtigung des Halleffektes. Cosmic Electrodynamics **2** (1971) 197—210

[3] Zum Halleffekt in der Elektrodynamik turbulent bewegter leitender Medien Beitr. Plasma Physik **11** (1971) 417—430

Herzenberg, A.:
[1] Geomagnetic dynamos. Phil. Trans. Roy. Soc. **A 250** (1958) 543—583

Ivanova, T. S. (Иванова, Т. С.):
[1] Метод решения задачи магнитогидродинамического динамо. Ж. Вычисл. Матем. Физ. **16** (1976) 958—968

Ivanova, T. S., and A. A. Ruzmaikin (Иванова, Т. С., и А. А. Рузмайкин):
[1] Магнитогидродинамическая динамо-модель солнечного цикла. Астрон. Ж. **53** (1976) 398—410
[2] Нелинейная стабилизация колебаний в мгд-динамо солнца. Препринт Ин-та прикл. Математики АН СССР, № 47, 1976
[3] Нелинейная магнитогидродинамическая модель динамо солнца. Астрон. Ж. **54** (1977) 846—858

Jepps, S. A.:
[1] Numerical models of hydromagnetic dynamos. J. Fluid Mech. **67** (1975) 629—646

Kazantsev, A. P. (Казанцев, А. П.):
[1] Об усленнии магнитного поля проводящей жидкостью. ЖЭТФ **53** (1967) 1806—1813 (Enhancement of a magnetic field by a conducting fluid. Sov. Phys.-JETP **26** (1968) 1031—1034)

Kippenhahn, R., and C. Möllenhoff:
[1] Elementare Plasmaphysik. Bibliographisches Institut Mannheim - Wien - Zürich 1975

Kit, L. G., and A. B. Tsinober (Кит, Л. Г., и А. Б. Цинобер):
[1] О возможности создания и исследования двумерной турбулентности в сильном магнитном поле. Магнитная Гидродинамика **1971**, **3**, 27—34

Köhler, H.:
[1] The solar dynamo and estimates of the magnetic diffusivity and the α-effect. Astron. & Astrophys. **25** (1973) 467—476

Kolesnikov, Yu. B. (Колесников, Ю. Б.):
[1] Двумерное турбулентное течение в канале с неоднородной электропроводностью стенок. Магнитная Гидродинамика **1972**, **3**, 32—36

Kolesnikov, Yu. B., and A. B. Tsinober (Колесников, Ю. Б., и А. Б. Цинобер):
[1] Двумерное турбулентное течение за круглым цилиндром. Магнитная Гидродинамика **1972**, **3**, 23—31
[2] Экспериментальное исследование двумерной турбулентности за решеткой. Изв. АН СССР, Механика жидкости и газа **1974**, **4**, 146—150

Kraichnan, R. H.:
[1] Inertial-range spectrum of hydrodmagnetic turbulence. Phys. Fluids **8** (1965) 1385—1387
[2] Helical turbulence and absolute equilibrium. J. Fluid Mech. **59** (1973) 745—752
[3] Diffusion of weak magnetic fields by isotropic turbulence. J. Fluid Mech. **75** (1976) 657—676
[4] Diffusion of passive-scalar and magnetic fields by helical turbulence. J. Fluid Mech. **77** (1976) 753—768

Kraichnan, R. H., and S. Nagarajan:
[1] Growth of turbulent magnetic fields. Phys. Fluids **10** (1967) 859—870

Krause, F.:
[1]* Eine Lösung des Dynamoproblems auf der Grundlage einer linearen Theorie der magnetohydrodynamischen Turbulenz. Habilitationsschrift, Univ. Jena 1967

[2] Zum Anfangswertproblem der magnetohydrodynamischen Induktionsgleichung. ZAMM **48** (1968) 333–343
[3] Explanation of stellar and planetary magnetic fields by dynamo action of turbulent motions. Acta Univ. Wratislaviensis **77** (1969) 157–170
[4] Zur Dynamotheorie magnetischer Sterne: Der „symmetrische Rotator" als Alternative zum „schiefen Rotator". Astron. Nachr. **293** (1971) 187–193
[5] Heat flow and magnetic field diffusion in turbulent fluids. Astron. Nachr. **294** (1972) 83–87
[6] The turbulent emf $\overline{\boldsymbol{u}' \times \boldsymbol{B}'}$ in the case of non-vanishing mean flow. ZAMM **53** (1973) 479–481
[7] The present status of mean-field magnetohydrodynamics and the turbulent dynamo problem. Ann. New York Acad. Sc. **257** (1975) 156–172
[8] Динамотеория. In: Магнитные Ар-Звёзды, Издательство „Элм", Baku 1975, 129–138
[9] Annotation on the paper "The turbulent emf $\overline{\boldsymbol{u}' \times \boldsymbol{B}'}$ in the case of non-vanishing mean flow". ZAMM **56** (1976) 172–173
[10] Mean-field magnetohydrodynamics of the solar convection zone. In: Basic Mechanism of Solar Activity, eds. V. Bumba and J. Kleczek, D. Reidel Publishing Company, Dordrecht-Holland 1976, 309–321
[11] Mean-field electrodynamics and dynamo theory of the Earth's magnetic field. Journ. Geophys. **43** (1977) 421–440

Krause, F., and H. Hiller:
[1] Zur Dynamotheorie stellarer und planetarer Magnetfelder. III. Über die Lösung des Eigenwertproblemes und die Berechnung der Feldgrößen. Astron. Nachr. **291** (1969) 271–286

Krause, F., and L. Oetken:
[1] On equatorially-symmetric models for magnetic stars as suggested by dynamo theory. In: Physics of Ap-stars, IAU-Colloqu. Nr. 32, eds. W. W. Weiss, H. Jenkner and H. J. Wood, Vienna 1976, 29–36

Krause, F., and K.-H. Rädler:
[1] Dynamo theory of the Sun's general magnetic field on the basis of a mean-field magnetohydrodynamics. In: Solar Magnetic Fields, ed. R. Howard, Reidel Publishing Company, Dordrecht-Holland 1971, 770–779
[2] Elektrodynamik der mittleren Felder in turbulenten leitenden Medien und Dynamotheorie. In: Ergebnisse der Plasmaphysik und der Gaselektronik. Bd. II, eds. R. Rompe and M. Steenbeck, Akademie-Verlag Berlin 1971, 1–154

Krause, F., K.-H. Rädler and G. Rüdiger:
[1] Über die Bedeutung des Nichols-Tolman-Effektes für das Zustandekommen kosmischer Magnetfelder. Gerlands Beitr. Geophys. **85** (1976) 26–34

Krause, F., and P. H. Roberts:
[1] Some problems of mean-field electrodynamics. Astrophys. J. **191** (1973) 977–992
[2] Bochner's theorem and mean-field electrodynamics. Mathematika **20** (1973) 24–33
[3] Comments on the paper "On the application of Cramer's theorem to axisymmetric incompressible turbulence" by I. Lerche. Astrophys. Space Sc. **22** (1973) 193–195
[4] The high-conductivity limit in mean-field electrodynamics. J. Math. Phys. **17** (1976) 1808–1809

KRAUSE, F., and G. RÜDIGER:
[1] On the turbulent decay of strong magnetic fields and the development of sunspots areas. Solar Physics **42** (1975) 107–119

KRAUSE, F., and M. STEENBECK:
[1] Models of magnetohydrodynamic dynamos for alternating fields. Czech. Slov. Acad. Sc. Astron. Inst. Publ. **51** (1964) 36–38
[2]* Untersuchung der Dynamowirkung einer nichtspiegelsymmetrischen Turbulenz an einfachen Modellen. Z. f. Naturf. **22a** (1960) 671–675

KROPATSHEV, E. P., S. N. GORSHKOV and P. M. SEREBRYANAYA (КРОПАЧЕВ. Э. П., С.Н. ГОРШКОВ и П. М. СЕРЕБРЯНАЯ):
[1] Модель кинематического динамо трёх сферических вихрей. I. Геомагн. и Аэрономия **17** (1977) 507–511
[2] Модель кинематического динамо трёх сферических вихрей. II. Геомагн. и Аэрономия **17** (1977) 927–929

KUMAR, S., and P. H. ROBERTS:
[1] A three-dimensional kinematic dynamo. Proc. Roy. Soc. **A 344** (1975) 239–258

LANDSTREET, J. D., and J. R. P. ANGEL:
[1] The polarisation spectrum and magnetic field strength of the White Dwarf Grw + 70° 8247. Astrophys. J. **196** (1974) 819–825

LARMOR, J.:
[1] How could a rotating body such as the Sun become a magnet? Rep. Brit. Assoc. Adv. Sc. 1919, 159–160

LEHNERT, B.:
[1] The decay of magnetoturbulence in the presence of a magnetic field and Coriolis force. Qu. J. Appl. Math. **12** (1955) 321–341

LEIGHTON, R. B.:
[1] Transport of magnetic fields on the Sun. Astrophys. J. **140** (1964) 1547–1562
[2] A magneto-kinematic model of the solar cycle. Astrophys. J. **156** (1969) 1–26

LÉORAT, J.:
[1] La turbulence magnétohydrodynamique hélicitaire et la géneration des champs magnétique a grande échelle. Thése doctoral, Univ. de Paris VII, 1975

LÉORAT, J., U. FRISCH and A. POUQUET:
[1] Helical magnetohydrodynamic turbulence and the non-linear dynamo problem. Ann. New York Acad. Sc. **257** (1975) 173–176

LERCHE, I.:
[1] Kinematic dynamo theory. Astrophys. J. **166** (1971) 627–638
[2] Kinematic dynamo theory. II. Dynamo action in infinite media with isotropic turbulence. Astrophys. J. **166** (1971) 639–649
[3] Kinematic dynamo theory. III. The effect of turbulent diffusivity in the dynamo equations. Astrophys. J. **168** (1971) 115–121
[4] Kinematic dynamo theory. IV. Dynamo action in non-rotating spheres with isotropic turbulence. Astrophys. J. **168** (1971) 123–129
[5] On the applications of Cramer's theorem to axi-symmetric, incompressible turbulence. Astrophys. Space Sc. **19** (1972) 189–193
[6] Kinematic dynamo theory. V. Comments on diverse matters including historical development, isotropic turbulence, and expansion techniques. Astrophys. J. **181** (1973) 993–1002
[7] Reply to foregoing comments. Astrophys. Space Sc. **22** (1973) 197

LERCHE, I., and B.-C. LOW:
[1] Kinematic dynamo action under incompressible, isotropic velocity turbulence. Astrophys. J. **168** (1971) 503–508
LERCHE, I., and E. N. PARKER
[1] The generation of magnetic fields in astrophysical bodies. VII. The internal small scale field. Astrophys. J. **168** (1971) 231–237
LEVY, E. H.:
[1] Effectiveness of cyclonic convection for producing the geomagnetic field. Astrophys. J. **171** (1972) 621–633
[2] Kinematic reversal schemes for the geomagnetic dipole. Astrophys. J. **171** (1972) 635–642
LIELAUSIS, O., A. KLYAVINYA and L. TIMANIS (ЛИЕЛАУСИС, О., А. КЛЯВИНЯ и Л. ТИМАНИС):
[1] О возможности специфических мгд-явлений при жидкометаллическом охлаждении мощных энергетических установок. Аннотация, АН ЛатвССР Инст. Физ. Рига, 1974
LILLEY, F. E. M.:
[1] On kinematic dynamos. Proc. Roy. Soc. **A 316** (1970) 153–167
LORTZ, D.:
[1] Impossibility of steady dynamos with certain symmetries. Phys. Fluids **11** (1968) 913–915
[2] Exact solutions of the hydromagnetic dynamo problem. Plasma Phys. **10** (1968) 967–972
[3] A simple stationary dynamo model. Z. Naturforsch. **27a** (1972) 1350–1354
LOW, B.-C.:
[1] Root mean square fluctuations of a weak magnetic field in an infinite medium of homogeneous stationary turbulence. Astrophys. J. **173** (1972) 549–555
[2] Errata. Astrophys. J. **178** (1972) 277
LOWES, F. J., and I. WILKINSON:
[1] Geomagnetic dynamo: a laboratory model. Nature **198** (1963) 1158–1160
[2] Geomagnetic dynamo: an improved laboratory model. Nature **219** (1968) 717–718
MALKUS, W. V. R., and M. R. E. PROCTOR:
[1] The macrodynamics of α-effect dynamos in rotating fluids. J. Fluid Mech. **67** (1975) 417–444
MEYER, F., and H. U. SCHMIDT:
[1] On the decay of sunspots. Mitt. Astron. Gesellsch. **32** (1973) 174–175
MEYER, F., H. U. SCHMIDT, N. O. WEISS and P. R. WILSON:
[1] The growth and decay of sunspots. Mon. Not. Roy. Astron. Soc. **169** (1974) 35–57
MOFFATT, H. K.:
[1] On the suppression of turbulence by a uniform magnetic field. J. Fluid Mech. **28** (1967) 571–592
[2] Turbulent dynamo action at low magnetic Reynolds number. J. Fluid Mech. **41** (1970) 435–452
[3] Dynamo action associated with random inertial waves in a rotating conducting fluid. J. Fluid Mech. **44** (1970) 705–719
[4] An approach to a dynamic theory of dynamo action in a rotating fluid. J. Fluid Mech. **53** (1972) 385–399
[5] The mean electromotive force generated by turbulence in the limit of perfect conductivity. J. Fluid Mech. **65** (1974) 1–10

[6] Appendix to the paper of Drobyshevski and Yuferev. "Topological pumping of magnetic flux by three-dimensional convection". J. Fluid Mech. **65** (1974) 41–44
[7] Generation of magnetic fields by fluid motion. Adv. Appl. Mech. **16** (1976) 119–181
[8] Magnetic field generation in electrically conducting fluids. Cambridge University Press. Cambridge - London - New York - Melbourne 1978

Moreau, R.:
[1] Homogeneous turbulence in the presence of a uniform magnetic field when the magnetic Reynolds' number is small. C. R. Acad. Sci. **A 263** (1966) 586–587
[2] Une solution simple pour le déclin de la turbulence homogene en présence d'un champ magnétique uniforme lorsque le nombre de Reynolds magnétique est petit. C. R. Acad. Sci. **A 264** (1967) 75–78
[3] On magnetohydrodynamic turbulence. Proc. Symp. on turbulence of fluids and plasmas, Brooklyn 1968, 359–372

Moss, D. L.:
[1] A numerical model of hydromagnetic turbulence. Mon. Not. R. astr. Soc. **148** (1970) 173–191

Nestlerode, I. A., and I. L. Lumley:
[1] Initial response of the spectrum of isotropic turbulence to the sudden application of a strong magnetic field. Phys. Fluids **6** (1963) 1260–1262

Oetken, L.:
[1] An equatorially-symmetric rotator model for magnetic stars. Astron. Nachr. **298** (1977) 197–207
[2] An equatorially-symmetric rotator model for magnetic stars. II. Inhomogeneous element distributions. Astron. Nachr. **300** (1979) 000–000

Parker, E. N.:
[1] The formation of sunspots from the solar toroidal field. Astrophys. J. **121** (1955) 491–507
[2] Hydromagnetic dynamo models. Astrophys. J. **122** (1955) 293–314
[3] The solar hydromagnetic dynamo. Proc. Nat. Acad. Sci. **43** (1957) 8–13
[4] The dynamical state of the interstellar gas and field. Astrophys. J. **145** (1966) 811–833
[5] The occasional reversal of the geomagnetic field. Astrophys. J. **158** (1969) 815–827
[6] The origin of magnetic fields. Astrophys. J. **160** (1970) 383–404
[7] The generation of magnetic fields in astrophysical bodies. I. The dynamo equation. Astrophys. J. **162** (1970) 665–673
[8] The generation of magnetic fields in astrophysical bodies. II. The galactic field. Astrophys. J. **163** (1971) 255–278
[9] The generation of magnetic fields in astrophysical bodies. III. Turbulent diffusion of fields and efficient dynamos. Astrophys. J. **163** (1971) 279–285
[10] The generation of magnetic fields in astrophysical bodies. IV. The solar and terrestrial dynamos. Astrophys. J. **164** (1971) 491–509
[11] The generation of magnetic fields in astrophysical bodies .V. Behaviour at large dynamo numbers. Astrophys. J. **165** (1971) 139–146
[12] The generation of magnetic fields in astrophysical bodies. VI. Periodic modes of the galactic field. Astrophys. J. **166** (1971) 295–300
[13] The generation of magnetic fields in astrophysical bodies. VIII. Dynamical considerations. Astrophys. J. **168** (1971) 239–249

[14] The generation and dissipation of solar and galactic magnetic fields. Astrophys. Sp. Sci. **22** (1973) 279–291

[15] The dynamo mechanism for the generation of large-scale magnetic fields. Ann. New York Acad. Sc. **257** (1975) 141–155

Pekeris, C. L., Y. Accad and B. Shkoller:

[1] Kinematic dynamos and the Earth's magnetic field. Phil. Trans. Roy. Soc. **A 275** (1973) 425–461

Pierson, E. S.:

[1] Electromagnetic self-excitation in the liquid-metal fast breeder reactor. Nucl. Sc. Engineering **57** (1975) 155–163

Pichakchi, L. D. (Пичахчи, Л. Д.):

[1] К теории гидромагнитного динамо. ЖЭТФ **50** (1966) 818–820

Ponomarenko, Yu. B. (Пономаренко, Ю. Б.):

[1] К теории гидромагнитного динамо. ПМТФ **1973**, **6**, 47–51

Pouquet, A., U. Frisch and J. Léorat:

[1] Strong mhd turbulence and the nonlinear dynamo effect. J. Fluid Mech. **77** (1976) 321–354

Pouquet, A., and G. S. Patterson:

[1] Numerical simulation of helical magnetohydrodynamic turbulence. J. Fluid Mech. **85** (1978) 305–323

Proctor, M. R. E.:

[1] Numerical solutions of the nonlinear α-effect dynamo equations. J. Fluid Mech. **80** (1977) 769–784

[2] On the eigenvalues of kinematic α-effect dynamos. Astron. Nachr. **298** (1977) 19–25

Rädler, K.-H.:

[1] Analogiebetrachtungen zu ebenen Problemen der Magnetohydrodynamik. Beitr. Plasmaphys. **2** (1964) 21–31

[2] Zur Elektrodynamik turbulent bewegter leitender Medien. Thesis, Univ. Jena 1966

[3]* Zur Elektrodynamik turbulent bewegter leitender Medien. I. Grundzüge der Elektrodynamik der mittleren Felder. Z. f. Naturforsch. **23a** (1968) 1841–1851

[4]* Zur Elektrodynamik turbulent bewegter leitender Medien. II. Turbulenzbedingte Leitfähigkeits- und Permeabilitätsänderungen. Z. f. Naturforsch. **23a** (1968) 1851–1860

[5] On some electromagnetic phenomena in electrically conducting turbulently moving matter, especially in the presence of Coriolis forces. Geod. Geoph. Veröff. Reihe II, H. 13 (1969) 131–135

[6]* Zur Elektrodynamik in turbulenten, Coriolis-Kräften unterworfenen leitenden Medien. Mber. dtsch. Akad. Wiss. Berlin **11** (1969) 194–201

[7]* Über eine neue Möglichkeit eines Dynamomechanismus in turbulenten leitenden Medien. Mber. dtsch. Akad. Wiss. Berlin **11** (1969) 272–279

[8]* Untersuchung eines Dynamomechanismus in turbulenten leitenden Medien. Mber. dtsch. Akad. Wiss. Berlin **12** (1970) 468–472

[9] Zur Dynamotheorie kosmischer Magnetfelder. I. Gleichungen für sphärische Dynamomodelle. Astron. Nachr. **294** (1973) 213–223

[10] Zur Dynamotheorie kosmischer Magnetfelder. II. Darstellung von Vektorfeldern als Summe aus einem poloidalen und einem toroidalen Anteil. Astron. Nachr. **295** (1974) 73–84

[11] Über den Korrelationstensor zweiter Stufe für ein inhomogenes turbulentes Geschwindigkeitsfeld. Astron. Nachr. **295** (1974) 85–92

[12] On the influence of a large-scale magnetic field on turbulent motions in an electrically conducting medium. Astron. Nachr. **295** (1974) 265–273

[13] Some new results on the generation of magnetic fields by dynamic action. Mem. Soc. Roy. Sc. Liege **VIII** (1975) 109–116

[14] Замечания о генерировании неосесимметричных магнитных полей при помощи динамомеханизма. In: Магнитные Ар-Звёзды, Издательство „Элм", Baku 1975, 139–140

[15] Mean-field magnetohydrodynamics as a basis of solar dynamo theory. In: Basic Mechanism of Solar Activity, eds. V. Bumba and J. Kleczek, D. Reidel Publishing Company, Dordrecht-Holland 1976, 323–344

[16] Mean-field approach to spherical dynamo models. Astron. Nachr. **301** (1980), in print

[17] Investigations of spherical mean-field dynamo models. Astron. Nachr. **301** (1980), in print

Roberts, G. O.:

[1] Spatially periodic dynamos. Phil. Trans. Roy. Soc. **A 266** (1970) 535–558

[2] Dynamo action of fluid motions with two-dimensional periodicity. Phil. Trans. Roy. Soc. **A 271** (1972) 411–454

Roberts, P. H.:

[1] An introduction to magnetohydrodynamics. Longmans, Green and Co. Ltd, London, 1967

[2] The dynamo problem. Woods Hole Oceanographic Inst. Rep. 67–54 (1967) 51–165

[3] Dynamo theory. In: Lectures on Applied Mathematics, ed. W. H. Reid, Amer. Mathematics Soc., Providence, R. I. **14** (1971) 129–206

[4] Kinematic dynamo models. Phil. Trans. Roy. Soc. **A 272** (1972) 663–703

Roberts, P. H., and A. M. Soward:

[1] Magnetohydrodynamics of the Earth's core. Ann. Rev. Fluid Mech. **4** (1972) 117–154

[2] On first order smoothing theory. J. Math. Phys. **16** (1975) 609–615

[3] A unified approach to mean-field electrodynamics. Astron. Nachr. **269** (1975) 49–64

Roberts, P. H., and M. Stix:

[1] The turbulent dynamo. A translation of a series of papers by F. Krause, K.-H. Rädler and M. Steenbeck. Techn. Note **60** (1971) NCRA Boulder, Colorado

[2] α-effect dynamos by the Bullard-Gellman formalism. Astron. & Astrophys. **18** (1972) 453–466

[3] On Vainshtein's simplest dynamo instability. Phys. Earth Plan. Int. **12** (1976) P19–P21

Rochester, M. G., J. A. Jacobs, D. E. Smylie and K. F. Chong:

[1] Can precession power the geomagnetic dynamo? Geophys. J. Roy. Astron. Soc. **43** (1975) 661–678

Rüdiger, G.:

[1] Behandlung eines einfachen hydromagnetischen Dynamos mittels Linearisierung. Astron. Nachr. **294** (1973) 183–186

[2] The influence of a uniform magnetic field of arbitrary strength on turbulence. Astron. Nachr. **295** (1974) 275–283

[3] Behandlung eines einfachen hydromagnetischen Dynamos mit Hilfe der Gitterpunktmethode. Publ. Astrophys. Obs. Potsdam **32** (1974) 25–29

[4] Die Wechselwirkung zwischen homogener Turbulenz und inhomogenem magnetischen Feld in der Umgebung neutraler Flächen. Astron. Nachr. **296** (1975) 133—141
[5] On α-effect for slow and fast rotation. Astron. Nachr. **299** (1978) 217—222

Schlüter, A., and L. Biermann:
[1] Interstellare Magnetfelder. Z. Naturforsch. **5a** (1950) 237—251

Schumann, U.:
[1] Numerical simulation of the transition from three- to two-dimensional turbulence under a uniform magnetic field. J. Fluid Mech. **74** (1976) 31—58

Soward, A. M.:
[1] Nearly symmetric kinematic and hydromagnetic dynamos. J. Math. Phys. **12** (1971) 1900—1906
[2] Nearly symmetric advection. J. Math. Phys. **12** (1971) 2052—2062
[3] A kinematic theory of large magnetic Reynolds number dynamos. Phil. Trans. Roy. Soc. **A 272** (1972) 431—462
[4] A convection-driven dynamo. I. The weak field case. Phil. Trans. Roy. Soc. **A 275** (1974) 611—651
[5] Random waves and dynamo action. J. Fluid Mech. **69** (1975) 145—177
[6] A thin disc model of the galactic dynamo. Astron. Nachr. **299** (1978) 25—33

Soward, A.M., and P. H. Roberts (Соуорд, А. М., и П. Х. Робертс):
[1] Современое состояние теории мгд-динамо. Магнитная Гидродинамика **1976**, **1**, 3—51

Steenbeck, M.:
[1] Probleme und Ergebnisse der Elektro- und Magnetohydrodynamik. Dtsch. Akad. Wiss. Berlin, Vorträge und Schriften Heft 73, Akademie-Verlag Berlin 1961
[2] Elementare magnetohydrodynamische Behandlung chaotisch turbulenter Bewegungen. Mber. dtsch. Akad. Wiss. Berlin **5** (1963) 625—629
[3] Warum hat die Erde ein Magnetfeld? Phys. Bl. **26** (1970) 158—168

Steenbeck, M., and G. Helmis:
[1] Zur Deutung der Neigung und der Westdrift des erdmagnetischen Hauptfeldes. Mber. dtsch. Akad. Wiss. Berlin **11** (1969) 723—734
[2] Rotation of the Earth's solid core as a possible cause of declination, drift and reversals of the Earth's magnetic field. Geophys. J. Roy. Astron. Soc. **41** (1975) 237—244

Steenbeck, M., I. M. Kirko, A. Gailitis, A. P. Klawina, F. Krause, I. J. Laumanis and O. A. Lielausis:
[1]* Der experimentelle Nachweis einer elektromotorischen Kraft längs eines äußeren Magnetfeldes, induziert durch eine Strömung flüssigen Metalls (α-Effekt). Mber. dtsch. Akad. Wiss. Berlin **9** (1967) 714—719
[2] (Штеенбек, М., И. М. Кирко, А. Гаилитис, А. П. Клявиня, Ф. Краузе, И. Я. Лауманис и О. А. Лиелаусис): Экспериментальное обнаружение электродвижущей силы вдоль внешнего магнитного поля, индуцированной течением жидкого металла (α-эффект). ДАН **180** (1968) 326—329 (Experimental discovery of the electromotive force along the external magnetic field induced by a flow of liquid metal (α-effect). Sov. Phys.-Dokl. **13** (1968) 443—445)

Steenbeck, M., and F. Krause:
[1] Elektromagnetische Rückkopplung durch Turbulenz unter der Einwirkung von Coriolis-Kräften. Mitt. Astron. Gesellsch. **19** (1965) 95

[2] Elektromagnetische Rückkopplung durch Turbulenz unter der Einwirkung von Coriolis-Kräften. Mber. dtsch. Akad. Wiss. Berlin **7** (1965) 900–906

[3]* Erklärung stellarer und planetarer Magnetfelder durch einen turbulenzbedingten Dynamomechanismus. Z. Naturforsch. **21a** (1966) 1285–1296

[4] Die Entstehung stellarer und planetarer Magnetfelder als Folge turbulenter Materiebewegungen. Gustav-Hertz-Festschrift, Akademie-Verlag Berlin 1967

[5] Возниковение магнитных полей звёзд и планет в результате турбулентного движения их вещества. Магнитная Гидродинамика **1967, 3,** 19–44

[6]* Zur Dynamotheorie stellarer und planetarer Magnetfelder. I. Berechnung sonnenähnlicher Wechselfeldgeneratoren. Astron. Nachr. **291** (1969) 49–84

[7]* Zur Dynamotheorie stellarer und planetarer Magnetfelder. II. Berechnung planetenähnlicher Gleichfeldgeneratoren. Astron. Nachr. **291** (1969) 271–286

STEENBECK, M., F. KRAUSE and K.-H. RÄDLER:

[1] Elektrodynamische Eigenschaften turbulenter Plasmen. Sitz.ber. dtsch. Akad. Wiss. Berlin, Klasse für Mathematik, Physik und Technik Nr. 1, Akademie-Verlag Berlin 1963

[2]* Berechnung der mittleren Lorentz-Feldstärke $\overline{\boldsymbol{v} \times \boldsymbol{B}}$ für ein elektrisch leitendes Medium in turbulenter, durch Coriolis-Kräfte beeinflußter Bewegung. Z. Naturforsch. **21a** (1966) 369–376

STIX, M.:

[1] A non-axisymmetric α-effect dynamo. Astron. & Astrophys. **13** (1971) 203–208

[2] Vergleich der Sonnenzyklen von Leighton und Steenbeck-Krause. preprint 1971

[3] Non-linear dynamo waves. Astron. & Astrophys. **20** (1972) 9–12

[4] Spherical $\alpha\omega$-dynamos, by a variational method. Astron. & Astrophys. **24** (1973) 275–281

[5] The galactic dynamo. Astron. & Astrophys. **42** (1975) 85–90

[6] Differential rotation and the solar dynamo. Astron. & Astrophys. **47** (1976) 243–254

[7] Dynamo theory and the solar cycle. In: Basic Mechanism of Solar Activity, eds. V. BUMBA and J. KLECZEK, D. Reidel Publishing Company. Dordrecht-Holland 1976, 367–388

SWEET, P. A.:

[1] The effect of turbulence on a magnetic field. Mon. Not. Roy. Astron. Soc. **110** (1950) 69–83

TOUGH, J. G.:

[1] Nearly symmetric dynamos. Geophys. J. Roy. Astron. Soc. **13** (1967) 393–406

[2] Corrigendum. Geophys. J. Roy. Astron. Soc. **15** (1900) 343

TOUGH, J. G., and R. D. GIBSON:

[1] The Braginskij dynamo. In: The application of Modern Physics to the Earth and Planetary Interiors, ed. S. K. RUNCORN, Wiley-interscience London - New York 1969, 55 –569

TOUGH, J. G., and P. H. ROBERTS:

[1] Nearly symmetric hydromagnetic dynamos. Phys. Earth Planet. Int. **1** (1968) 288–296

TSINOBER, A. B. (ЦИНОБЕР, А. Б.):

[1] Магнитогидродинамическая турбулентность. Магнитная Гидродинамика **1975, 1,** 7–22

TVERSKOJ, B. A. (ТВЕРСКОЙ, Б. А.):

[1] К теории гидродинамического самовозбуждения регулярных магнитных полей. Геомагн. и Аэрономия **6** (1966) 11—18 (Theory of hydrodynamical self-excitation of regular magnetic fields. Geomagn. i Aeron. **7** (1966) 7—12)

VAINSHTEIN, S. I. (ВАЙНШТЕЙН, С. И.):

[1] О генерации крупномасштабного магнитного поля турбулентной жидкостью. ЖЕТФ **58** (1970) 153—159 (Generation of a large scale magnetic field by a turbulent liquid. Sov. Phys.-JETP **31** (1970) 87)

[2] Задача о генерации магнитного поля при наличии акустической турбулентности. ДАН **195** (1970) 793—796

[3] Задача о магнитном поле в неоднородном турбулентном потоке. ПМТФ, № **1** (1971) 12—18

[4] О нелинейной задаче турбулентного динамо. ЖЕТФ **61** (1971) 612—620 (Non-linear problem of the turbulent dynamo. Sov. Phys.-JETP **34** (1972) 327—331)

[5] Функциональный подход в теории турбулентного динамо. ЖЕТФ **62** (1972) 1376—1385 (Functional approach to the turbulent dynamo. Sov. Phys.-JETP **35** (1972) 725)

[6] ,,Антидинамо" — возможный механизм явлений происходящих нейтральних слоях магнитного поля. ЖЕТФ **65** (1973) 550—561 ("Antidynamo"—a possible mechanism of phenomena occurring in neutral layers of a magnetic field. Sov. Phys.-JETP **38** (1973) 270—275)

[7] Перестройка магнитного поля плазменными механизмами. УФН **120** (1976) 613—645 (Modification of a magnetic field by plasma mechanism. Sov. Phys.-Usp. **19** (1976) 987—1006)

[8] Возбуждение магнитного поля ленгмюровскими колебаниями и ионным звуком (α-эффект в плазме). Gerlands Beitr. Geophys., Leipzig **85** (1976) 93—102

[9] Простейшая динамо-неустойчивость. ЖЭТФ **68** (1975) 997—1004 (Simplest dynamo-instability. Sov. Phys.-JETP **41** (1975) 494—000)

VAINSHTEIN, S. I., and YA. B. ZEL'DOVICH (ВАЙНШТЕЙН, С. И., и Я. Б. ЗЕЛЬДОВИЧ):

[1] О происхождении магнитних полей в астрофизике. УФН **106** (1972) 431—457 (Origin of magnetic fields in astrophys. Sov. Phys.-Usp. **15** (1972) 159—172)

VOIGTMANN, L.:

[1] Über das zeitliche Verhalten zylindersymmetrischer globaler Magnetfelder unter dem Einfluß einer nicht spiegelsymmetrischen homogenen isotropen Turbulenz. Diplomarbeit, Univ. Jena 1968

VOTSISH, A. D., and YU. B. KOLESNIKOV (Воциш, А., Ди Ю. Б. Колесников):

[1] Пространственная и завихренность в двумерной однородной турбулентности. Магнитная Гидродинамика **1976, 3,** 25—28

[2] Аномальный перенос импулса в сдвиговом мгд-течении с двумерной структурной турбулентности. Магнитная Гидродинамика **1976, 4,** 47—52

WEISS, N. O.:

[1] The dynamo problem. Q. J. Roy. Astron. Soc. **12** (1971) 432—446

YOSHIMURA, H.:

[1] Complexes of activity of the solar cycle and very large scale convection. Solar Physics **18** (1971) 417—433

[2] On the dynamo action of the global convection in the solar convection zone. Astrophys. J. **3** (1972) 863–886
[3] Solar-Cycle dynamo wave propagation. Astrophys. J. **201** (1975) 740–748
[4] A model of the solar cycle driven by the dynamo action of the global convection in the solar convection zone. Astrophys. J. Suppl. Series **294** (1975) 467–494
[5] Phase relation between the poloidal and toroidal solar-cycle general magnetic fields and location of the origin of the surface magnetic fields. Solar Physics **50** (1976) 3–23

ZEL'DOVICH, YA. B. (Зельдович, Я. Б.):
[1] Магнитное поле в проводящей турбулентной жидкости при двумерном движении. ЖЭТФ **31** (1956) 154–155 (The magnetic field in the two-dimensional motion of a conducting turbulent fluid. Sov. Phys.-JETP **4** (1957) 460–462)

INDEX